HUMAN EVOLUTION

Readings in Physical Anthropology

THIRD EDITION

NOEL KORN, Editor
Los Angeles Valley College

HOLT, RINEHART AND WINSTON, INC.
New York   Chicago   San Francisco   Atlanta   Dallas
Montreal   Toronto   London   Sydney

To the memory of
Harry Reece Smith
and
Fred W. Thompson

Copyright © 1959, 1967, 1973 by Holt, Rinehart and Winston, Inc.
All rights reserved
Library of Congress Catalog Card Number: 73-236
ISBN: 0–03–091288–1
Printed in the United States of America
3 4 5 6   090   9 8 7 6 5 4 3 2 1

PREFACE

This collection of readings for beginning students in physical anthropology is intended to accompany assignments in whatever textbook is required by the instructor. The selections have been made from representative works illustrating the different lines of evidence that provide a unified view of human evolution—its processes, results, and implications. Thus, the reader will find here studies in genetics and ecology, field studies of primate behavior, description and interpretations of the fossil record, experimental studies, and descriptions of research in ongoing evolution among human populations. These form the subject matter of modern evolutionary research. I have tried to provide examples which are not only significant, but also accessible to college students with minimal biological backgrounds.

Scientists employ the experimental method to establish generalizations concerning the relationships among phenomena. At first it might be thought that physical anthropologists cannot utilize the experimental method in the study of human evolution. Indeed, population studies have to be undertaken with "free-ranging" subjects responding in real ways to their natural environment. Fossils must be recovered in the field and conclusions drawn about their significance. Human populations cannot be induced to take up residence in a laboratory.

Nevertheless, the processes of evolution—the "how" of evolution—can be studied experimentally. In fact, the synthesis of knowledge about evolutionary processes—mutation, recombination, selection, drift, and hybridization—has been gained largely through laboratory studies; descriptive studies of fossils have also proved fruitful in providing our present picture of human evolution. Such studies have succeeded in integrating the study of human culture and of human evolutionary biology into a single discipline: anthropology.

Thus the readings assembled here have as much relevance for so-called "cultural anthropology" as they do for "physical anthropology." I have attempted to select papers which illustrate how anthropologists treat the relation between human culture and human evolution. I have also chosen papers by nonanthropologists that consider areas of research from which anthropologists can profit in constructing a generalizing "science of man."

In this collection, Part One deals with evolution, genetics, and natural selection. The population concept in biology is discussed in the context of the history of Western thought. It is a typological way of thinking that multiplies the difficulty for students in understanding the modern concept of evolutionary units—populations. Here the processes of organic evolution are viewed, together with some of the major kinds of evidence that support today's "synthetic theory" of evolution.

Microevolutionary processes, such as those that differentiate clusters of population from each other at individual gene loci, are described in concrete examples drawn from current research. Some of the more interesting chromosomal abnormalities and their phenotypic expressions are described. The purpose of this section is to help the student obtain some insight into the principles and agencies of biological evolution, as geneticists and other evolutionary biologists have been able to establish them.

Part Two deals with the evolutionary significance of primate studies. The interest of anthropologists in the evolution of human behavior has led to a need to know more about the behavior of monkeys, apes, and other nonhuman primates. The genetic capacity for behavior is as much the result of evolution as the genetic capacity to grow body hair or produce blood antigen B. Behavioral evolution has left only a fragmentary fossil record, and the archeological record fills in only some of the blanks. Thus the evolution of behavior must be reconstructed using all available evidence, including that of its prehuman base.

Defined biologically, man is characterized, not simply by the body structures that differentiate him from other forms, but also by his behavior. "Man is a cultural animal" sums it up. The evolutionary anthropologist is interested in tracing the ways in which particular populations of animals came, in the course of time, to be arrays of human bodies behaving in human ways. The dynamics of human evolution, the part that the capacity for culture has played in human evolution, are discussed in Part Three, "The Dynamics of Hominid Evolution."

It used to be thought that if only there were enough fossils, the course of human evolution would become clear. Thus, for a long time, anthropologists relied primarily on fossil-collecting and the comparison of fossils with living forms of man and his closest relatives, the apes. Yet it is now evident that the fossil record itself is simply not enough; it must be interpreted according to rules and principles that are not at all apparent in the study of the fossils themselves. In this section, the student can become acquainted with some of the problems of describing and interpreting the fossil record of human evolution. Natural selection as the prime evolutionary force is the major principle enabling paleontologists and archeologists to make sense out of the welter of bone fragments and broken artifacts that make up their respective raw data.

Similarly, the articles in Part Four, "Human Populations and their Genetic Diversity," reflect the evolutionary approach to human differences. Evolution is a continuing process, and natural selection did not simply occur in the past as a series of biohistorical events.

In the same way that natural selection and the other evolutionary processes have produced man as a separate species, they have created differences among modern human populations. It used to be thought that the small-scale changes differentiating populations developed through the operation of one set of rules, while the large-scale changes that differentiated man from other animals operated according to another. Here again, the synthesis of genetic evolutionary research with the long-standing interest of physical anthropologists in racial differences has produced a new understanding of how populations within our species have come to differ from one another. Kelso's term for this, *heterography*, is an apt one.

Part Five, "Is Man Still Evolving?" suggests some of the problems and possibilities of ameliorating the "human condition" through a knowledge of the evolutionary dynamics of human populations.

The task of editing a collection of articles such as this for the use of beginning students is as much a question of what not to include as of what to choose. In looking over the possibilities among the publications of the last decade or so, I was surprised at how many papers originally written for scientific audiences lent themselves to reprinting. Other editors might easily have chosen the same number of other equally useful selections.

Regretfully, I have had to omit examples of some important and dramatic directions of research. Recent work in molecular genetics is distinguished as much by its inaccessibility to students lacking biochemical background as by its research possibilities and the intensity of the controversies it has generated. The reader will also find unrepresented the valuable work being done by archeologists in adding to our knowledge of the evolution of man. It seems to me that all introductory texts in general use deal with the archeological record at a depth perfectly adequate for most students.

Years ago I took a graduate course with a distinguished anthropologist. The course was conducted in the time-honored format of a seminar; each week a student would read an original paper to the assembled group. The professor would then take the lead in making comments, suggestions, and criticisms. Invariably he would open up on a trembling student scientist in the same way: "Your paper," he would intone, "raises more questions than it answers." The way he said it left no doubt that he intended a negative judgment. To raise more questions than one answered was apparently some kind of anthropological heresy—not a serious one, to be sure, but worthy of pointing out.

In the interviewing years it has been brought home to me, over and over again, that a paper that raises more questions than it answers may be a very good paper. It is often one that students can read with profit, that can motivate their thinking and assist them in their own learning. Possession of this quality is one of the criteria that have governed the selection of materials for this third edition of *Human Evolution: Readings in Physical Anthropology*.

Each edition has functioned in a different way for students and teachers. The first edition (1959) appeared at a time when few of the available textbooks had integrated the "new physical anthropology" with the traditional concerns of the discipline. That edition, made up of articles, book chapters, and snippets from here and there, was designed to help teachers bridge the gap between the textbook and the new concerns of ecology, the population approach, and evolutionary processes. The selections were primarily content-laden, for the convenience of many teachers of physical anthropology who were actually specialists in other fields.

The second edition, published in 1967, reflected the new directions in the anthropological study of free-ranging monkeys or apes, functional anatomy, and reconstruction studies based on the fossil hominid record. Rather than reprinting papers that presented data, this edition provided a selection of theoretical articles to illuminate areas of disagreement and to reflect on the promising beginnings of the new synthesis in anthropological inquiry.

This third edition makes its appearance in a different climate of anthropological culture. For one thing, there are now many up-to-date textbooks from which the teacher may select to serve as the basic reading for his course. For another, the explosion of graduate training in physical anthropology and primate studies has resulted in a wide distribution of specialists in the "new" physical anthropology in American colleges and universities. The great accomplishments of the 1960s have been carried out, moreover, against the background of an ideological ferment in all the sciences of man. Phylogenetic explanations of what used to be considered as cultural phenomena have proliferated. The reading public outside of the classroom has come to perceive anthropology as capable of providing an evolutionary and ethnological rationale for many of the events that have brought man to his present crisis. Aggression, "law and order," war, the revolution in sex attitudes, as well as other social developments, have been accompanied by a shake-up in the public's confidence in rationality, planning, and decision-making. People both in the colleges and outside have embraced such "Original Sin" hypotheses of behavior as Ardrey's popular *African Genesis*.

Indeed, many primate studies have lent themselves easily to interpretations of genetic programming of behavior. It is really only in the last few years that an ecological emphasis on primate behavior has begun to dominate the field, with an insistence on the flexibility of species-specific behavioral potential, and an exploration of ecological contributions to behavioral phenotypes. Thus, a major function of this collection is corrective. Some of the selections represent a kind of "revisionism," in recognizing the role played by ecological factors in theorizing about primate social behavior.

Another American cultural trait has reappeared: the social myth of the intellectual inequality of those populations that anthropologists a generation ago used to identify with such glibness as "races." The resurgence of this idea, armed now with protocols of thousands of test situations and reinforced also by a crisis in our political attempts to solve ethnic minority problems in America, cannot be ignored by teachers of anthropology. Some of the papers selected for this edition have a bearing on the issues intrinsic to this controversy.

I have also attempted to achieve a better balance between two emphases in contemporary anthropological research: microevolutionary studies of contemporary populations on the one hand, and paleoanthropology on the other. While this twin emphasis is apparent in the professional literature, it has been less apparent in textbook materials. Articles have been chosen with this disparity in mind.

Earlier editions have deleted the bibliographies provided by the authors, in the belief that students in introductory courses could dispense with the references to the literature. However, in response to requests from many instructors, this edition reprints the bibliographies as furnished by the authors.

The reader will also find introductions to the articles that direct the student to specific questions raised by each selection. Moreover, the number of selections has been reduced to permit a more intensive treatment of a smaller number of topics.

In preparing this edition I have had frequent occasions to recall with fondness my collaborators on the earlier editions: Harry Reese Smith and Fred W. Thompson. Katherine Franklin, Fred's daughter, has assisted with the many tasks involved in editing this collection of readings for student use. My wife, June, has supported me throughout with affectionate thoughtfulness and good humor.

And lastly, I should like to thank all those students who have bravely and candidly asked "So what?" so many times in so many classes, and thus have played a major role in directing my attention to the relevant and humane aspects of anthropology as a science of man.

Van Nuys, California   N.K.

CONTENTS

PART ONE

EVOLUTION, GENETICS, AND NATURAL SELECTION

Our present-day understanding of evolution has developed from many different kinds of research in the biological sciences. Advances both in theory and in the accumulation of information have made clearer the nature of the processes by which changes in hereditary material are distributed and redistributed within populations, and by which new kinds of populations can come into being. Close study of the fossil record by paleontologists has been enlarged by insights from the field of ecology: the study of evolution has benefited from understanding how animal and plant populations are related to one another in space.

Several principles may be said to be the product of contemporary research into evolution. All have implications for the study of human evolution, implications that are now being pursued and tested. First is the realization that *evolution is what happens to populations*. The meaningful unit

1

of study in evolutionary biology is the population, the naturally interbreeding group of organisms more or less isolated reproductively from other similar groups.

Second, the study of evolution concerns the study of *dynamics* or of *processes*. Evolutionary anthropology has swung away from research as "sorting out the results of evolution," in Sherwood Washburn's felicitous phrase, toward attempts to understand the nature of the relationships of populations to each other and to the environment over spans of time. Paleontology also has been vastly enriched by this shift in emphasis.

Furthermore, evolutionary processes did not simply occur in the past, but operate continuously through the mechanism of heredity. Natural selection, for example, the major agency of evolutionary change, does not merely act to redistribute genetic material in the gene pools of populations as a response to alterations in their ecological niches; it acts to maintain an equilibrium between the genetic potential of a population and its conditions of life. There are probably no such things as structures that are "evolutionarily neutral." The hereditary characteristics of a population and their distribution at any given time are expressions of the ecological relation between that population and its ecological niche at that time. (In the face of this realization, we might question the validity of the interpretation of modern "racial differences" as simply being adaptations to conditions of 700 generations ago—that is, the late Pleistocene.)

We are, moreover, becoming increasingly aware that the human fossil record itself is comprehensible in evolutionary terms only if we try to understand the changes in behavior that correspond to the changes in structure of the hominids through the last two million years.

In the selections that follow, the major cornerstones of the modern synthetic theory of evolution are interpreted. Dobzhansky specifies what geneticists and evolutionary biologists mean by their use of the term "population." Stebbins reviews the principal agencies of evolutionary change and discusses the evidence on which our understanding of them is based. Livingstone describes studies of evolutionary dynamics of human populations. Heller describes some of the chromosomal abnormalities related to dysfunction, which constitute a part of man's "genetic load."

1

ON TYPES, GENOTYPES, AND THE GENETIC DIVERSITY IN POPULATIONS

Theodosius Dobzhansky

A basic precept of the theory of evolution is that the unit of evolutionary change is the population, *the array of individuals who make up a reproductive community. Individuals are born, contributing their genes to the gene pool that constitutes the behavioral and structural potential of the next generation, and individuals die. The potential development of each generation is limited by the genetic constitutions of precursor generations. The mix—the variety of genotypes that characterize the gene pool of a population— may be stable, or under the pressure of natural selection, the proportions of genotypes within the gene pool may change. The population is the unit we study for evidences of stability and change.*

Darwin took variation, or the diversity of a population, as his starting point for considering the action of natural selection. But as Dobzhansky shows in the following article, the intellectual climate of the last 2500 years has made it difficult for students of evolution to integrate the concept of population into the ways we view the world. Every chart of the "characteristics" of a group of organisms, human or infrahuman, implies a fixed type, to which individuals may be compared, and against which they fit, as "examples" or "exceptions." If we think this way about evolutionary change, then it is very difficult to deal with the problems of raciation, speciation, or even variation within populations.

In this article Dr. Dobzhansky traces the roots of "typological" thinking in the intellectual history of the Western world, and shows why the concept of "type" is inadequate in dealing with the shifting data of life. In reading the article, note particularly how

Reprinted from J. N. Spuhler, editor, *Genetic Diversity and Human Behavior* (Chicago: Aldine Publishing Company 1967). Copyright © 1967 by the Wenner-Gren Foundation for Anthropological Research, Inc. By permission of the author and the publisher.

*the implications of population thinking for human evolution are
developed: the significance of the concept for understanding the
original divergence of hominid populations in the evolution
of early hominids, the evolution of man's capacity for culture,
and finally, contemporary human diversity.*

*Theodosius Dobzhansky is Professor at the Rockefeller University
in New York. A life-long student of genetics and evolution, he
has in recent years sought to apply his background to an understanding of the heredity and evolution of man.*

Genetic differences in man, like those between individuals of any animal or plant species and those between species, are all products of the
evolutionary development of the living world. These differences, with their
behavioral consequences, can only be understood in the light of evolution,
sub specie evolutionis. Our understanding of evolution, however, is itself
evolving. The Darwin-Wallace theory of evolution appeared more than a
century ago. The development of the evolutionary thought has gone
through several stages. It was necessary, first of all, to demonstrate beyond reasonable doubt that evolution had, in fact, taken place. Roughly
the first four decades after Darwin, until about 1900, were dominated by
studies in comparative anatomy, embryology, systematics, zoogeography,
phytogeography, and paleontology, designed to discover and examine the
evidences of evolution. During the same period, the first sketchy attempts
were made to trace the paths that the evolution of the living world actually
followed. The evidences of evolution are now so well known that they are
part of the basic college and even high school courses of biology. On the
contrary, the phylogenies of the animal and plant kingdoms, and particularly the phylogeny of man, are far from completely known, and they are
actively studied today.

The study of the causes that bring evolution about came to the forefront
of attention at the turn of the century. The first three decades of the
current century were mainly a period of data-gathering. This was the time
of the rapid growth of basic genetics, and genetics was more often than
not studied for its own sake, rather than because of its bearing on evolutionary problems. Nevertheless, although a causal analysis of evolution
grounded in genetics may well be wrong, one flying in the face of the
established principles of genetics cannot possibly be right. The next twenty
years or so, roughly from 1930 to 1950, saw the gradual formulation of
the modern biological, also called the synthetic, theory of evolution.
Adumbrated by Chetverikov in 1926, the basic principles of the biological
theory were arrived at almost simultaneously by Fisher, Haldane, and
Wright around 1930. For the first time in the history of biology a biological theory was deduced mathematically from, ultimately, a single fundamental premise—Mendel's law of segregation. The theory was soon

shown to possess a great explanatory power in biology as a whole. Within two decades Mayr, Simpson, Rensch, Schmalhausen, Stebbins, Darlington, White, and others demonstrated that the theory makes sense of experimental genetics, paleontology, zoological and botanical systematics, cytology, comparative morphology, and embryology. More recently, the theory has been applied successfully in anthropology, ecology, physiology, and biochemistry. It has been strengthened by the spectacular developments in molecular genetics.

Historical perspectives tend to become blurred when one approaches the present. Where are we now and where are we going? Has the theory of evolution congealed into a dogma? Far from that, evolutionary thought is now in a period of rapid development and change. I agree with Mayr (1963) that "the replacement of typological thinking by population thinking is perhaps the greatest conceptual revolution that has taken place in biology." My only cavil is that Mayr's "has taken place" is overoptimistic; I do believe, however, that this conceptual revolution is under way.

The differences between typological and populational thinking are often as subtle as they may be profound. Many biologists are completely unaware of these differences, and yet Mayr is right that "virtually every major controversy in the field of evolution has been between a typologist and a populationist." To a typologist, the diversity of living individuals is illusory. Individuality is an accidental deviation from the one real, constant, and unchangeable type of the species to which the individual belongs. It is the basic species type that a scientist should endeavor to discover and to understand. To a populationist, concretely existing individuals, not types, are the biological entities. Individuals may even be the ontological entities. The differences between individuals must be investigated, not ignored, and explained, not explained away.

Averages are statistical abstractions, and types are at best convenient models or devices that an observer utilizes to facilitate the description and communication of the data he has gathered. The use of "types" for this purpose is, of course, a legitimate procedure. A textbook of zoology can hardly avoid using schemes, or "body plans," of a worm, an insect, a vertebrate animal, and so forth. Such schemes or "types" are abstractions that represent the common features of a group of organisms and omit the characteristics by which they differ. In a sense, any classification involves some kind of abstraction, and in some fields of study (for example, archeology and linguistics), the word "typology" is used in a way that makes it a near synonym of what in biology is called "classification." Classification and schematization, however, cease to be legitimate, and become specimens of typological thinking, when the classifier or schematizer forgets that the animals, plants, and people exist before and independently of any schemes. Typological thinking is what in philosophy is called "reification" of abstractions. Populational thinking, especially when applied to man, becomes closely related to individualism. It could also be called existential thinking, if it were not that the word "existentialism" in recent years has been used in so many different senses that it has become utterly ambiguous.

PHILOSOPHICAL ANTECEDENTS

The roots of both the typological and the populational approaches go deep in the history of human thought. As Bronowski (1956) beautifully put it, "Science is nothing else than the search to discover unity in the wild variety of nature—or more exactly, in the variety of our experience. Poetry, painting, the arts are in the same search, in Coleridge's phrase, for unity in variety. Each in its own way looks for likeness under the variety of human experience." The oldest and most important instrument whereby the runaway variety is bridled is in symbolic language. Words like "man," "cat," or "dog" refer not to particular persons or animals but to representatives of mankind, cat-kind, and dog-kind. Human language is a spontaneous typologist.

The diversity of persons, cats, and dogs is nevertheless inevasible. Moreover, individual persons and animals change and grow old. It is tempting to declare the diversity and change to be false fronts or spurious appearances. Parmenides did so about 24 centuries ago. A similar result was achieved by Plato more gently, by his famous theory of ideas. God has created the eternal, unchangeable, and inconceivably beautiful prototypes, or ideas, of Man, of Horse, and even of such mundane and inanimate objects as Bed and Table. Individual persons, horses, beds, and tables are only pale shadows of their respective ideas. Acquisition of wisdom is a way to catch a glimpse of the ideas that one formerly saw only as shadows.

Platonic philosophy, of course greatly modified in form but still true to its spirit, has survived the centuries and millennia. It is a living influence today. Aristotle did not like the Platonic ideas, but assumed one cosmic idea which manifests itself in the visible world. The realization of the ideal form is the purpose of nature, most clearly observable in the living world. All animals are variants of a single architectonic plan. For more than two millennia, until the time of Darwin, the organic world was viewed as either Plato or Aristotle saw it. The living, and even the nonliving, bodies formed a "Great Chain of Being" (Lovejoy, 1960), ranging from a lesser to greater perfection. Leibniz in the seventeenth and Bonnet in the eighteenth centuries felt satisfied that the Chain is single and uninterrupted. Bonnet saw it starting with fire and "finer matters," and extending through air, water, minerals, corals, truffles, plants, sea anemones, birds, ostriches, bats, quadrupeds, monkeys, and so to man.

To Lamarck and to Geoffroy St. Hilaire, but not to their predecessors and contemporaries, the Great Chain implied evolution. Has the world actually evolved through the stages corresponding to the links of the Chain? In the famous debate between St. Hilaire and Cuvier in 1830, the issue of evolution was stultified by being mixed up with the Great Chain. Cuvier was, of course, absolutely right that there is no discernible single plan of body structure common to all animals. The hypothesis that living organisms are manifestations of a limited number of basic types or ideas of structure nevertheless proved to be very useful. It inspired studies on comparative anatomy and classification from Belon, through Buffon, Daubenton, Goethe, Oken, and the school of German *Naturphilosophie*, to St. Hilaire, Karl von Baer, and Richard Owen.

There is no doubt that the data collected by these investigators furnished a base for the doctrine of evolution. It is, however, questionable whether they can validly be regarded as pioneer evolutionists. Goethe's *Urpflanze* was not claimed to have actually grown and flowered anywhere in particular; still more important, the existing plants were not claimed to have the *Urpflanze* as their actual ancestor; the *Urpflanze* was merely a blueprint according to which the "creative power" (*schopfende Gewalt*) designed the existing plants. That Baer and Owen strenuously opposed Darwin should come as no surprise. As late as 1876, Baer insisted that "one must clearly distinguish the ideal from the genetic or genealogical kinship," and, while favoring the former, denied the latter. Owen was even more explicit; his morphological archetypes were expressions of Plato's eternal ideas (see Zimmermann, 1953, and Kanaev, 1963, for anthologies and discussions of the history of evolutionary thought).

The roots of the populational thought are less easily traced than those of typology. The former arises primarily in connection with observations on human beings, rather than with studies on animals and plants. Human individuality is stressed in most religions, and is one of the tenets of the Christian world outlook. The situation is complicated because that outlook has also a typological component—the doctrine of Original Sin, by which human nature is held to be ineluctably debauched. A geneticist is tempted to imagine that the fruit that Adam and Eve so imprudently ate contained some powerful mutagen, which induced a heavy genetic load in all their descendants. It is, however, an individual person who commits sins; we possess free wills, and are accountable for our actions. Pelagius stated this most clearly (too clearly, it would appear, since he was declared a heretic):

> Everything good and everything evil, in respect of which we are either worthy of praise or of blame, is done by us, not born with us. We are not born in our full development, but with a capacity for good and evil; we are begotten as well without virtue as without vice (quoted after Dunham, 1964).

The formal philosophical codification of human individuality came only in 1785, in Kant's doctrine that every human being is an end in himself.

This doctrine is, however, implicit in the belief that every man has certain inalienable rights; this belief became popular during the Age of Enlightenment, and it serves as the foundation of the theory of democracy. In general, the typological view of man goes most easily with conservative, and the view that every man is an end in himself with liberal political persuasions.

But man's efforts to know himself are often frustrated by his remarkable capacity to deceive himself. Quite inconsistently, the conservatives also favor the view that different people are genetically different, while the liberals prefer the belief that man is at birth a *tabula rasa*, a blank slate, and that all differences are the products of upbringing and education.

EVOLUTION VS. TYPOLOGY

Darwin wrote that his great work, *The Origin of Species*, was "one long argument." The heart of this long argument was "that species are only strongly marked and permanent varieties, and that each species first existed as a variety." Varieties, hence, are not merely diverse manifestations of the same archetype, of the Platonic idea of their species; they are themselves incipient species, and the existing species are aggregations of varieties. Now, Darwin used the word "variety" in a sense so inclusive that it became ambiguous. His "varieties" ranged from what we would call subspecies or races—that is, genetically distinct populations occupying definite geographic territories to the exclusion of other races of the same species—to breeds of domestic animals and cultivated plants, and down to polymorphs and aberrant individuals found within an interbreeding population. This ambiguity proved so confusing that "variety" is no longer used as a technical term. In Darwin's "long argument," however, his use of the word was not unreasonable; what Darwin set out to demonstrate was that there existed within a species genetic raw materials from which natural selection could compound new species.

Platonic types exist in some realm above the world of human senses; they are intrinsically static and unchangeable. The theory of evolution has shattered the typological statics, and replaced it with a populational dynamism. The living world is changeable and capable of transformations; some of the transformations may represent progress and others degradation. Moreover, the changeability is not a false front or an illusion, as the followers of Parmenides and Plato thought; it may produce real novelties. An individual changes with age, but this change follows, within limits, a fixed pattern. A population may change in a variety of directions; it owes its plasticity to the genetic diversity within it. Genetic differences in man signify that the human species has changed, and may change in the future.

Typological thinking is un-evolutionary or anti-evolutionary. With the general acceptance in biology of evolutionism, typological thinking should have been, it would seem, replaced by populational thinking. This has not happened everywhere, because established habits of thought are often refractory to change. Not all biologists are aware of the profound incompatibility of evolution with typology. As pointed out above, biological classification, and perhaps any classification, goes easily with typological thinking. If one strives to delineate species or races, the presence of intermediate forms is a nuisance that one is tempted to shrug off whenever possible. Borderline cases between race and species are a godsend to an evolutionist but an annoyance to a museum man, whose business is to write a label for every specimen in his collection. Although the usefulness of classification is unquestionable, a maximal neatness of the taxonomic pigeonholes is not the ultimate aim of systematics. A modern systematist classifies the living world in order to understand it, and he is forced to pay more than lip service to evolution.

To illustrate the subtle but pervasive differences between the typological and the populational approaches, two examples may be considered briefly. The current civil rights movement in the United States has prompted the racists to devise a curious quasi-scientific justification of their social and political biases. It is alleged that, compared with whites, Negroes have a smaller average brain size and lower average scores on various intelligence and achievement tests. That these differences are questionable, and if real may be of environmental origin, need not concern us here. The point is that not even racists deny that all these measurements vary among whites as well as among Negroes. The variation ranges overlap so broadly that the large brains and the high scores in Negroes are decidedly above the white averages, and the small brains and the low scores in whites are much below the Negro averages.

Now, the typological reasoning goes about as follows. All Negroes are but manifestations of the same Negro archetype, and all whites of the same white archetype; no matter what his brain size or his intelligence score may be, an individual is either a Negro or a white, and should be treated accordingly. Some racists go so far as to claim, apparently in all sincerity, that it is unkind to Negroes to treat them as though they were white. The populational approach is simply that an individual is himself, not a pale reflection of some archetype. Every person has a genotype and a life history different from any other person, be that person a member of his family, clan, race, or mankind. Beyond the universal rights of all human beings (which may be a typological notion!), a person ought to be evaluated on his own merits.

One aspect of the controversy started by the recent book of Coon (1962) may be considered as our second example. According to this author, the species *Homo erectus* consisted of five subspecies, each of which was transformed into a subspecies of *Homo sapiens*, the transformation taking place in different places and at different times. Just what can this claim possibly mean? In a symposium held at Burg Wartenstein in 1962, and in a book published before Coon's (Dobzhansky, 1962, 1963), I argued that, at least from mid-Pleistocene on, there was but a single hominid species; that anagenesis (transformation) has been more important in the evolution of the human stock than cladogenesis (splitting); and that the species *Homo erectus* gradually evolved into *Homo sapiens*. Coon believes, however, that *sapiens* is an evolutionary "grade," which can be achieved independently, and earlier by one race than by another.

A species is an inclusive Mendelian population, a reproductive community, not a type or a grade. Becoming a new species is not instantaneous, like getting a new degree or being admitted into an exclusive club. The population of a species gradually changes in genetic composition, sometimes more rapidly and at other times more slowly. Eventually the accumulated genetic alterations become so clearly expressed in the fossil remains that a paleontologist is forced to apply a new label, in other words, a new species name. A typologist is then prone to assume that an individual to which a certain species label is applied is always more like any other individual that goes under the same species label than it is like an individual of any other species. This may be true only of species living contemporaneously, and only in a sense that an individual does not as a rule belong to more than one biological reproductive community. Even so, with respect to some genes an individual may resemble a representative of another species more than he resembles his sibling. For example, some humans and some chimpanzees have the A blood type and others have the O blood type. When an ancestral species is transformed in time into a descendant species, the criterion of reproductive community is no longer applicable. One obviously cannot tell whether individuals who lived many generations apart could or could not interbreed. They most certainly did not do so. An individual or a population of *Homo erectus* at a certain time level may then be more like one called *Homo sapiens* on the next higher level, than a more ancient population labelled *Homo erectus*. They were living populations, not typological "grades."

THE *TABULA RASA* THEORY

Racism is a form of typological misjudgment which assumes that an individual is a manifestation of a racial type. Mankind consists of several races or subspecies; the differences between the races are important, the differences among individuals within a race are trivial. However, another form of typology recognizes but a single type—that of the species, *Homo sapiens*. The classical statement of this view is John Locke's *tabula rasa* theory. This is frequently misrepresented as claiming that all people are similar at birth. Now, Locke was one of the most levelheaded among philosophers, and he did not claim anything so rash as that all infants are born alike. What he maintained is, rather, that there are no inborn "ideas." In his words,

> Let us then suppose the mind to be, as we say, white paper, void of all characters, without any ideas; how comes it to be furnished? . . . All that are born into the world being surrounded with bodies that perpetually and diversely affect them, variety of ideas, whether care be taken about it or not, are imprinted on the minds of children.

The *tabula rasa* theory has had a great and abiding influence on the climate of "ideas" in which the civilized part of mankind lives. It has been adopted by the thinkers of the Age of Enlightenment, from Voltaire and Helvetius, Rousseau and Condorcet, to Bentham and Jefferson. It has been, and continues to be, an important ingredient in the philosophies of democracy. Its attractiveness lies in that it seems to uphold an optimistic view of human nature. A lot of people behave stupidly, wretchedly, and viciously. This is, we are assured, because they were badly brought up, and spoiled by corrupting influences of badly organized society. A better-organized society, better care, guidance, and education will, however, make everybody behave in accordance with the natural goodness and reasonableness that are the normal and universal endowment of every representative of the human species. Genetic differences among people have no behavioral consequences, at least not in a democratically organized society. I believe that it can be demonstrated that, on the contrary, the assumption of an innate uniformity of all human beings makes nonsense of democracy. Though this problem lies outside the frame of reference of the present discussion, I shall make some brief comments about it below.

It is no exaggeration to say that different variants of the *tabula rasa* theory are entertained, explicitly or implicitly, by most social scientists, and by some influential schools of psychology, especially those with psychoanalytical learnings. Ruth Munroe lists as "the fourth basic concept accepted by all psychoanalytic schools" that

> Early childhood is the time when the malleable, adaptable, flexible human psyche takes on the essential direction it will pursue. Because the human infant is so unformed by nature, because these experiences are the first, the events of infancy have a psychological importance very naturally overlooked by the adult philosopher in his study or the adult scientist among his instruments.

An idea carried to extremes may suddenly turn into its opposite. So it has happened with the *tabula rasa* concept. We have seen that some early Christian philosophers, particularly Pelagius, used an idea very similar to the *tabula rasa* notion to affirm human individuality, in opposition to those who maintained that all human natures are equally depraved by Adam's original sin. The same idea in its modern guise makes the human nature uniform, and reduces individuality to the status of a veneer applied by the infant-rearing practices and by circumstances of a person's biography.

The trouble is that the *tabula rasa* theory starts with a valid premise but goes on to draw an erroneous conclusion. The valid premise is that the behavioral development of *Homo sapiens* is remarkably malleable by external circumstances. Man's paramount adaptive trait is his educability. In the broad sense in which it is used here, *educability* means that man is able to adjust his behavior to circumstances in the light of experience. The genotype of the human species has been so formed in the process of evolution that the educability is a universal property of all nonpathological individuals. This universality, or near-universality, is a product of natural selection. It has made possible the human cultural development. In all cultures, primitive and advanced, the vital ability is, and always was, to be able to learn whatever is necessary to become a competent member of the group or society in which the individual happens to be born or is placed by circumstances.

Environmental plasticity, however, is not incompatible with genetic diversity. Genetic differences do have behavioral consequences. To a geneticist this consideration is trivial, but he is constantly forced to remind his colleagues, especially those in the social sciences. It cannot be stressed too often that what is inherited is not this or that "trait" or "character" but the way the development of the organism responds to its environment. This is particularly important when what is being considered is the genetic determination of the diversity in behavioral traits. The sex cells obviously cannot transmit "behavior"; what they do transmit are genes, or, if you wish, chain molecules of deoxyribonucleic acids, which determine the pattern of the development of the organism, including the pattern of its behavioral development. The key word in the preceding sentence is "determine," and it must be understood correctly.

Genes determine the pattern of the development in the sense that, given a certain sequence of environmental influences, the development follows a certain path. The carriers of other genetic endowments in the same environmental sequence might well develop differently. But the development of the carrier of a given genotype might also follow different paths in different environments. The observed diversity of individual phenotypes would be more limited than it actually is if all people were as similar genetically as identical twins, but the environments were as manifold as they actually are; the diversity would be also more limited than it is, given the existing genetic variability but a uniform environment. To put it in another way, the observed phenotypic variance has both genetic and environmental components.

SOME PARAMETERS OF GENETIC DIVERSITY

Individuals of which a species is composed may be regarded as more or less faithful or defective incarnations of the species archetype. This is the typological approach. In modern biology it usually takes a disguise. For example, it is postulated that there exists somewhere, or can be obtained by breeding, the optimal genotype (= archetype), composed entirely of genes adaptively most favorable in a given environment, or even in all environments in which the species can live. Individuals observed in reality deviate more or less noticeably from the postulated possessors of the optimal genotype; they carry "genetic loads," which, in theory, can be measured by their deviations from the optimum genotype. The populational approach envisages a biological species as a Mendelian population composed of carriers of diverse genotypes. No genotype can really be "optimal," because the population faces a great variety of environments, inconstant in space and in time and with different adaptive requirements. How great is the genetic diversity within a given species is at best only sketchily known. In particular, the genetic diversity in the human species is explored quite inadequately. The matter needs more investigation, research, and analysis. This is one of the pivotal problems in the present book, and what I am trying to do in this introductory discussion is to place it in its proper perspective.

According to Ford's (1940) original definition, an interbreeding population is said to be genetically polymorphic if it consists of two or more clearly distinct and genetically different kinds of individuals. Some species and populations are obviously highly polymorphic, others less so, and still others seem to be monomorphic. The matter is, however, beset with complications. Our ability to recognize polymorphisms depends upon the techniques used. Before 1900 there was little reason to suppose that human populations are polymorphic for blood groups, because techniques were not available for diagnosing them. New biochemical polymorphisms are now being discovered in man almost annually, and the end is not yet in sight.

The immense field of polygenic variability is explicitly excluded by Ford's definition of polymorphisms, because polygenes, also by definition, do not produce discontinuous and phenotypically discrete variants. Yet at the biochemical level there is presumably no difference between the major genes and the polygenes. Some geneticists are loath to venture from the solid ground of discrete gene effects to the apparent morass of polygenes. The classical methods of genetics, recording the Mendelian segregation ratios in the progenies of hybrids, are not easily applicable to polygenic inheritance. Biometrical methods, selection experiments, and heritability determinations have to be used. Some geneticists have even recommended that a kind of moratorium be declared on studies of polygenic inheritance, until such time as it can be reduced to molecular terms and then presumably handled by the classical methods. I do not think this recommendation stands a chance of being adopted, for the simple reason that a majority of the differences between normal, nonpathological humans happen to be manifestations of polygenic differences. This field of study is too important to be relegated to an indefinite future time. Most relevant to our discussion, the nonpathological genetic variations in behavior traits in man seem to be mostly under polygenic control.

Compared to the number of persons who ever lived, the diversity of potentially possible genotypes in man is so vast that no more than a minute fraction of this diversity could be realized. The number of the genes in a human sex cell is not known. It could hardly be less than 10,000, which is the estimate, also not too reliable, usually given for the number of genes in *Drosophila*. With only two variant alleles per gene, this makes $2^{10,000}$ potentially possible genetically different gametes, and $3^{10,000}$ potentially possible zygotes. Many, or most, gene loci may be capable of giving more than two alleles per locus. Although not all potentially possible combinations have equal probabilities of actually being formed, the possible diversity is practically infinite. With every person (identical twins excepted) being genotypically different from every other, an immense majority of potentially possible genotypes will never be formed in reality.

The existing genotypes are, obviously, not just a random sample of the potentially possible ones. Even assuming that the mutation rates in some genes are as low as one per billion per generation, most genes will have actually mutated somewhere in the human species. However, the mutant genes will not necessarily persist in the populations. The dominant lethal mutations arising in a given generation will be extinguished in that same generation. Some of the spontaneous abortions, stillbirths, and neonatal deaths represent eliminations of the dominant, or semidominant, lethal mutant genes. A dominant mutation that would make its carrier unable to reproduce would not technically be called a lethal, but it will likewise be eliminated within a generation after its origin. This may be of interest to students of behavioral abnormalities which interfere with the reproduction of their carriers. Hutchinson (1959) called such abnormalities paraphilias; whether any of them are expressions of dominant mutant genes remains to be discovered.

Complete lethality, absolute sterility, and complete dominance are limiting cases. Semilethal and subvital genotypes, various degrees of lowered reproductive capacity, and semidominant and incompletely recessive genetic conditions are much more numerous in reality. Mutational changes giving rise to such conditions are of great interest and importance. The mutant genes persist in populations for an average number of generations, which is greater or smaller depending on the degree of the reduction of viability, reproductive capacity, and dominance. Completely recessive mutants, which according to some geneticists are very rare, will, of course, persist longer than incompletely recessive ones of the same degree of harmfulness. Simple mathematical formulas have been worked out relating these variables. It is not necessary for us to consider them here; what is important is that all living populations carry genetic loads consisting of more or less drastically deleterious mutants not yet eliminated by natural selection. The part of the genetic load maintained by the mutation pressure against selection is termed the mutational load.

The genetic variants that compose the mutational load are important components of the variability we observe in populations. This is because the genetic variants that cause slight or minute decreases of the adaptive values of their carriers, as measured by their reproductive performance, may be retained for quite considerable numbers of generations. Suppose that having a nose of an esthetically unappealing shape reduces the reproductive value of its carrier very slightly, say by one-thousandth, 0.001. If this nose shape is due to a single dominant gene arising by mutation once in 100,000 gametes (0.00001), then about 2 per cent of the population will have this nose shape. If a single recessive gene is responsible, then 10 per cent of the gametes will carry the gene but only 1 per cent of the population will have unattractive noses.

It is evidently difficult to obtain evidence for or against the hypothesis that the morphological, physiological, and behavioral variants found in human or other populations confer on their possessors adaptive advantages or disadvantages of the order of, say, 10 per cent and lower. This difficulty explains the persistence in population genetics of conflicting theories of population dynamics and particularly of genetic loads. This matter is one of crucial importance for understanding the nature of the genetic variation in mankind and in other biological species, and I shall discuss it as concisely and as objectively as I can.

The "classical" theory assumes that most of the unfixed, variable, genes are represented in populations by two or more alleles, one of which is normal, typical, and adaptively superior to the others; these latter make their carriers, if ever so slightly, inferior in fitness. Heterotic alleles, which make the heterozygotes superior to the homozygotes, are assumed to be rare. A genotype homozygous for all the normal alleles would evidently have the optimal fitness. The process of mutation, however, keeps injecting into the gene pool of the population deleterious mutant alleles; the deleterious alleles as yet uneliminated by the normalizing natural selection form the genetic load of the population; and the genetic load consequently is a mutational load. Ingenious statistical methods have actually been devised to measure the deviation of the genotypes from this theoretical optimal genotype.

The "balance" theory regards the optimal genotype as typological fiction. The populations of sexually reproducing and outbreeding species are arrays of diverse genotypes for several reasons other than the mutation process interfering with the achievement of the optimal genotype. In the first place, it is unrealistic to imagine that any population exists in an absolutely uniform environment; since a population has to face a variety of environments, some genotypes highly fit in an environment A may be inferior in B or in C, and vice versa. For example, some genotypes may be resistant to certain infectious diseases, others may be able to get along on less food, still others give brave warriors, or peaceful citizens, or obedient slaves. A polymorphic, or genetically diversified, population which contains all these genotypes may have an adaptive advantage over a genotypically uniform one. The diversifying, also called "disruptive" (in my opinion, an unsuitable name), natural selection will maintain or increase the adaptive polymorphism. However, some or all of the genotypes from time to time may be confronted with environments unfavorable for them, such environmentally misplaced genotypes will act as constituents of the genetic load. This form of genetic load will be maintained in the population not by the pressure of recurrent mutation but by a balancing natural selection.

Another form of balancing natural selection can maintain a genetic diversity in a population even in a uniform environment. Suppose that the heterozygote, A_1A_2, for two gene alleles or gene complexes, A_1 and A_2, is heterotic, that is, adaptively superior to both homozygotes, A_1A_1 and A_2A_2. The heterotic form of natural selection will maintain both A_1 and A_2 in the population. This will be true even if one or both homozygotes are severely handicapped or lethal. The relative frequencies of A_1 and A_2 in a constant environment will depend upon the relative fitness of the two homozygotes; the speed with which these equilibrium frequencies will be attained will depend upon the degree of heterosis in the heterozygote. If the heterozygous and homozygous genotypes are clearly distinct phenotypically, we have the simplest form of balanced polymorphism.

This introductory article cannot describe several other forms of natural selection that maintain the genetic diversity in populations irrespective of mutation pressure. The major unsettled and controversial problem in population genetics is whether most genetic diversity in populations is maintained by the mutation pressure and counteracted by the normalizing natural selection, or whether an appreciable, and perhaps a major, part of this diversity is sustained by various forms of balancing natural selection.

The classical and the balance theories of population structure have rather different implications for studies on the genetic variables of human behavior. Suppose that the genetic diversity in human populations is principally mutational in origin as contended by at least some partisans of the classical theory. Most individuals in a population would then be expected to be homozygous for "normal" alleles at most gene loci. The genetic diversity would be limited to a minority of loci, which in some individuals would be represented by at least one allele being a more or less deleterious mutant, waiting to be cast out by the normalizing natural selection. Although the mutant alleles might be numerous in the aggregate, the mutants at any one gene locus would usually be present in a small minority of individuals. It then would not be inconceivable that one would somehow apprehend the characteristics, including the behavior features, of that elusive entity, the carrier of the normal, typical, and optimal genotype of the species. All deviations from these typical features would then be accidental imperfections, to be eliminated as far as possible.

If the bulk of the genetic diversity in a species is balanced, the optimal genotype becomes a will-o'-the-wisp. At best, it is an abstraction convenient for some kinds of mathematical calculations. The genetic diversity will then be expected to be not only greater than the classical theory would suggest, but its character will also be different. No one genotype being the "optimal" or "normal" one, there is only an array of genotypes that make their carriers able to live and to reproduce successfully in the environments inhabited by the species. This array of genotypes may be designated as the adapative norm of the species or population. It must be stressed that the "adaptive norm" is an array of forms not identical in their fitness in any one environment; some may be fitter than others in a certain environment, but less fit in other environments.

The ill-adapted, pathological variants are the expressed genetic load of the population, and the exceptionally vigorous or outstandingly successful or valuable variants, its genetic elite. Again, the role of the environment must not be lost sight of; a genotype belonging to the genetic load in one environment may shift into the adaptive norm in another environment. The same is true, of course, of the genetic elite. It is, then, the diversity of the genotypes and phenotypes, including the diversity of behaviors, and the factors that maintain the genetic variance in the populations, and not the "typical" nor even the statistically average behavior, that is most important to study.

It would be out of place to attempt here an evaluation of the evidence at present available for and against the classical and the balance theories of population structure. As stated above, the issue is a controversial one. It is, however, safe to say that the classical and the balance theories are not alternatives, in the sense that one of them must be wholly right and the other wrong. The problem is to what extent each of the two theories describes a real situation. It is undeniable that some genetic variants are maintained in populations by balancing selection, particularly by its heterotic form; it is likewise undeniable that some mutant genes are deleterious in all existing environments, both when homozygous and when heterozygous. The occurrence of such mutants in populations must be due to a mutation pressure.

Furthermore, no single solution of the problem can be valid for all living species. If, for example, it were demonstrated that most of the genetic variability found in populations of *Drosophila* is balanced, it would not follow that the same is true in human populations, or vice versa. Not only different species but even ecologically different populations of the same species may well contain different proportions of mutational and balanced variability. Comparative genetics is a line of study logically as well justified as comparative anatomy, comparative physiology, and comparative psychology.

GENETIC DIVERSITY IN EVOLUTIONARY PERSPECTIVE

Much of the foregoing discussion has dealt with genetic diversity in general, rather than with the genetic diversity in man. Genetic differences cannot be divided rigorously into purely morphological, purely physiological, and exclusively behavioral. It is a platitude that any genetic difference is necessarily a physiological and hence biochemical difference, because genes can act in no way other than through chemical messengers. The converse is not necessarily true—not every physiological difference is reflected in morphological or behavioral traits. Many, though not quite all, genetic differences produce what are known as pleiotropic, or manifold, effects; they alter two, or several, traits that do not have any obvious developmental interdependence. Darwin's classic example of pleiotropism (not called, of course, by this name) was that white cats with blue eyes are deaf. Many hereditary diseases, in man and elsewhere, produce complex syndromes of symptoms, and these syndromes often include more or less dramatic changes of behavior. Others are reflected in behavioral traits only very indirectly. For example, the ideal of feminine beauty has varied in different places and at different times all the way from linear to obese. That body build is genetically conditioned is not subject to doubt, but what behavioral consequences a given sort of body build will have depend on culturally developed tastes.

A developmental interdependence of traits may not be easily apparent, but it may exist nevertheless. There must be a reason why white fur and blue eyes go together with the deafness in cats. On the other hand, is the gene, or genes, that modify both the color of human hair and the color of the iris of the eye pleiotropic? The old idea that some forms of behavior in man go together with certain facial features and other "stigmata" has given rise to numerous superstitions, but the possibility of such correlations based on pleiotropic effects of genes should not be dismissed out of hand. The main point here is the need to be on one's guard against the hoary fallacy that there must be one gene for every trait and one trait for every gene. If you are puzzled why natural selection has established as species characters, and even as characters of higher categories, some apparently useless morphological traits, remember that the useless traits may be parts of pleiotropic syndromes that also include less conspicuous but more vitally important components.

The phenomenon of pleiotropism must be kept in mind in connection with the controversial problem of the so-called constitutional types, somatotypes, and racial types. As a geneticist sees it, the kernel of this problem is this: Are human populations polymorphic for genes, or for linked groups of genes sometimes called supergenes, that determine complexes of traits inherited as units? Sheldon and his followers claim that these complexes consist not only of constellations of morphological but also conjoined psychic traits. One could then imagine that there exist three genes, or gene complexes, forming multiple allelic series, which determine respectively the so-called endomorphic, mesomorphic, and ectomorphic "components." This hypothesis was submitted to a test only once, by Osborne and DeGeorge (1959). They somatotyped 59 pairs of identical and 53 pairs of fraternal twins, and had their typing confirmed by Sheldon. The heritability of the "components" proved to be low and, in addition, not equal in females and in males. This negative result may only mean, of course, that the criteria for the "components" have not been chosen correctly. Even more dubious are the claims of a Polish school of anthropologists (for example, Czekanowski, 1962, and Wiercinski, 1962), that they can distinguish in Polish and other European, and also in African, populations, a limited number of racial types inherited as units. No attempt to test the validity of this claim by using the twin method has been made, as far as I know.

As stated above, the genetic diversity that underlies the behavioral variations is not in a class by itself; it is similar in principle to that underlying the physiological and the morphological variations. On the other hand, it should always be kept in mind that the genetic and the evolutionary processes that control behavior have, in the human species, certain important peculiarities that make them different from those in any other species. It has already been pointed out above that one of the diagnostic attributes of the human species is genetically conditioned educability. The hallmark of humanity is the ability to learn, to profit by experience, and to adjust one's behavior accordingly. This developmental plasticity of human behavior is also man's most basic adaptive feature; without it, human social and cultural evolution would have been impossible. An individual whose genotype deprives him of this plasticity is an obvious misfit and belongs to the genetic load of the population in which he has appeared. The plasticity of learned behavior in man contrasts with the relative fixity of preponderantly instinctual behavior in other animals, especially insects. This does not mean that human behavior is independent of genetics; it does mean that the genetic system in man is constructed in a way that is unique in the living world.

The developmental fixity or plasticity of a trait is genetically conditioned. In general, a fixity is a property characteristic of traits or qualities, the presence and the precise form of which is indispensable for survival and reproduction. The developmental processes that give rise to such traits are said to be homeostatically buffered, or canalized; in other words, they can take place in about the same way over the entire range of the environments that the species normally encounters in its habitats. The environmental variations within that normal range do not influence the phenotype of the organism to any appreciable extent. With very few exceptions, babies are born with two eyes, a four-chambered heart, a suckling instinct, physiological mechanisms maintaining constant body temperature, and so forth. The genetics of developmental canalization has been studied experimentally in suitable materials, chiefly *Drosophila* flies. We can form a fairly good idea about how it is achieved in the evolutionary process.

There are, however, characters in which fixity would be disadvantageous, and a plasticity is beneficial. If a *Drosophila* larva is given ample food, it develops into a fly of a certain size. If the food is scarce, although above a certain minimum, the starving larva does not die; it pupates and gives an adult insect of a diminutive size. The number of eggs deposited by a *Drosophila* female on abundant food and at favorable temperatures may easily be ten times greater than with scarce food and an unfavorable temperature. A fixity of the body size and of the number of eggs produced would evidently be disadvantageous; instead of stabilizing these traits at certain values, natural selection has destabilized them, made them contingent on the environment in which an individual finds itself. Experimental evidence on how such a destabilizing selection may work is limited, but a good paper was recently published by Prout (1962).

The genetic endowment of *Homo sapiens* guarantees the development in all individuals belonging to the adaptive norm of the species of such capacities as educability, symbolic thinking, and communication by symbolic language. It is perhaps justified to say that human evolution has been dominated by a stabilizing selection for these capacities, and for a consequent destabilization of the overt observable behavior. This is responsible for the illusion that man at birth is a *tabula rasa* as far as his prospective behavioral development is concerned. In reality, the educability goes hand in hand with a genetic diversity. Any society has diverse functions and callings. With the development of civilizations the diversity of vocations has not grown smaller; on the contrary, it has increased, and continues to increase, enormously. The division of labor in a primitive society is chiefly between sexes and among different age groups; in a civilized society it is between individuals, occupational groups, guilds, social classes, or castes.

The trend of cultural evolution is obviously not toward making everybody have identical occupations, but toward an increasingly differentiated occupational structure. What is the adaptively valid response to this trend? First of all, the human genetically secured educability makes most individuals trainable for most occupations. Almost everybody, if brought up and properly trained, could become a fairly competent farmer, a craftsman of some sort, a soldier, a sailor, a tradesman, a teacher, or a priest. It is, however, eminently probable that some individuals, because of their genetic endowments, are more easily trained to be soldiers and others to be teachers. It is even more probable that only some individuals have the genetic wherewithal for certain specialized professions, such as a musician or a singer or a poet, or for peak achievement in sports or wisdom or leadership.

Human educability is traditionally emphasized by those who espouse liberal political views, while the genetic differences are harped upon by conservatives. As pointed out above, this is sheer confusion. The main tenet of liberalism, it seems to me, is that every human being, every individual of the species *Homo sapiens* is entitled to equal opportunity to achieve the realization of his potentialities, at least insofar as they are compatible with the realization of the potentialities of other people. Now, an equality of opportunity is needed not because people are all alike but because they are different. If they were all alike in their capacities and tastes, it would not really matter who is embarking on what career; the careers and vocations could as well be distributed by drawing lots. The diversifying form of natural selection, however, made the human populations polymorphic for tastes and capacities. A class or caste society leads unavoidably to the environmental misplacement of carriers of different genotypes. The biological justification of equality of opportunity is that it minimizes the chance of loss to the society of valuable human resources, as well as the chance of personal misery resulting from environmental misplacement.

BIBLIOGRAPHY

Bronowski, J. 1956 *Science and human values.* New York: Julian Messner.
Coon, C. S. 1962 *The origin of races.* New York: Alfred A. Knopf.
Czekanowski. J. 1962 The theoretical assumptions of Polish anthropology. *Current Anthrop.,* 3:481–494.
Dobzhansky, Th. 1962 Genetic entities in hominid evolution. In S. L.

Washburn (Ed.), *Classification and human evolution*. Chicago: Aldine.
1963 *Mankind evolving*. New Haven, Conn.: Yale University Press.
1964 *Heredity and the nature of man*. New York: Harcourt Brace.
Dunham, B. 1964 *Heroes and heretics*. New York: Alfred A. Knopf.
Ford, E. B. 1940 Polymorphism and taxonomy. In J. Huxley (Ed.), *The new systematics*. Oxford: Clarendon Press.
———— 1964 *Ecological genetics*. London: Methuen–John Wiley.
Hutchinson, G. E. 1959 A speculative consideration of certain possible forms of sexual selection in man. *Amer. Natur.*, 93:81–91.
Kanaev, I. I. 1963 Essays on the history of comparative anatomy before Darwin. (In Russian) Moscow-Leningrad: Academy of Sciences U.S.S.R.
Lovejoy, A. O. 1960 *The great chain of being*. New York: Harper.
Mayr, E. 1963 *Animal species and evolution*. Cambridge: Belknap Press.
Osborne, R. H., and F. V. DeGeorge 1959 *Genetic basis of morphological variation*. Cambridge, Mass.: Harvard University Press.
Prout, T. 1962 The effects of stabilizing selection on the time of development in *Drosophila melanogaster*. *Genet. Res. Camb.*, 3:364–382.
Wiercinski, A. 1962 The racial analysis of human populations in relation to their ethnogenesis. *Current Anthrop.*, 3:2–46.
Zimmerman, W. 1953 *Evolution, die Geschichte ihrer Probleme und Erkenntnisse*. Frieburg-München: K. Alber.

2

THE DYNAMICS OF EVOLUTIONARY CHANGE

G. Ledyard Stebbins

Bisexual reproduction carries on the processes of heredity, ensuring a general similarity from one generation to the next. Cats have kittens, pumpkins produce pumpkins, dinosaurs hatch dinosaurs. But on the other hand, these same processes of heredity provide opportunities for change; built into the replication of DNA, the

Reprinted from J. Ives Townsend, editor, *Lectures in Biological Sciences* (Knoxville: The University of Tennessee Press, 1963). Copyright © 1963 by the University of Tennessee Press. By permission of the author and the publisher.

duplication of chromosomes, cell division, meiosis, and fertilization are the possibilities for variation and change that may be exploited by natural selection and the other forces of evolution.

Population geneticists, zoologists, and other biological scientists have put together the picture we now possess of how biological evolution occurs: the "synthetic theory" of evolution. The major outlines of that picture are now pretty clear; continuing research is filling in the details of how variation, natural selection, random genetic drift, and gene flow act to transform the hereditary properties of populations through time.

To summarize the ways in which evolution operates, Dr. Stebbins here reduces the outlines of evolutionary theory to seven postulates, and then goes on to review experimental evidence, and evidence from outside the laboratory, to demonstrate how we know what we know.

In granting permission to reprint this paper, Dr. Stebbins recently wrote: ". . . I find that during the twelve years since it was prepared, my views have not changed at all, and it expresses well my present point of view. In fact, a considerable series of experiments and observations made during this interval have provided additional support for it."

Thus, while new research may fill in the details and the background of the picture of biological evolution, it is unlikely that its major outlines will be significantly altered by the acquisition of new information.

G. Ledyard Stebbins is Professor of Botany at the University of California, Davis.

Since the time of Charles Darwin, the concept of evolutionary change has permeated the entire field of biology. It is accepted without question by all working biologists, because of the immense body of facts which now demonstrates its existence. Furthermore, change through time is now recognized as one of the guiding principles of the universe, from stellar galaxies to human societies. Biologists of modern times are relatively little concerned with showing that change has taken place. Even the directions which evolutionary changes have taken in particular groups of plants and animals have proven to be a less rewarding field of study than the mechanism of evolutionary change. The challenging question which most modern evolutionists are trying to answer is: "What forces or processes can bring about the evolution of a population of organisms into a new type, with a different appearance and a changed way of life?"

This was, of course, the main question which Darwin himself was trying to answer. To do this, he developed the theory of natural selection, for which he was able to gather together a body of factual evidence, much of it experimental, which in his day appeared impressive enough to convince most biologists that his views were correct.

But in the period after Darwin's death, and particularly in the early part of the twentieth century, experimental biology made great strides, both in the facts about living things which it uncovered and the techniques of investigation which it developed. The biologists of this time, moreover, were concerned chiefly with the mechanism of life. This mechanism can be understood best by intensive studies of individual organisms, while evolution is best understood by means of the comparative study of populations. Consequently, biological techniques and standards of experimentation far outstripped those which were used by Darwin and which were adequate in his day, without contributing much to our understanding of evolution.

This fact explains in part the low opinion which many biologists of a generation ago held in regard to Darwin's theory of natural selection. In addition, the rediscovery of Mendel's laws of inheritance focused attention upon unit characters, or conspicuous differences between organisms which are inherited as single units. When the origin of such unit characters by genetic mutation was demonstrated in the evening primrose (*Oenothera*) by De Vries and in the fly *Drosophila* by Morgan, this seemed to many biologists to spell the doom of the natural selection theory, based as it was on the accumulation through selection of small differences between individuals.

The past thirty years have seen a great revival of interest in Darwinism, and an added confidence not only in the correctness of the evolutionary theory itself, but also in Darwin's explanation of natural selection. This revival has been brought about by scientists who have applied modern quantitative and experimental methods to the sphere of action in which the evolutionary processes are actually at work. This is the genetics of natural populations. Population genetics has erected three great pillars which can support the edifice of evolutionary theory. First, it has shown that multiple-factor or polygenic inheritance is the way in which most differences between natural populations are inherited. This leads to the corollary that, as Fisher and many others have maintained, mutations with relatively small effects, which contribute to the almost imperceptible differences between individuals upon which Darwin relied, play a larger role in evolution than the conspicuous mutations observed by De Vries and Morgan. Second, population genetics has accumulated a great body of evidence showing that the selective value of any single gene depends partly upon the other genes with which it is associated in any individual. This fact disposes simultaneously of any theories which would make either the individual organism or the gene the basic unit of evolutionary change. Mendelian segregation tells us that the individual is merely the temporary home of a collection of genes which is partly scattered and assembled anew with each successive generation of individuals. The rarity of mutations makes each gene a relatively permanent entity, but as an evolutionary unit its significance is reduced by its changing adaptive values. Because of this fact, Dobzhansky and others have correctly recognized the "Mendelian population" in sexually reproducing organisms as the basic unit of evolutionary change. The third pillar erected by population genetics is the quantitative, experimental approach to the study of selection itself. In the last century, evolutionists tried to establish the validity of natural selection chiefly by showing how a particular characteristic, such as body color, could have a selective advantage in a given environment. They failed to realize that the interactions between an organism and its environment are extremely complex, so that the way in which a characteristic is advantageous is often indirect and very hard to demonstrate conclusively. Population genetics has taught us that the best way to show the action of natural selection is to demonstrate statistically that a given gene, or gene combination, has an adaptive advantage in a known, controlled environment. This demonstration is relatively easy in organisms which are well known genetically, and can sometimes lead to a further study of how selection acts.

Supported by these three pillars, evolutionists have erected a theory of how evolution works which has often been called the "synthetic theory," and sometimes the "Neo-Darwinian theory." Natural selection still serves as its cornerstone, but the mutation theory, Mendelian genetics, and the statistics of population dynamics are almost equally important components of its foundations. The symposia which were held in various parts of the world during the year of the Darwinian centennial, 1959, impressed most of their participants with the degree to which biologists actively working in the field of evolution are agreed on the soundness of this theory. A generation ago, many leading biologists felt that none of the processes which had been invoked to explain evolution was satisfactory and that the true explanation would come with the discovery of some hidden "cause" not then known. Now most biologists agree in believing that the major causes of evolution are known, but that no one of these causes is sufficient in itself. Mutation, genetic recombination, and natural selection are complementary to each other. The task of the future lies chiefly in showing how these processes act in relation to each other and to the changing environments through which the earth has passed and is passing.

SEVEN BASIC POSTULATES

At the Cold Spring Harbor Symposium of 1959, the present writer became so strongly impressed by the solidity and implications of this widespread agreement about the major processes of evolution that he undertook to formulate a series of seven basic postulates which seem to express the extent of the agreement and may serve as solid foundations for future working hypotheses about how the processes of evolution operate. The discussion presented there served as a summary of the symposium, which it concluded, and was therefore illustrated solely with examples taken from that symposium. In the present article, the same postulates will be discussed in a broader fashion, with examples selected from all of the evolutionary literature as most clearly illustrative of the postulate being discussed. One must remember, however, that, in a discussion as brief as the present one must be, only a tiny fraction of the available evidence can be reviewed.

The first basic postulate can serve as a summary of the present introduction, or of the entire discussion. Its validity depends upon that of the other six, so that no particular evidence needs to be cited in its favor. It is stated as follows:

At least in higher animals and plants, evolution proceeds principally as the result of the interaction between four indispensable processes: mutation, gene recombination, natural selection, and isolation.

THE ROLE OF MUTATION AND GENE RECOMBINATION

The next two postulates are stated as follows:

Second, mutation neither directs evolution, as the early evolutionists believed, nor even serves as the immediate source of variability upon which selection may act. It is, rather, a reserve or potential source of variability which serves to replenish the gene pool as it becomes depleted through the action of selection.

Third, the mutations which are most likely to be accepted by selection and so to form the basis of new types of organisms are those which individually have relatively slight effects on the phenotype, and collectively form the basis of polygenic or multiple-factor inheritance.

These two postulates are complementary to each other. Together, they express the conclusion that in sexually reproducing organisms both mutation and genetic recombination are essential contributors to the gene pool. This can be defined as the supply of hereditary variation in a population on which natural selection can act. The existence of a great store of variability in the gene pool is best shown by two types of experimental evidence: first, from experiments with artificial selection and second, from experiments which have uncovered concealed genetic variability by obtaining certain chromosomes or chromosome segments in the homozygous condition.

To understand the meaning of recent experiments with artificial selection, we should first make predictions as to what the limits of selection would be on the basis of two contrasting hypotheses. One hypothesis assumes that the gene pool is very small, so that selection changes populations by increasing the frequency of favorable mutations just as soon as they occur, and cannot be effective unless favorable mutations are constantly occurring at a reasonably high rate. We may call this the hypothesis of *direct selection* of favorable mutations. Either by direct statement or by implication, we can recognize this hypothesis as basic to the thinking of De Vries, Johannsen, Bateson, Morgan, and most geneticists of a generation ago.

The second hypothesis assumes that the gene pool is very large, so that, if a population is placed into a new environment and subjected to selection in a new direction, new gene combinations are sorted out from the supply of genes already present in the gene pool, and selection can be effective for many generations even if no new mutations occur. We can call this the hypothesis of *indirect action,* since it assumes an indirect rather than a direct connection between mutation and selection.

If evolution usually proceeds by direct selection of new mutations, then the limits of selection would be set by the rate of favorable mutations, which is always very low. Selection would change the population for only a very few generations. This assumption was made by De Vries, Bateson, and particularly Johannsen, and was apparently borne out by Johannsen's well-known experiments with the garden bean. On the other hand, if the gene pool of most populations is large and diverse, then selection can be expected to change a population in the same direction for many generations, even if no new mutations occur at all.

Recent selection experiments with corn, *Drosophila,* mice, chickens, and other cross-fertilizing populations of higher plants have consistently shown that such populations are capable of responding to selection for many generations on the basis of the genetic variability stored in them. Two good examples are the Illinois corn experiment and the selection experiments by Mather and his associates for the number of abdominal bristles in *Drosophila.*

The Illinois experiment, described by Woodworth and Jugenheimer, was as follows: In 1895 the agronomists at the University of Illinois decided to find out the number of generations in which they could produce a change in a characteristic by continuous artificial selection, using a cross-fertilizing population of field corn. They selected for four different characteristics—high protein of the kernels, low protein, high oil content of the kernels, and low oil. This experiment was continued for fifty generations; the most recently published results were obtained in 1959 (the experiment was briefly interrupted during the last war). In all four lines the population responded to selection for at least thirty-five generations. In the case of high oil and low protein, a significant change took place in the populations even between the forty-fifth and fiftieth generations. During the experiment, the protein content was more than doubled in the high-protein line and reduced to less than half of the original concentration in the low-protein line. Even greater results were obtained by selecting for high and low oil content. The kernels of the original population contained 4.7 per cent of oil. After fifty generations of selection, the mean oil content in the high line was raised to 15.4 per cent, and in the low line it was lowered to 1.0 per cent.

The facts which we know about mutation rates in corn tell us definitely that the Illinois agronomists must have been sorting out genetic differences which existed in their original population. Since in each line of corn being selected the number of plants raised per generation was between 200 and 300, the total number of plants raised during fifty generations in each line was between 10,000 and 15,000. We do not know the actual rates of mutation for changes in oil or protein content. But the data of Stadler on rates of mutation for other characteristics in corn tell us that for a particular characteristic the occurrence of one mutation in 50,000 plants is a relatively rapid rate. Hence the occurrence of even one mutation in the desired direction during the course of the experiment is rather unlikely. The slow, steady way in which the populations responded to selection shows that many genetic differences were being sorted out. Since these differences could not have arisen by mutation during the course of the experiment, they must have existed in the gene pool of the original, unselected, but cross-fertilizing population.

The experiments of Mather and his associates on selection for high and low number of abdominal bristles in *Drosophila* gave similar results. In these experiments the response to selection was less regular than in corn, and the undesirable secondary effects, particularly the appearance of a high degree of sterility, were more marked. The actual response was in a series of short bursts, lasting five to twenty generations, and separated by intervals of several generations during which no response to selection occurred. Nevertheless, selection was effective even after one hundred generations, and the calculations of Mather and Harrison, like those given above for the corn experiment, show that the response to selection was based upon genetic differences already present in the gene pool of the initial population, rather than mutations occurring during the course of the experiment. Furthermore, Breese and Mather tested six different marked regions of one chromosome for genes affecting the difference between the line selected for high and that for low bristle number. They found such genes in every region, which suggested that the gene pool contained a very large supply of genes affecting bristle number, at many different loci. Indirect evidence leading to the same conclusion was obtained by Clayton and Robertson, who found that an inbred line of *Drosophila*, which would be expected to contain a smaller gene pool than the crossbred population used by Mather and Harrison, responded very slowly to selection for increased bristle number.

The experiments of Goodale and MacArthur on selection for body size in mice, and those of Lerner on shank length in chickens have yielded similar results, indicating that large gene pools for various characteristics are normally present in populations of cross-fertilizing higher organisms. The experiments of King on selection of *Drosophila* for resistance to DDT showed that the gene pool of this fly contains so many genes which affect DDT resistance that, if two lines derived from the same original wild population are kept apart and subjected to simultaneous but independent selection for resistance, they will each acquire a different combination of genes for DDT resistance.

A second type of experiment which has explored the nature of the gene pool is the analysis of hidden variability by artificial inbreeding, usually accompanied by special ways of manipulating the chromosomes so that their role in determining this variability can be understood. Most of these experiments have been conducted on various species of *Drosophila* by Dobzhansky and his associates. They have shown us that natural populations of this fly not only possess a great store of hidden genetic variability, but also that this variability is highly organized. Genetic linkage serves to bind together adaptive combinations of genes on particular chromosome segments. Often the adaptiveness of a particular linked combination depends not entirely upon its own properties but on its interaction with another combination in a different, but homologous, chromosome. Such chromosome segments are said to be co-adapted. A fly containing such pairs of interacting homologous chromosomes has heterosis or hybrid vigor. In many species, the integrity of such combinations is preserved by means of chromosomal rearrangements, particularly inversions of chromosomal segments. The fact that natural selection can increase the frequency of chromosomal rearrangements because of their value in promoting hybrid vigor has been demonstrated experimentally by Dobzhansky and Pavlovsky, and Levine. Furthermore, Levene, Pavlovsky, and Dobzhansky have shown that this complexity of the gene pool strongly affects the adaptive value of genes. The adaptive value of a gene can increase or decrease not only through changes in the external environment, including the other living organisms with which the population is associated. Even if these factors remain constant, a gene can change markedly in adaptiveness as it becomes associated with different genes during the continuous process of segregation and recombination which goes on in any cross-fertilizing population.

These facts have led Dobzhansky to the "balance hypothesis" of population structure. According to this hypothesis, adaptiveness is normally maintained through a combination of genes and chromosomal segments kept in the heterozygous condition. Individual genes do not usually respond directly to selection as separate units, but rather their adaptive value depends upon the way in which they contribute to the gene pool.

Large numbers of genes are thus kept in the germ plasm because their effects are neutral or actually beneficial in heterozygous combinations with other alleles, when these same genes would be strongly disadvantageous or actually lethal if they were present in the homozygous, or "pure," condition.

The experiments of modern population genetics, therefore, tell us that natural selection acts indirectly upon a spectrum of variability generated largely by genetic recombination which takes place in a very complex fashion. Mutation serves to replenish the store of variability as it becomes depleted through selection. Most of the genes which enable a population to change in response to a new environment are not newcomers which have recently appeared through mutation; they are "old-timers" which presumably originated by mutation many generations ago and have been preserved either as hidden recessives or as contributors to some past adaptive complex.

This principle does not mean that genes with large effects on the appearance of the individual never play an important role in evolution. On the contrary, some of the most striking demonstrations of the action of natural selection which have been made in recent years have involved such genes. The most striking example is industrial melanism. In the industrial areas of Europe, populations of moths have become completely transformed in a few years from light- to dark-colored types through selection of dominant mutations for dark color. The experiments of Kettlewell have shown conclusively that these changes were the result of natural selection and have told us much about how selection has acted. In other examples, resistance of insects to insecticides has been shown to result largely from the selection of single mutations with large effects.

In plants, genetic information recently obtained about the columbine, *Aquilegia*, suggests that the distinctive features of its flower were originally acquired through the occurrence and establishment of a single gene with large effects. The columbine differs from all of its relatives in that its petals bear long spurs containing nectar. This serves to attract animal pollinators, which may be bumblebees, hawk moths, or hummingbirds, depending upon the species of columbine. The related genera all have white flowers without spurs and are pollinated by various kinds of insects. A single species of *Aquilegia* closely related to the columbine, but without spurs on its petals, is found in eastern Asia. When this species was crossed with the common European columbine, the spur of the latter was found to be determined by a single dominant gene, although many different genes affected its length. We can suppose, therefore, that the new direction of evolution, which started long ago in the columbine genus and evolved species having flowers with a variety of shapes, sizes, colors, and spur lengths, was originally triggered off by the establishment in a population of a single mutation with a conspicuous and highly adaptive effect.

The Positive Effect of Natural Selection

The fourth basic postulate is stated as follows:

The role of natural selection is much more than the purely negative one of eliminating unfit types. By greatly increasing the frequency of gene combinations which otherwise have a very low chance of appearing, selection has an essentially creative and progressive effect.

The experiments on the genetic structure of populations which have already been reviewed can themselves be regarded as strong evidence in favor of this postulate. Even more striking evidence has recently been obtained from experimental studies of mimicry in butterflies. The phenomenon of mimicry, or resemblance in outward appearance between a distasteful or harmful species and an unrelated, harmless species, was used even in Darwin's day as evidence for the power of natural selection. Although some biologists in more recent times have been skeptical of the existence of this phenomenon, the recent experiments of Brower have established its validity beyond reasonable doubt. She offered specimens of the distasteful monarch butterfly to jays under carefully controlled conditions and showed that birds "trained" to avoid the monarch butterflies would also refuse to eat its mimic, the viceroy, although this butterfly was shown to be completely palatable to birds.

Sheppard has recently studied the genetic basis of mimicry in African butterflies of the swallowtail genus (*Papilio*) by extensive hybridization between the different mimicking forms which occur in a single species (*P. dardanus*), as well as crosses between mimetic and non-mimetic forms. These studies have shown clearly that mimetic forms will become more common and their mimicry will become more perfect in regions where the distasteful models are the most frequent. Where models are rare, mimicry is imperfect and variable within the same population, while in regions where the distasteful species is absent, the species consists entirely of non-mimetic forms. Crosses between mimetic and non-mimetic forms show that the general features of the mimetic pattern are determined by a single gene with a large effect, but that perfection of the mimicry is brought about by the action of many modifying genes, each with a small effect. In *Papilio dardanus* of central Africa there exist half a dozen or more different mimetic forms, each of which mimics a different species or subspecies of distasteful butterfly, with three different genera represented among the models. The different mimetic races are all determined by different allelomorphic genes at a single locus, with each of these major or "switch" genes giving the butterfly a superficial resemblance to a particular model. In regions where the model is abundant, the "switch" gene responsible for mimicry of that model has gathered around itself a collection of modifiers at many different loci, which have perfected the resemblance between mimic and model to a remarkable degree. The evolution of mimicry in these butterflies has, therefore, been brought about by natural selection of many different genetic changes, some with large and others with small effects. These have all been brought together into a harmonious combination which adapts the population to a special situation (i.e., the presence of a particular distasteful model). There is good reason to believe that this type of progressive action of natural selection on a number of highly specific genetic changes has been responsible for all of the extraordinary adaptations found in nature, such as the remarkable shapes and color patterns of such flowers as orchids and milkweeds, the unbelievably complex instincts of such animals as spiders and solitary wasps, and the social behavior of bees, ants, and primitive men.

The Origin of Species

The fifth and sixth basic postulates of the modern synthetic theory of the causes of evolution concern the problem of the origin of species:

Fifth, the continued separation of new adaptive lines of evolution from related lines with different adaptations requires the origin of barriers of reproductive isolation, preventing or greatly restricting gene flow between them. This separation is essential for maintaining the diversity of adaptations which exists in any one habitat, and so should be regarded as the basis of species formation.

Sixth, the origin of reproductive isolation, like that of new adaptive types, requires the establishment of many new genetic changes, including structural alterations of the chromosomes and cytoplasmic changes as well as gene mutations.

These two postulates place the origin of species on a different level from the origin of different adaptations or races of a single species. Both processes are based upon natural selection of genetic differences, but for the origin of species natural selection must sort out differences of a very special kind. These affect the behavior of a population toward certain specific related populations rather than its adaptation to the environment as a whole.

A recent symposium on the species problem has shown us that the problem of the nature and origin of species is by no means solved and that divergent opinions exist both as to what species are and how they came into being. Nevertheless, a large array of facts obtained from a great variety of experiments tells us that species are distinct from each other chiefly because they are separated by barriers which restrict or prevent gene exchange between populations.

The basic importance of reproductive isolation in maintaining adaptive diversity within a community of organisms is evident from the widespread occurrence of two phenomena which have been much studied in recent years, hybridization and character displacement. If two related populations which have been isolated from each other in different habitats are permitted by environmental changes to come together, then their subsequent evolution will depend upon the degree of reproductive isolation which has developed between them. If they are still able to cross and produce fertile hybrids, the result of their contact will be the production of an intermediate, hybrid swarm, in which the identity of the original populations will become lost to a greater or lesser degree. Numerous examples of this can be cited, particularly in higher plants. On the other hand, if their ability to intercross and produce fertile hybrids is so much reduced that each population can maintain intact its own adaptive properties, then the populations will tend to compete with each other. Since direct competition tends to destroy reproductive capacity, natural selection will favor genotypes of each population which are as differently adapted as possible from the norm of the competing population and so will cause the two populations to diverge from each other. Lack has shown that in the finches of the Galapagos Islands, a race of a particular species, if it is the sole occupant of a particular island, will have a rather wide and generalized range of variation. Another race of the same species, which on a different island is sharing its habitat with a distinct but related species, will have a narrower and more specialized range of variation. This phenomenon, known as character displacement, is also described in ants by Brown and Wilson, who review additional examples in birds.

The diversity and genetic complexity of reproductive isolating barriers are evident from a great wealth of observational and experimental evidence, which is summarized in part by Stebbins, Dobzhansky, and Mayr. If all of the barriers separating two related species are studied, they are found to be of various sorts, and any two species are usually isolated by many different kinds of barriers. Sometimes the species can be easily hybridized artificially, and fertile offspring can be raised under human supervision, but they nevertheless fail to hybridize in nature because of differences in their breeding seasons, mating instincts, or similar factors. The distinctness of various species of pines in California is of this nature, as is also the separation of the mallard from the pintail duck. In other instances, as in the wild rye (*Elymus*) genus of grasses and its relatives, natural hybrids are common, but they are so sterile that they rarely, or never, reproduce. More common, however, are examples in which two related species are separated from each other by a variety of barriers. A classic example is *Drosophila pseudoobscura* and *D. persimilis*, which are separated by different temperature requirements for maximum sexual activity, an instinctive tendency for females to mate with males of their own species, sterility of F_1 hybrids (which is complete in the male sex), and weakness or sterility of backcross progeny from the partly fertile F_1 females.

We can also see how diverse and complex the reproductive isolating barriers between two species can be by crossing two related species many different times, using as parents different races of the same pair of species. Such experiments have been performed in *Drosophila* and various groups of higher plants, such as the tarweeds, *Madiae*; the phlox family, *Polemoniaceae*; and the grasses. In each example, the same kind of result has been obtained. Both the ease of crossing and the fertility of the F_1 hybrid differ greatly depending upon the particular parental strains used. Any widespread species contains a great store of genetic variability, including not only genes which affect the adaptation of the population to its environment, but also those which help or hinder the ability of a species to cross and exchange genes with individuals of another species.

The genetic complexity of reproductive isolating mechanisms is shown by studies of segregation for fertility in later generations of hybrids between partly interfertile species. Two good examples are the hybrid between *Galeopsis getrahit* and *G. bifida*, of the mint family, analyzed by Müntzing, and that between *Drosophila pseudoobscura* and *D. persimilis*. In both examples segregation was very complex, and indicated that many different gene pairs were contributing to the sterility barrier. Less complete data from many other crosses point in the same direction. The inability of two species to cross and form fertile hybrids is not acquired through the appearance of one or two mutations with profound effects. Like other differences between races and species, it is built up gradually through the accumulation of many genetic differences, each one with a small effect.

The ways in which natural selection can build up reproductive isolating barriers are by no means fully understood, but evidence concerning them is accumulating. For instance, in both frogs and fishes, races of the same species which are adapted to growth under different optimal temperature conditions may produce abnormal embryos when crossed with each other. This is apparently the result of disharmony in interaction between genes controlling different rates and temperature optima of embryonic growth.

In plants, the sterility of interspecific hybrids is often due largely to differences between the parental species in chromosomal structure. Within certain species, such as *Trillium kamschaticum*, special cytological techniques have revealed many differences between both individuals and races in respect to small details of chromosome structure. These differences are partly correlated with both the climatic features of the present environment in northern Japan, where the species is common, and with the geological history of that region. Since chromosomal diversity in this species runs parallel with diversity in morphological characteristics, we can suspect that natural selection has operated in similar ways to bring about both kinds of differences. Although the chromosomal differences between the races of this particular species of *Trillium* are not great enough to cause hybrids between races to be sterile, the related species of *Trillium* differ from *T. kamschaticum* in respect to more numerous chromosomal differences of the same kind. Furthermore, although hybrids between diploid species of *Trillium* are not known, a wealth of evidence from hybrids between species in other plant genera shows that chromosomal differences such as those which exist between the diploid species of *Trillium* can be responsible for hybrid sterility. Furthermore, this sterility can be overcome by doubling the chromosome number. This places together in the same hybrid nucleus duplicate sets of chromosomes, which are consequently able to pair normally and produce viable gametes.

Once reproductive isolating barriers have arisen, they can be strengthened and reinforced by natural selection which favors those individuals of a species having an instinctive tendency to mate with others of their own species. This process has been fully discussed by Dobzhansky.

A final way in which species-separating barriers can be formed is as a secondary result of hybridization between pre-existing species. The process of amphiploidy, or the production of fertile, true-breeding species by doubling the chromosome number of sterile hybrids, is now well known to geneticists, and numerous examples have been described, both in the garden and in nature. In addition, more recent experiments have shown that highly sterile, though slightly fertile interspecific hybrids can produce more fertile offspring in later generations without change in chromosome number. Furthermore, some of the fully fertile strains which can be bred from such offspring may form partly sterile hybrids when crossed with either of the original parental species. This is apparently brought about by the effects of genetic recombination, which in rare instances may build up a new harmonious recombination of those genetic differences which were responsible for the original sterility barrier.

Half a century ago, many evolutionists believed that the critical experiment which would demonstrate our understanding of evolutionary processes would be the production under controlled conditions of a new species. This they defined as a population which would breed true and would be reproductively isolated by hybrid inviability or sterility from all other pre-existing populations. Amphiploids, or doubled hybrids, fulfill this qualification in every respect, and the partially isolated segregates with unchanged chromosome number which have been produced in *Elymus, Nicotiana,* and *Delphinium* come close to it. While the former constitute a special type, rare or lacking in the animal kingdom, segregates of the latter type might be expected in all types of organisms. Furthermore, since partial sterility is found in many hybrids between different races of the same animal species, the origin of new species from such hybrids would require the addition of only a few mutations to the extreme segregants which gene recombination can produce in the progeny of such hybrids. The origin of species, one of the most crucial steps in evolution, is well on the way to becoming understood.

THE MAJOR TRENDS OF EVOLUTION

The seventh and final basic postulate about evolutionary processes is as follows: The origin of genera and other higher categories, as well as the longtime trends which have given rise to increasingly complex and highly organized forms of life, results from the continuation into geologic spans of time of the processes responsible for evolution on the racial and species level. The only new element which must be considered is the increasingly evident extinction of populations intermediate between the successful lines.

The first type of evidence in favor of this postulate is the fact that some of the same kinds of differences which in one group of organisms may form the distinction between genera, or even families, can in a related group exist as differences between species of the same genus or even between races of the same species. This has been illustrated elsewhere for the grass family.

Even more convincing evidence has been obtained from careful studies of fossil lineages, particularly in mammals. This evidence is carefully reviewed by Simpson. He has shown that the earliest representatives of lines which led eventually to very different kinds of animals were so much alike that they would unquestionably be placed in the same group if their descendants were not known. For instance, ancestral lineages of the modern horse, rhinoceros, and the Biblical cony (*Hyrax*) can all be traced back through a succession of fossil forms to the Eocene period, fifty to sixty million years ago. At this period, these lineages have converged to such an extent that the forms representing them are much alike in size, shape, head form, and tooth structure. Furthermore, the different representatives of the same lineage which followed immediately after each other in time were usually so much alike that they could easily be visualized as having evolved through the accumulation of many relatively small genetic differences.

To be sure, many examples are known in which a new type of animal or plant appears suddenly and seems to be completely separate in respect to many large differences from any earlier fossil form. To explain these apparent saltations Simpson assumes that the fossil record contains many highly significant gaps. Furthermore, both his evidence and logical arguments suggest strongly that those conditions which would be most likely to bring about the origin of a new major adaptive complex and hence a new higher category would also be most likely to produce gaps in the fossil record. Organisms which exist as large populations in stable environments have the greatest chance of being preserved as fossils, but are the least likely to give rise to new adaptive types. New departures in evolution are most likely to occur when a system of relatively small populations, partly isolated from each other, is evolving in a rapidly changing environment. This combination of conditions is perhaps more unfavorable than any other for preserving such forms as fossils.

If, as Simpson believes, apparent saltations are produced by a combination of rapid evolution plus unfavorable conditions for fossilization, then those groups with the poorest fossil record should have the largest number of apparent saltations, and the improvement of our knowledge of the fossil record should progressively fill in the gaps. Recent progress in fossil discovery has shown this to be true in a striking fashion. In particular, the fossil history of man, which a generation ago had to be interpreted on the basis of fragments which could almost have been counted on the fingers of one's hands, now is illustrated by a variety of prehuman types, many of which are represented by a considerable number of individual fossils. As Le Gros Clark has emphasized, these new finds have definitely filled in some of the gaps. They have made it highly probable that the fossil primates which are waiting to be unearthed will eventually give us a continuous sequence extending from the ape- or monkey-like common ancestor which existed thirty or forty million years ago up to modern man.

SOME THOUGHTS ON AN
EVOLUTIONARY PHILOSOPHY

This discussion will close with some thoughts on how our present knowledge of evolution could affect our philosphy of life. The point of view adopted here is entirely personal, and I make no apologies for it. The facts upon which it is based are derived from a variety of sources, and most of the ideas have already been expressed by a number of other writers.

When we apply the concepts of evolution to our own past, we quickly realize that man is the product of two different kinds of evolution. Our bodies have evolved in the same way as those of other mammals, particularly primates. As mentioned above, the fossil record of man's ancestors is gradually being laid bare. We now can reconstruct with some assurance the way in which our ancestors first started to walk erect, to use tools and fire, and to hunt game in groups or primitive societies. We are sure that the size of the human brain increased gradually, as did also man's ability to make better tools. If, therefore, we consider only our bodies, we must conclude that we are no more than large apes that walk erect and have unusually large brains.

But our present way of life does not depend only upon our bodies and our brain power. Our minds and our social organization contribute far more to human nature than our bodies. Furthermore, our minds and social behavior, although based upon the foundation of our biological, genetic heredity, must nevertheless be reconstructed in each generation by learning. From our parents, our teachers, and from the leaders of our society we acquire a vast store of cultural heredity, which has been built up slowly and carefully by the thousands of generations of men who have preceded us. It is upon this heritage that our present way of life depends. When men first began to make tools, wear clothes, build shelters, and talk to each other, they set in motion a new kind of evolution, which we call socio-cultural evolution. Although built upon the foundations of organic evolution, socio-cultural evolution follows new directions and is governed by new principles. The familiar biological processes of mutation and genetic recombination are replaced by invention, learning, and cultural spread, or diffusion. Selection exists in socio-cultural evolution, but it makes progress through differential survival of customs and inventions rather than of men. It is thus radically different from the natural selection which guides organic evolution.

Finally, cultural evolution produces its effects in an entirely new way. Through organic evolution, organisms became modified to suit their environment; socio-cultural evolution enables man to modify the environment to suit his own needs. Animals became adapted to cold climates by developing fur; man, by building furnaces or by borrowing fur from animals. Birds became able to fly by growing wings and profoundly modifying their bodies; we fly infinitely higher and faster by elaborately designed machines, which even transport a bit of low-level, warm, temperate climate many miles above the earth.

Most important, organic evolution is opportunistic in direction. It is governed by the chance combination of environmental factors and the types of organisms which happen to exist at any one time. Socio-cultural evolution, on the other hand, is determined at least in part by man's own foresight and his ability to conceive of a better way of life for himself and his descendants.

We cannot overemphasize the fact that socio-cultural evolution is totally new in quality and has made man qualitatively different from all animals. This is true in spite of the fact that its beginning depended largely upon quantitative increases in brain power, and that in particular mental characteristics, like the ability to learn, memorize, and communicate with each other, we differ only in degree from the more intelligent kinds of apes. One of the most important facts about all of evolution is that from time to time new qualities emerge through more complex organization of simpler substances and systems. On the chemical level, we see this in many compounds which have properties very different from the chemical elements and simpler compounds of which they are built. The properties of salt and the elasticity of rubber are examples. Life itself differs from non-living matter only in having a special type of very complex organization. Since the dominant theme of cultural evolution has also been increasing complexity of organization, one need not be surprised that it has generated entirely new qualities. Our minds, our foresight, and our social structure, although they are the products of evolution, are nevertheless completely real and new. They set us apart from animals just as truly as if they had been specially created.

Another important fact about socio-cultural evolution is that it progressed for a very long time through traditions and learning which were passed down to each successive generation by word of mouth, without benefit of writing. Men have been able to speak to each other for at least five hundred thousand years; they have had brains as highly developed as ours for at least seventy-five thousand years; they have had such spiritual beliefs as that in an afterlife for at least fifty thousand years, as witnessed by ancient graves which include implements for use in the world to come. But writing as a means of perpetuating tradition is barely six thousand years old. Now everything we know about modern peoples who are not, or were not, able to write leads us to believe that among them reason dominates only the immediate events of their lives. Their social structure and their plans for the future are bound up in their emotions and are passed on from generation to generation by spoken rules, stories, chants, poems, and incantations, surrounded with the symbols of religious worship. We do not know how religion began, but we can be sure that it has guided man's evolution for at least a hundred thousand years. Before the advent of writing, the stability of society depended upon the ability of children to learn from their elders the spoken word, and this was developed largely through the force which the symbols of religion gave to certain essential moral precepts. The ability to receive these words and to accept these precepts must, therefore, have had as high a selective value in primitive society as any other characteristic. Spiritual qualities must have been essential to the earliest rational men. Consequently, we must think of man as basically spiritual, regardless of whether we believe that religion was given to him by a supernatural supreme being, or whether, as I believe, we consider that it evolved through the socio-cultural process. The ability of men to put their ideas into writing, thus rendering them much more precise and constant, has enabled us to substitute rational thinking for many of the superstitions of the older religions, and we have not reached the end of this process. Nevertheless, the ties which bind us to our traditional heritage, which enable us to work together, and which stimulate our dreams for the future are still made up largely of emotional and spiritual attachments, and the experience of those nations which have attempted to sever them and substitute purportedly rationalistic philosophies like Marxism has emphasized sharply for us the dangers of such a course.

Hence to the questions, "Why am I here?" and "What is the meaning of life?" I give these answers: Whatever mind or spirit that I possess, as well as the comforts of the civilization in which I live, has been given to me by the work, care, and ideals of my own parents and teachers, their parents and teachers, and so on back through the ages. I owe to them an immense debt, which I can repay only through following their examples, and, like them, learning how to work with my fellow men and to develop ideals and dreams which I can pass on to future generations. And because I believe that human progress has been shaped in the past not by the unalterable will of an inscrutable supreme being, but by the hopes, ideals, and working together of men and women like ourselves, I can hope that whatever I do that is of worth will make the world better for future generations than it would have been if I had not made the effort. This is the greatest satisfaction for which I can hope in either the present life or any conceivable future existence.

LITERATURE CITED

Breese, E. L., and K. Mather 1957 The organisation of polygenic activity within a chromosome in *Drosophila*. Heredity 11:373–395.

Brower, Jane Van Zandt 1958a Experimental studies of mimicry in some North American butterflies. Part I. The monarch, *Danaus plexippus*, and viceroy, *Limenitis archippus archippus*. Evolution 7:32–48.

———— 1958b Experimental studies of mimicry in some North American butterflies. II. *Battus philenor*, and *Papilio troilus*, *P. polyxenes*, and *P. glaucus*. Evolution 12:123–136.

———— 1958c Experimental studies of mimicry in some North American butterflies. III. *Danaus gilippus berenice* and *Limenitis archippus floridensis*. Evolution 12:273–286.

Brown, W. L., Jr., and E. O. Wilson 1956 Character displacement. Systematic Zool. 5:49–64.

Clark, W. Le Gros 1959 The crucial evidence for human evolution. Proc. Am. Phil. Soc. 102 (2):159–172.

Clausen, J. 1951 Stages in the evolution of plant species. Cornell Univ. Press, Ithaca. 206 p.

Clayton, G., and A. Robertson 1955 Mutation and quantitative variation. Am. Naturalist 89:151–158.

Dobzhansky, Th. 1951 Genetics and the origin of species. 3rd ed. Columbia Univ. Press, New York. 364 p.

———— 1955 A review of some fundamental concepts and problems of population genetics. Cold Spring Harbor Symp. Quant. Biol. 20:1–15.

———— 1958 Species after Darwin, p. 19–55. *In* S. A. Barnett, [ed.], A century of Darwin. Heinemann, London.

———— and O. Pavlovsky 1955 An extreme case of heterosis. Proc. Nat. Acad. Sci 41:289–295.

Fisher, R. A. 1930 The genetic theory of natural selection. Clarendon Press, Oxford. 272 p.

Goodall, H. D. 1937 Can artificial selection produce unlimited change? Am. Naturalist 71:433–459.

——— 1942 Further progress on artificial selection. Am. Naturalist 76: 515–519.

Grant, V. 1957 The plant species in theory and practice, p. 39–79. *In* E. Mayr, [ed.], The species problem. Am. Assn. Adv. Sci. Publ. 50.

Kettlewell, H. B. D. 1955 How industrialisation can alter species. Discovery 16:507–511.

King, J. C. 1955 Integration of the gene pool as demonstrated by resistance to DDT. Am. Naturalist 89:39–46.

Kurabayashi, M. 1958 Evolution and variation in Japanese species of *Trillium*. Evolution 12:286–310.

Lack, D. 1917 Darwin's finches. Cambridge Univ. Press, Cambridge. 208 p.

Lerner, I. M. 1958 The genetic basis of selection. John Wiley, New York. 298 p.

Levene, H., O. Pavlovsky, and Th. Dobzhansky 1958 Dependence of the adaptive values of certain genotypes in *Drosophila pseudoobscura* on the composition of the gene pool. Evolution 12:18–24.

Levine, L. 1955 Genotypic background and heterosis in *Drosophila pseudoobscura*. Genetics 40:832–849.

MacArthur, J. W. 1944 Genetics of body size and related characters. II. Satellite characters associated with body size in mice. Am. Naturalist 78: 224–237.

Mather, K. 1955 Response to selection. Cold Spring Harbor Symp. Quant. Biol. 20: 158–165.

——— and B. J. Harrison 1949 The manifold effect of selection. Heredity 3:1–52.

Mayr, E. 1957 Difficulties and importance of the biological species concept, p. 371–388. *In* E. Mayr, [ed.], The species problem. Am. Assn. Adv. Sci. Publ. 50.

Minamori, S. 1957 Physiological isolation in Cobitidae. VI. Temperature adaptation and hybrid inviability. J. Sci. Hiroshima Univ. 17:1–65.

Moore, J. A. 1957 An embryologist's view of the species concept, p. 325–338. *In* E. Mayr, [ed.], The species problem. Am. Assn. Adv. Sci Publ. 50.

Prazmo, W. 1960 Genetic studies on the genus *Aquilegia* L. I. Crosses between *Aquilegia vulgaris* L. and *Aquilegia ecalcarata* Maxim. Acta Societatis Botanicorum Poloniae 29 (1):57–77.

Sheppard, P. M. 1959 The evolution of mimicry; a problem in ecology and genetics. Cold Spring Harbor Symp. Quant. Biol. 24:131–140.

Simpson, G. G. 1953 The major features of evolution. Columbia Univ. Press, New York. 434 p.

——— 1960 History of life, p. 117–180. *In* S. Tax, [ed.], Evolution after Darwin, v. 1. Univ. Chicago Press, Chicago.

Stebbins, G. I. 1950 Variation and evolution in plants. Columbia Univ. Press, New York. 643 p.

———— 1956 Taxonomy and the evolution of genera, with special reference to the family Gramineae. Evolution 10:235–245.

———— 1958 The inviability, weakness and sterility of interspecific hybrids. Ad. Genet. 9:147–215.

———— 1959a The role of hybridization in evolution. Proc. Am. Phil. Soc. 103 (2):231–251.

———— 1959b The synthetic approach to problems of organic evolution. Cold Spring Harbor Symp. Quant. Biol. 24:305–311.

———— and A. Vaarama 1954 Artificial and natural hybrids in the Gramineae, tribe Hordeae. VII. Hybrids and allopolyploids between *Elymus glaucus* and *Sitanion* spp. Genetics 39:379–395.

Woodworth, C. M., and R. W. Jugenheimer 1948 Breeding and genetics of high protein corn. What's new in the production, storage, and utilization of hybrid seed corn. Rep. 3rd Ann. Industry-Research Conf. Chicago, p. 75–83.

3

THE FOUNDER EFFECT AND DELETERIOUS GENES

Frank B. Livingstone

How do populations within a species come to differ from one another? What triggers two or more arrays onto different evolutionary pathways? An obvious answer is adaptation, working through the mechanism of natural selection. If given the opportunity by way of appropriate gene variants and number of generations, populations will become gradually differentiated from each other as they become more closely fitted to the requirements of their respective ecological niches. The niche to which a human population may be adapted might include diseases, climate, or other natural environmental phenomena, the abundance or rarity of nutrients in the food supply, and perhaps other as yet unknown selective agents.

Natural selection, originally identified and described by Darwin

Reprinted from the *American Journal of Physical Anthropology*, 30:1, January 1969, pages 55–60. By permission of the author and the publisher.

*in 1858, is the most dramatic mechanism in differentiating popula-
tions. But it is not the only agency. Random genetic drift or sampling
error may result in an unpredictable increase of nonadaptive
genes, particularly within the gene pools of small populations.
Conversely, neutral or favorable genes may decline in frequency
or even disappear entirely from one generation to the next.*

*Another related phenomenon may take place when a new
population is established by a small number of individuals. Obviously the only genes available to descendants of such a founders'
group will be those of the original ancestors. The frequencies of
genes in the gene pool of succeeding generations may be vastly
affected by the distribution of the genes among the founders of
the population.*

*It is not difficult to perceive how this might happen for a genetic
polymorphism which is less than vital for survival. But in the
following article the author describes a situation in which the gene
is actually deleterious, or harmful: the celebrated gene for hemoglobin S, better known as the sickle-cell gene. The study he
describes here is of interest on a number of counts:*

*1. The use of modern computer technology to test a hypothesis
concerning the possible gene frequencies among the founders
of a genetic isolate;*

*2. The illustration of the ways in which selection together with
other evolutionary mechanisms may operate simultaneously upon
a population;*

*3. The way in which the question of what we mean by a
deleterious gene is illuminated.*

*Dr. Livingstone scored a major breakthrough in his 1958 study
on sickle-cell anemia. In it he demonstrated the probability that
balanced polymorphism of the sickle-cell gene in West Africa,
with its high percentage of malaria-resistant heterozygotes, was
tied to the origin of agriculture in the region; he thus showed
how a cultural event could play a significant role in the evolution
of a human population.*

*Frank B. Livingstone is a member of the Department of
Anthropology at the University of Michigan.*

Many distinctive human populations are characterized by the presence of
one or more lethal or severely deleterious genes in frequencies which
would be defined as polymorphic according to Ford's ('40) famous definition. The particular genetic disorder, however, varies. The Old Order
Amish of Lancaster County, Pennsylvania have a gene frequency of 0.07
for the recessive Ellis-van Creveld syndrome, while the Amish as a whole

have a frequency of about 0.05 of the recessive cartilage-hair hypoplasia syndrome (McKusick et al., '64). Many of the tri-racial isolates of the Eastern United States also have a high frequency of a deleterious gene (Witkop et al., '66). Although such populations are frequently defined by religious or ethnic criteria, there are others not so defined. Several island populations in the Åland archipelago have a gene frequency of greater than 0.1 for von Willebrand's disease (Eriksson, '61), and the Boer population of South Africa and some populations of Northern Sweden have frequencies of porphyria much greater than those of other populations (Dean, '63; Waldenström and Haeger-Aronsen, '67). However, these conditions are dominant and do not have the very severe effects of other hereditary disorders found in high frequencies. On the other hand the population of the Chicoutimi District of Quebec has recently been found to have a gene frequency of about 0.02 for tyrosinemia, which is a lethal recessive (Laberge and Dallaire, '67).

In most of these cases the population in question has undergone a rapid increase in recent years, and the question arises· as to whether this rapid expansion and the original small size of the isolate could account for the high frequency of the deleterious gene. Such an explanation by the founder effect seems obviously to apply to most of the cases cited above, but the founder effect may well be a more general explanation of human gene frequency differences. It is now becoming apparent that the major populations of mankind vary significantly in their frequencies of deleterious genes and that many large populations such as Eastern European Jews have high frequencies of deleterious genes which are found in low frequencies in other populations (McKusick, '66). There have been many attempts to determine how such genes could be polymorphic, for example, Anderson et al. ('67) and Knudson et al. ('67) have discussed cystic fibrosis and Myrianthopoulos and Aronson ('66), Tay-Sachs disease. The purpose of this paper is to attempt to determine the extent to which the founder effect can cause high frequencies of deleterious genes with various models of population expansion.

The occurrence which initiated this research is the gene for sickle cell hemoglobin in the Brandywine isolate of Southeast Maryland. At present the sickle cell gene frequency in this isolate is about 0.1 (Rucknagel, '64). The high frequencies of this gene in many parts of Africa, India, and the Middle East are now well accepted as being due to a relative resistance of the sickle cell heterozygote to falciparum malaria. The high frequency in the Brandywine isolate may have a similar explanation, but the surrounding Negro population does not have such a high frequency. And although the endemicity of falciparum malaria in Southeast Maryland in the last century is not known in any detail, it would not appear to have been great enough to explain the high sickle cell frequency in the Brandywine isolate. The isolate also has many other deleterious genes in high frequency (Witkop et al., '66).

The Brandywine isolate seems to have had its beginning in the early Eighteenth Century when laws were passed to prohibit co-habitation and marriage among races, which prior to then were presumably frequent or at least known. Up to 1720 there were several prosecutions under these laws of individuals with surnames currently present in the isolate (Harte, '63). Harte ('63) has maintained that the Brandywine isolate is derived from these illegal unions, and Witkop et al. ('66) show that the most common surname came from such a union. In 1790 the first United States Census recorded 190 persons with the group's surnames as "other free people," and since then over 90% of the recorded marriages have been endogamous or between individuals with surnames within the group (Harte, '59). According to Harte ('59) there are six "core" surnames which have been associated with the group since its founding and comprise 66% of the population and another ten surnames which entered the group after the Civil War, but Witkop et al. ('66) list seven core surnames and eight marginal ones. The total population of the isolate is now estimated to be 5,128 (Witkop et al., '66), and the statistics do indicate rapid, if erratic, growth (Gilbert, '45; Harte, '63).

In order to simulate gene dynamics the population has been assumed to have doubled itself in early generations, and then after slower growth to have approached a doubling in recent generations. The simulation was run for 10 generations with the following numbers in succeeding generations: 20, 40, 80, 160, 320, 640, 664, 728, 856, 1,112. This approximates the early demographic history of the Brandywine isolate, but the isolate is much larger today. However, gene frequency change in later generations with a large population is very small.

The simulation model randomly selects two parents from the initial population which has been assumed to have either one or two sickle cell heterozygotes among the founders. A family size is randomly determined, the offspring generated and then selected out or stored with no compensation for those not surviving (a copy of the program is available on request). Since the population is increasing rapidly, the family size distribution approximates that recorded by Roberts ('65) for a population in Tanzania which has about 4.0 surviving offspring per female. The founder population can actually vary in size, however. The size of the offspring generation is the number which is set; but with an average of 4.0 offspring per marriage and 40 offspring, the founder population would be expected to consist of ten marriages or 20 individuals.

Figure 3–1 shows the distribution of the deleterious gene frequencies after ten generations for two sets of 50 runs each with different initial conditions and different fitnesses for the genotypes. With a gene frequency of 0.05, which is comparable to having two sickle cell heterozygotes among the founders, the gene is present at a frequency of greater than 0.04 in almost 40% of the populations, while for a starting gene frequency of 0.025 or one founder with the sickle cell trait, 16% of the populations have the gene at a frequency of greater than 0.04. With two founders there were runs which resulted in a gene frequency as high as that of the sickle cell gene in the Brandywine isolate, but with one founder there were none as high. However, there were many frequencies close to it, so that such an outcome is possible if not probable. Hence there seems to be no necessity to postulate a selective advantage for the sickle cell in the Brandywine isolate. It should be pointed out that this simulation and further ones assume the population is closed. Gene flow

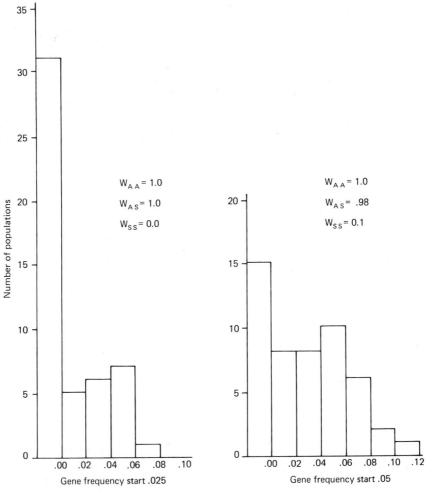

FIGURE 3–1—The distributions of the frequency of the sickle cell gene (S) after ten generations of expansion for two sets of 50 runs each with different initial gene frequencies and different fitnesses for the genotypes (W's). Note: the populations in which the S gene has been completely eliminated are separated in the left column from those in which it is still present.

from other populations would tend to decrease the frequency of the deleterious gene, but if most of the population's expansion is due to natural
increase, then the founder effect would be most important.

In order to determine whether such high frequencies could occur in a
population with a greater number of founders, a similar program was run
with 40 and 80 founders. The results are shown in Figure 3–2. These runs
were started with one carrier of the lethal gene and the population doubled itself for five generations, so that it ended with 1,280 and 2,560
individuals, respectively. Although the lethal gene is not present in high
frequencies in as many populations, it is still present in about 5% in a
frequency greater than 0.04. The fact that populations begun with a few
founders should have such high frequencies of lethal genes seems to
indicate that they can contribute to the problem of the genetic load.
According to Morton ('60), the average individual has the equivalent of
four recessive lethals in the heterozygous state. For a population with 40
founders this would imply over 100 lethal or deleterious genes among the
founders, so that several would be expected to attain high frequencies.
The fact that the number of deleterious genes in small populations started
by a few founders seems to average much less may indicate a lower
genetic load, but the Brandywine isolate, the Amish, and the Eastern
European Jews do have several deleterious genes in high frequencies. In
any case, it seems to be a possible way to study the problem.

The most recent population expansion which seems to have increased
the frequency of a lethal gene is the peopling of the Saguenay River and
Lake St. John region by French Canadians. Settlement of the Upper
Saguenay did not begin until the 1830's, and Chicoutimi was founded in
1840 by 220 individuals from La Malbaie, 66 from Eboulements, and 37
from Baie St. Paul (Buies, 1896). In 1861 the population of Chicoutimi
was 10,478 and in 1871 it rose to 17,483. Much of this increase was
undoubtedly due to immigration, but given the enormous rate of increase
of the French Canadian population as a whole (Henripin, '54), the expansion of the population of Chicoutimi was due to a great extent to
natural increase. Under these conditions a lethal gene frequency of 0.02
would not be unlikely and seems to agree with our simulation model. The
fact that the entire French Canadian population stems from about 10,000
original settlers (Henripin, '54) may have led to this population having
its own set of lethal genes.

The high frequencies of deleterious genes in the Eastern European
Jewish populations of Lithuania and Eastern Poland may have a similar
explanation, although there is disgreement about this possibility (Mc-
Kusick, '66). Myrianthopoulos and Aronson ('66) do not consider such
an explanation likely for the Tay-Sachs gene. Instead they propose a
slight selective advantage for the heterozygote. They have to postulate the
operation of this selective advantage for 50 generations which is longer

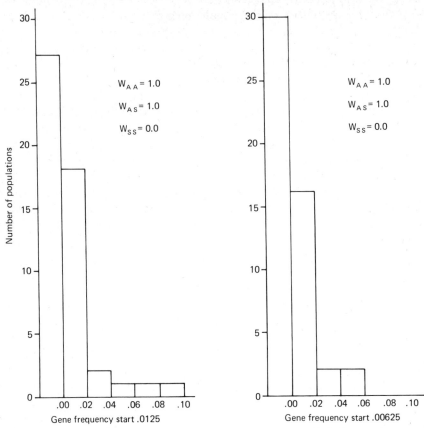

FIGURE 3–2—The distributions of the frequency of a lethal recessive gene (S) after five generations of expansion for two sets of 50 runs each with different initial gene frequencies which represent one carrier of the lethal gene in a founder population of 40(0.0125) and one carrier in a population of 80(0.00625). Note: the populations in which the lethal gene has been completely eliminated are separated in the left column from those in which it is still present.

than the population has been there. It is also much longer than the factor thought to confer the selective advantage, typhoid fever in the ghettos (Aronson, '64), seems to have been present as a serious disease. The Tay-Sachs gene attains its highest frequencies in the Jewish populations of Southern Lithuania and Northeast Poland, which were founded in the Twelfth Century after the Crusades led to the persecution of the Jews in Germany. Although the Jewish settlements in Lithuania were founded by refugees from the west, according to Herzog ('65) they preceded by two or three centuries the Jewish settlements in Mazovia to the west in Poland. Thus, these colonies were isolated for some time and were actually expanding to the west into Northern Poland when the Jews were expelled from Lithuania in 1495. Most moved to adjacent territories but then moved back to Lithuania in 1503. Hence, the population history of these Jewish groups seems to be one of expansion from a few founders. In any case by the time of the flowering of Eastern Jewish culture in the Sixteenth Century, the population was very large and continued to expand up to the Twentieth Century.

When a population size of 1,000 or more is attained, the change in gene frequency is approximated by deterministic equations. For a lethal recessive the frequency after n generations is:

$$q_n = \frac{q_0}{1 + nq_0},$$

where q_0 is the initial gene frequency. A lethal gene in a large population is thus eliminated at a very slow rate, particularly when it occurs in a very low frequency. If the Tay-Sachs gene increased to 0.05 in the early generations before the population became large, in the approximately 30 generations since then, the gene would have decreased to

$$\frac{0.05}{1 + 30(0.05)} = 0.02,$$

which is about the frequency today in Eastern European Jews.

The fact that lethal genes are eliminated at such a slow rate in large populations would make it possible for them to have "polymorphic" frequencies long after the original expansion. Since most of the world's populations have expanded rapidly in the last 1,000 years, much of the variability in the frequencies of lethal genes (or non-lethals for that matter) could be a consequence of the original expansions of the major populations. As an example, it is suggested that this effect may explain the high frequencies of cystic fibrosis in the populations of Europe, which range around a gene frequency of 0.02. For the non-Caucasian populations on Hawaii Wright and Morton ('68) have estimated the gene frequency for cystic fibrosis to be 0.003, which presumably is close to the equilibrium frequency due to a balance of selection and mutation. Given this frequency, over 50% of a set of founder populations of size 100 would be expected to have a carrier of this lethal. With the sudden expansion of such a set of founder populations it seems possible that such a lethal could attain a frequency of 0.05 for the entire population. A more precise mathematical expression of the problem seems possible and could perhaps yield a solution.

LITERATURE CITED

Anderson, C. M., J. Allan and P. G. Johansen 1967 Comments on the possible existence and nature of a heterozygote advantage in cystic fibrosis. In: Cystic Fibrosis. E. Rossi and E. Stoll, eds. Bibliotheca Paediatrica, No. 86. S. Karger, New York, pp. 381–387.

Aronson, S. M. 1964 Epidemiology. In: Tay-Sachs' Disease. B. W. Volk, ed. Grune and Stratton, New York, pp. 118–154.

Buies, A. 1896 Le Saguenay et le Bassin du Lac Saint-Jean. Léger Brousseau, Québec, 3rd edition.

Dean, G. 1963 The prevalence of the porphyrias. So. Afr. J. Lab. Clin. Med., 9:145–151.

Eriksson, A. W. 1961 Eine neue Blutersippe mit v. Willebrand-Jürgens' scher Krankheit (erbliche Thrompathie) auf Åland (Finnland). Acta Genet. Med. Gemell., 10:157–180.

Ford, E. B. 1940 Polymorphism and taxonomy. In: The New Systematics. J. S. Huxley, ed. Clarendon Press, Oxford, pp. 493–513.

Gilbert, W. H. 1945 The wesorts of Maryland; an outcasted group. Jour. Wash. Acad. Sci., 35:237–246.

Harte, T. J. 1959 Trends in mate selection in a tri-racial isolate. Social Forces, 37:215–221.

——— 1963 Social origins of the Brandywine population. Phylon, 24: 369–378.

Henripin, J. 1954 La Population Canadienne au Début du XVIIIe Siècle. Institut national d'études démographiques, Paris, Cahier No. 22.

Herzog, M. I. 1965 The Yiddish Language in Northern Poland: Its Geography and History. Indiana University Research Center in Anthropology, Folklore and Linguistics, Pub. No. 37.

Knudson, A. G., L. Wayne and W. Y. Hallett 1967 On the selective advantage of cystic fibrosis heterozygotes. Amer. J. Hum. Genet., 19:388–392.

Laberge, C., and L. Dallaire 1967 Genetic aspects of tyrosinemia in the Chicoutimi Region. Canad. Med. Assoc. J., 97:1099–1100.

McKusick, V. 1966 Clinical genetics at a population level. The ethnicity of disease in the United States. Alabama J. Med. Sci., 3:408–424.

McKusick, V., J. A. Hostetler, J. A. Egeland and R. Eldridge 1964 The distribution of certain genes in the Old Order Amish. Cold Spring Harbor Symp. Quant. Biol., 29:99–114.

Morton, N. E. 1960 The mutational load due to detrimental genes in man. Amer. J. Hum. Genet., 12:348–364.

Myrianthopoulos, N. C., and S. M. Aronson 1966 Population dynamics of Tay-Sachs Disease. I. Reproductive fitness and selection. Amer. J. Hum. Genet., 18:313–327.

Roberts, D. F. 1965 Assumption and fact in anthropological genetics. Jour. Roy. Anthrop. Inst., 95:87–103.

Rucknagel, D. L. 1964 The Gene for Sickle Cell Hemoglobin in the Wesorts. Thesis, University of Michigan, Ann Arbor.

Waldenström, J., and B. Haeger-Aronsen 1967 The porphyrias: a genetic problem. *In:* Progress in Medical Genetics. Volume V. A. G. Steinberg and A. G. Bearn, eds. Grune and Stratton, New York, pp. 58–101.

Witkop, C. J., C. J. MacLean, P. J. Schmidt and J. L. Henry 1966 Medical and dental findings in the Brandywine isolate. Alabama J. Med. Sci., 3: 382–403.

4

HUMAN CHROMOSOME ABNORMALITIES AS RELATED TO PHYSICAL AND MENTAL DYSFUNCTION

John H. Heller

The "raw material" of evolution is the array of variations that naturally characterize any population. Individual organisms, whose phenotypes demonstrate the effects of their genotypes, contribute in different proportions to the genotype array of succeeding generations, depending on the adaptiveness of the genotypes they carry. Thus, a population at any time consists of a range of genetically based adaptations to the requirements of its total environment.

A particular feature of the environment may impose very stringent limits of variation relevant to the trait. In such a case the population will demonstrate only a narrow degree of variation in regard to that trait. In other instances, the limits imposed by the environment may be relatively broad. Thus, with respect to the trait adaptive to that aspect of the environment, there may be a considerable range of variation. In contemporary human populations, height may be such a trait; every human population today is characterized by a considerable degree of polymorphism and

Reprinted from the *Journal of Heredity*, 60:5, September-October 1969, pages 239–248. By permission of the author, and the publisher, the American Genetic Association.

continuous variation in height. The same may hold true for nasal shapes and other facial features.

The range of variation respecting a given trait, then, may be relatively wide, or the environment may permit little or no variation. Additionally, there are abnormal traits that dramatically inhibit the adaptiveness of organisms, resulting in early death or failure to reproduce, traits with "negative evolutionary impact."

Many of the traits best known to human geneticists are of this latter kind. The genetics of abnormality is a field of research that continues to be extremely fruitful in illuminating the hereditary basis of human variation. It has been estimated that of 1900 known gene loci on the human chromosomes, fully 400 are sites of aberrant pathological conditions of human hemoglobin.

It would be useful if we had as much data about the "normal" variations in human genotypes. But it is usually an abnormality in body function, or an abnormal phenotype, which inspires genetic investigation; consequently it need not be surprising that most of the array of "normal" genotypes continues to elude genetic study. Meanwhile, the list of known human genetic abnormalities increases significantly from year to year.

In recent years the marked technical advances in karyotyping have made possible the detailed analysis of the gross shape, structure, and number of the chromosomes themselves. As a result a number of human phenotype defects have been correlated with chromosome abnormalities (aberrant numbers, shapes, and so on). In the following article, a review is provided of some of the more dramatic physical and mental dysfunctions with known chromosomal basis.

Dr. Heller is the President of the New England Institute in Ridgefield, Connecticut.

The relationship of human disease syndromes to chromosome aberrations is assuming an increasingly greater role in the detection, diagnosis, treatment and prediction of mental and physical defects in man. By means of karyotype analysis one is enabled to recognize previously unknown syndromes and to differentiate between separate but phenotypically similar entities. Proper diagnosis permits suitable therapuetic measures to be undertaken and enables genetic counselors to assess correct risks in many instances. Recent refinements in sampling embryonic cells by amniocentesis make it feasible to determine, in high risk cases, whether the embryo has a chromosome abnormality or whether it is a male, which has a high risk of sex-linked genetic defect. Termination of pregnancy can be recommended on the basis of this knowledge.

CLASSES OF CHROMOSOME ABNORMALITIES

Chromosome abnormalities have been known in plant and animal species for a very long time. They occur firstly as variations in the number of chromosomes per cell deviating from the normal two sets (maternal and paternal), existing either as complete multiples of sets, a condition called polyploidy (triploidy, tetraploidy, etc.), or as addition or loss of chromosomes within a set, a situation known as aneuploidy (monosomy, trisomy, tetrasomy, etc.). The origin of deviations in chromosome number is known to be through nondisjunction, either during the meiotic divisions in the maturation of the germ cells or during mitotic divisions in the developing individual, or through lagging of chromosomes at anaphase of cell division.

Secondly, chromosome aberrations occur as structural modifications such as duplications, deficiencies, translocations, inversions, isochromosomes, ring chromosomes, etc. These aberrations result from chromosome breakage and reunion in various patterns different from the normal sequence of loci. In most cases, especially the "spontaneous" instances, the cause of chromosome breaks is unknown, but many extraneous agents have been demonstrated experimentally to be efficacious in inducing fragmentation. Foremost among these agents is ionizing radiation but many chemical substances (alkylating agents, nitroso-compounds, antibiotics, DNA precursors, etc.) and viruses have been implicated.

GENETIC EFFECTS OF CHROMOSOME ABERRATIONS

The striking genetic alterations accompanying chromosome aberrations were brilliantly analyzed by Blakeslee and coworkers on *Datura*, and by the *Drosophila* workers (Morgan, Bridges, Muller, Sturtevant, Painter, Patterson and many others). The task was greatly facilitated in *Drosophila* by the fortunate circumstance in the larval salivary glands where the giant polytene chromosomes exhibit intimate somatic pairing as well as characteristic banding patterns that permit identification of specific gene loci.

Particularly illuminating were Bridges' analyses of sex chromosomes and sex determination in *Drosophila*, utilizing the phenomenon of nondisjunction of the sex chromosomes and culminating in the genic balance theory of sex determination. In this insect the female normally has two X chromosomes plus the autosomes, and the male has one X and one Y. Two X chromosomes and one Y chromosome results in a female, whereas a chromosome constitution of XO produces a sterile male.

In contrast, the Y chromosome in mammals has a strongly masculinizing influence. The presence of a single Y is sufficient to induce differentiation into a male phenotype in the presence of one to five X chromosomes. The XO constitution differentiates into a female phenotype in both mouse and man.

MAMMALIAN CHROMOSOME STUDIES

The first reported instance of chromosome aberration in mammals was discovered by genetic methods in the waltzing mouse by William H. Gates in 1927[37], and analyzed cytologically by T. S. Painter.[68] Many difficulties in techniques prevented accurate counting and analysis of mammalian chromosomes—large number and relative small size of chromosomes, tendency to clump on fixation, cutting of chromosomes in sectioned material, etc. Even the somatic chromosome number in man was accepted erroneously as 48 until 1956 when Tjio and Levan[88] established the correct count of 46. This count was quickly confirmed by Ford and Hamerton[28], and in 1959 the first positive correlation of a chromosome abnormality and human disease syndrome was made by Lejeune *et al.*[54] (also Jacobs *et al.*[44])—the trisomic number 21 chromosome, and Down's syndrome or mongolism. Shortly thereafter Klinefelter's[46] and Turner's[30] syndromes were identified with XXY and XO sex chromosome constitutions respectively, and in rapid succession reports of many other human chromosome abnormalities appeared, such as trisomy 17, trisomy 18, partial trisomy, ring X chromosome, sex chromosome mosaics, cri-du-chat syndrome, et cetera[9,26].

This sudden explosion of human chromosome studies, in contrast to the long delay of confirmation in human cells of chromosome abnormalities long known in plants and other animals, was made possible by new techniques of preparation. The accumulation of many cells in the metaphase stage of mitosis with colchicine, the use of hypotonic solution to swell the cells and separate chromosomes on the spindle, the discovery that phytohemagglutinin stimulates mammalian peripheral lymphocytes to undergo mitosis, and the method of squashing or spreading on slides of loose cells taken from bone marrow or tissue culture, all contributed to the rapid and accurate analysis of mammalian and human chromosome number and structure.

Karyotype analysis involves the careful comparison of chromosomes in a particular individual to the standard pattern for human cells, including precise measurements of lengths, arm ratios and other morphological features. Special attention is given to comparison of homologous chromosomes where differences may indicate abnormalities. An idiogram is a diagrammatic representation of the entire standard chromosome complement, showing their relative lengths, position of centromeres, arm ratios, satellites, secondary constrictions and other features. Figure 4–1 shows an idiogram of a normal human male with 22 pairs of autosomes and XY sex chromosome constitution. A karyotype is constructed from photographs of chromosomes which are arranged in pairs similar to the idiogram. Figure 4–2 shows a karyotype of a normal human female.

INCIDENCE OF HUMAN CHROMOSOME ANOMALIES

Chromosome anomalies are relatively frequent events. They have been estimated to occur in 0.48 percent of all newborn infants (one in 208)[81]. At least 25 percent of all spontaneous miscarriages result from gross chromosomal errors.[13] The general incidence of chromosome abnormalities in abortuses is more than fifty times the incidence at birth.

Although it is impossible to obtain an accurate total of victims suffering from effects of chromosome aberrations, one can make rough calculations on the basis of their estimated frequencies in the population of the United States, assuming that there is no appreciable difference in life expectancy between these individuals and those with normal chromosome complements. Although this assumption probably is unjustifiable, it suf-

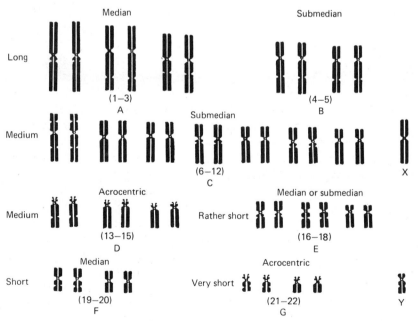

FIGURE 4–1—Idiogram of a normal male with 22 pairs of autosomes and XY chromosome constitution. (Modified from Patau,[69] Sohval,[84] Ferguson-Smith *et al.*, [27] and Palmer and Funderburk.[68a])

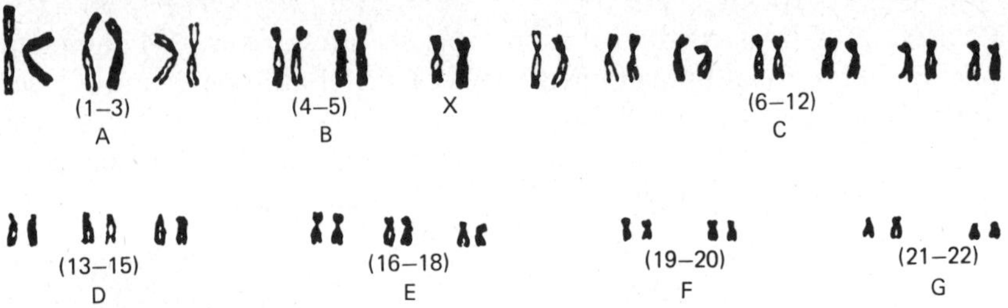

FIGURE 4-2—Karyotype of a normal female with 22 pairs of autosomes and two X chromosomes.

fices for this rough calculation. Among the current population of 202 million we arrive at a figure of 1,136,971 total afflicted with chromosome abnormalities. This total probably represents an underestimate since it does not include all types of chromosome aberrations. Table 4-1 indicates totals for a number of specific syndromes.

SYNDROMES RELATED TO AUTOSOME ABNORMALITIES

Down's Syndrome

This defect results from duplication of all or part of autosome 21, either in the trisomic state or as a translocation to another chromosome, usually a 13-15 (D group) or 16-18 (E group) but may be to another G group chromosome. The overall incidence is about 1 in 700 live births[71], but the trisomic type is correlated with age of the mother, having a frequency of about 1 in 2000 in mothers under 30 years of age, and increasing to 1 in 40 in mothers aged 45 or over. The translocation type constitutes about 3.6 percent of cases and is unrelated to the mother's age, but is transmitted in a predictable manner. Among mental retardates mongoloids represent 16.7 percent.

Clinical features include physical peculiarities ranging from slight anomalies to severe malformations in almost every tissue of the body. Typical appearance of a mongoloid shows slanting eyes, saddle nose, often a large ridged tongue that rolls over a protruding lip, a broad, short skull and thick, short hands, feet and trunk. Frequent complications occur: cataract or crossed eyes, congenital heart trouble, hernias, and a marked susceptibility to respiratory infections. They exhibit characteristic dermatoglyphic patterns on palms and soles. Also they have many biochemical deviations from normal, such as decreased blood-calcium levels and diminished excretion of tryptophane metabolites. Early aging is common.

All mongoloids are mentally retarded; they usually are 3 to 7 years old mentally. Among the relatively intelligent patients, abstract reasoning is exceptionally retarded.

TABLE 4–1

*Total frequencies in the United States of various types of chromo-
somal abnormalities, calculated on the basis of 202 million current
population and the estimated frequency of each abnormality. (It
must be noted that the grand total does not include all types of
chromosome aberrations, therefore must be lower than the real value.)*

SYNDROME	CHROMO-SOME NUMBER	ESTIMATED INCIDENCE	CALCULATED NUMBER IN U.S.
Down's trisomy 21	47	1 in 700	288,571
Trisomy D	47	1 in 10,000	20,200
Trisomy E	47	1 in 4000	50,500
Trisomy X	47	1 in 10,000 females	101,000
Turner's XO	45	1 in 5000 females	20,200
Klinefelter's XXY	47	1 in 400 males	252,500
Double Y XYY	47	1 in 250 males	404,000
		Total	1,136,971

Female mongoloids are fertile and recorded pregnancies have yielded
approximately 50 percent mongoloid offspring. Fortunately male mongoloids are sterile. Examination of their testes reveals varying degrees of
spermatogenic arrest correlated with the abnormal chromosome features.

Among mongoloids there is a prevalence of leukemia in childhood; the
incidence is some twenty times greater than in the general population.
Simultaneous occurrence with other syndromes, such as Klinefelter's, also
is found, and many cases of mosaicism have been described.

E Trisomy Syndrome

This is another autosomal anomaly, which involves chromosomes 16,
17 and 18, and is estimated to occur at a frequency of 1 in 4000 live
births[20]. Many others die before birth, thus contributing to the large
number of miscarriages and stillbirths. These individuals survive only a
short time, from one-half day to 1460 days, with an average of 239 days,
but females live significantly longer than males.

Trisomy 17 Syndrome

Many serious defects usually are present in afflicted individuals[25]: odd
shaped skulls, low-set and malformed ears, triangular mouth with reced-
ing chin, webbing of neck, shield-like chest, short stubby fingers, and toes

with short nails, webbing of toes, ventricular septal defect and mental retardation, as well as abnormal facies, micrognathia and high arched palate.

TRISOMY 18 SYNDROME

This anomaly[70,82] is characterized by multiple congenital defects of which the most prominent clinical features are: mental retardation with moderate hypertonicity, low-set malformed ears, small mandible, flexion of fingers with the index finger overlying the third, and severe failure to thrive. It generally results in death in early infancy. Its frequency increases with advanced maternal age. Three times as many females as males have been observed; one would expect that more males with this syndrome will be found among stillbirths and fetal deaths.

D SYNDROME

This trisomy[19,56,70,83] involves chromosomes 13, 14 and 15, and has an estimated frequency of about 1 in 10,000 live births. Many others die *in utero*. Survival time has been reported from 0 to 1000 days, with an average of 131 days.

Clinical features include: microcephaly, eye anomalies (corneal opacities, colobomata, microphthalmia, anophthalmia), cleft lip, cleft palate, brain anomalies (particularly arrhinencephaly), supernumerary digits, renal anomalies (especially cortical microcysts), and heart anomalies.

TRISOMY 22 SYNDROME

This syndrome produces mentally retarded, schizoid individuals. Reports of its occurrence are too few to permit an estimate of its frequency in the population.

CRI-DU-CHAT SYNDROME

Lejeune *et al.*[55] first described this anomaly in 1965, which involves a deficiency of the short arm of a B group chromosome, number 5. Translocations appear to be a common cause of the defect, an estimated 13 percent of cases being associated with translocations; described cases have had B/C, B/G, and B/D translocations[23]. The high proportion emphasizes the importance of unbalanced gamete formation in translocation heterozygotes as a cause of this syndrome. Among parents the frequency of male and female carriers is approximately the same, a situation that contrasts with the much greater frequency of female carriers of a D/G translocation among parents of translocation mongoloids.

Typical clinical features of cri-du-chat individuals are: low birth weight, severe mental retardation, microcephaly, hypertelorism, retrognathism, downward slanting eyes, epicanthal folds, divergent strabismus, growth retardation, narrow ear canals, pes planus and short metacarpals and metatarsals. About 25 to 30 percent of them have congenital heart disorders. A characteristic cat-like cry in infancy is responsible for the name of the syndrome. The cry is due to a small epiglottis and larynx and an atrophic vestibule. However, this major diagnostic sign disappears after infancy, making identification of older cases difficult.

An estimate of the frequency of this syndrome is given as over 1 percent but less than 10 percent of the severely mentally retarded patients. Many have IQ scores below 10, and most are institutionalized.

PHILADELPHIA CHROMOSOME

Finally, among autosomal aberrations, a deleted chromosome 21 occurs in blood-forming stem cells in red bone marrow. This deletion, which shows up long after birth, appears to be the primary event causing chronic granulocytic leukemia. This aberration was discussed in 1960 by Nowell and Hungerford[67] (also Baikie *et al.*[3]).

SYNDROMES RELATED TO SEX CHROMOSOME ABERRATIONS

The great majority of known chromosomal abnormalities in man involve the sex chromosomes. In one survey (that excluded XYY) it was estimated that abnormalities occurred in 1 out of every 450 births; if the recent estimate of XYY[81] is correct, the frequency actually is much higher. Increased knowledge about sex chromosome aberrations is probably related to the greater concentration of attention on patients with sexual disorders, but is due in part to the ability to detect carriers of an extra X chromosome by the so-called sex chromatin body or Barr body[6]. This structure is a stainable granule at the periphery of a resting nucleus and, according to the Lyon hypothesis[57], is considered to be an inactivated X chromosome. A normal female cell has one Barr body, since it has two X chromosomes, and is said to be sex chromatin positive (or one positive). A normal male cell has no Barr body and is said to be sex chromatin negative.

KLINEFELTER'S SYNDROME

The first sex chromosome anomaly described in 1959 by Jacobs and Strong[46] and also by Ford *et al.*[29] was the XXY constitution that is typical of Klinefelter's syndrome. Buccal smears from these patients are sex chromatin positive. They can be tentatively diagnosed by this test along with clinical symptoms. Final confirmation of diagnosis can be achieved by karyotype analysis using either bone marrow aspiration or peripheral blood culture.

Victims of Klinefelter's syndrome are always male but they are generally underdeveloped, eunachoid in build, with small external genitalia, very small testes and prostate glands, with underdevelopment of hair on the body, pubic hair and facial hair, frequently with enlarged breasts (gynecomastia), and many have a low IQ.

The classical type with two X chromosomes and one Y chromosome was the first case discovered, but subsequently, chromosome compositions of XXXY, XXXXY, XXYY[66] and XXXYY[7,8,63,77] have been reported. In addition, numerous mosaics have been described, including double, triple and quadruple numeric mosaics, as well as combinations of numeric and structural mosaics. These conditions are summarized in Table 4–2. They all resemble the XXY Klinefelter's phenotypically and are considered modified Klinefelter's syndromes. The classical XXY type may have low normal mental development or may be retarded, but other types show increasingly greater mental retardation.

The incidence of Klinefelter's syndrome is estimated to be 1 in 400 male live births, which represents from 1 to 3 percent of mentally deficient patients. This condition also has been correlated with age of the mother: the older the mother, the greater the risk of having such a child. These individuals usually are sterile. Spermatogenesis is generally totally absent. Hyalinization of the semeniferous tubules begins shortly before puberty. Congenital malformations are rare. Mental retardation is present in approximately 25 percent of affected individuals, and mental illness may be more common than in the general population.

TURNER'S SYNDROME

Female gonadal dysgenesis was described by Turner in 1938 as a syndrome of primary amenorrhea, webbing of the neck, cubitas valgus and short stature, coarctation of aorta, failure of ovarian development and hormonal abnormalities. Patients exhibit sexual infantilism; their breasts are usually underdeveloped, nipples often widely spaced, particularly in those subjects who have a shield or funnel chest deformity. Usually sexual hair is scanty; external genitalia are infantile; labia small or unapparent; clitoris usually normal, although may be enlarged. The uterus is infantile; the tubes long and narrow; the gonads represented by long, narrow, white streaks of connective tissue in normal position of ovary. They are almost always sterile. Hormonal secretions usually are abnormal. Shortness of stature is characteristic and many other skeletal abnormalities occur. Peculiar facies include small mandible, anti-mongolian slant of eyes, depressed corners of mouth, low-set ears, auricles some-

TABLE 4–2

Reported sex chromosomal constitutions in Klinefelter's syndrome.
(Modified from Reitalu.[77])

		SEX CHROMOSOMAL CONSTITUTION			
Only one karyotype observed per individual		XXY			
		XXYY			
		XXXY			
		XXXYY			
		XXXXY			
Numeric mosaics	Double	XX	XXY		
		XY	XXY		
		XY	XXXY		
		XXY	XXYY		
		XXXY	XXXXY		
		XXXX	XXXXY		
	Triple	XY	XXY	XXYY	
		XX	XXY	XXXY	
		XY	XXY	XXXY	
		XO	XY	XXY	
		XX	XY	XXY	
		XXXY	XXXXY	XXXXYY	
		XXXY	XXXXY	XXXXXY	
	Quad-ruple	XXY	XY	XX	XO
Numeric and structural mosaics	Double	XXY	XXxY		
	Triple	XY	XXY	XXxY	
		XxY	Xx	XY	

times deformed. Cardiovascular defects are frequent, the most common being coarctation of the aorta. Slight intellectual impairment is found in some patients, particularly those with webbing of the neck.

In 1954 it was discovered that many patients with ovarian agenesis were sex chromatin negative, and in 1959 Ford and colleagues[30] gave the first chromosome analysis showing that Turner's syndrome has the sex chromosome abnormality of only one X chromosome (XO) rather than two X's. It was quickly confirmed by Jacobs and Keay[45] and by Fraccaro et al[33].

Mosaicism is known to exist—both 45 chromosome cells and 46 chromosome cells occur side by side in tissues of the individual—and can result from nondisjunction in early embryonic development. Isochromosomes sometimes are involved, e.g., creating a situation with 3 long arms of the X chromosome but only 1 short arm.

The incidence of XO Turner's syndrome is estimated as 1 in approximately 5000 women; many die *in utero*.

Large scale screening of newborn babies by buccal smears can permit detection of chromatin negative females, chromatin positive males, and double, triple, quadruple and quintuple positive cases of either sex[58]. Table 4–3 shows the relationship between sex chromosome complements and sex chromatin pattern.

TRIPLO-X SYNDROME

Females containing three[47], four and five X chromosomes are known[4,47,63]. The triplo-X syndrome is thought to have an incidence of about 1 in 800 live female births. This syndrome was first described by Jacobs *et al.*[47] in 1959. Although it has no distinctive clinical picture, menstrual irregularities may be present, secondary amenorrhea or premature menopause. Most cases have no sexual abnormalities and many are known to have children. The most characteristic feature of 3X females is mental retardation. Quadruple-X[14] and quintuple-X[50] syndromes are much rarer. These individuals are mentally retarded; usually the more X chromosomes present, the more severe the retardation. Frequently these individuals are fertile.

An extra X chromosome confers twice the usual risk of being admitted to a hospital with some form of mental illness. The loss of an X, on the other hand, has no association with mental illness; thus the chance of mental hospital admission is not raised for an XO female. An extra X chromosome also predisposes to mental subnormality. The prevalence of psychosis among patients in hospitals for the subnormal is unusually high in males with two or more X chromosomes.

Numerous other sex chromosome anomalies occur[38], many involving mosaics and structural chromosome aberrations. For example, occasionally an XY embryo will differentiate into a female, a situation referred to as testicular feminization male pseudohermaphrodite (Morris syndrome)[60]. These individuals have only streak gonads and vestigial internal genital organs. They usually have undeveloped breasts and do not menstruate. They are invariably sterile[76].

Still other sexual abnormalities are intersexes and true hermaphrodites, many of which have an XX sex chromosome constitution or are mosaics for sex chromosomes such as XO/XY or XX/XY or more complicated mixtures[31]. Sex chromosome mosaicism is very common. Almost every sex chromosome combination found alone has been found in association with one or more cell lines with a different sex chromosome constitution. These mosaics exhibit quite a variable expression; for example in an

XO/XY mosaic the external genitalia can appear female, male or inter-sexual[85].

THE YY SYNDROME

The male with an extra Y chromosome (XYY) has attracted much attention in the public press as well as in scientific circles because of his reputed antisocial, aggressive and criminal tendencies[1, 2, 64]. Although this abnormality belongs in the above category of syndromes related to sex chromosome aberrations, it has been singled out for special discussion because of its social and legal implications.

Evidence supporting the existence of a double Y syndrome has accumulated within the last six years. Studies in Sweden[32] showed an unusually large number of XXYY and XYY men among hard-to-manage patients in mental hospitals. These observations received impressive confirmation in studies of maximum security prisons and hospitals for the criminally insane in Scotland where an astonishingly high frequency (2.9 percent) of XYY males was found[48]. This was over fifty times higher than the then current estimate of 1 in 2000 in the general population. Subsequently many additional studies on the YY syndrome have appeared and a composite picture of the XYY male emerged[5, 19, 15, 21a, 34, 35, 41, 72-75, 78, 80, 92].

The principal features of the extra Y syndrome appear to be exceptional height and a serious personality disorder leading to behavioral disturbances. It seems likely it is the behavior disorder rather than their intellectual incompetence that prevents them from functioning adequately in society[18].

Clinically the XYY males are invariably tall (usually six feet or over) and frequently of below-average intelligence. They are likely to have

TABLE 4–3

The relationship between sex, sex chromosome complement and sex chromatin pattern (modified from Miller[63])

SEX CHROMATIN PATTERN	SEXUAL PHENOTYPE	
	Female	*Male*
−	XO	XY
−	XY	XYY
	(testicular feminization)	
+	XX	XXY
		XXYY
+ +	XXX	XXXY
		XXXYY
+ + +	XXXX	XXXXY
+ + + +	XXXXX	

unusual sexual tastes, often including homosexuality. A history of anti-social behavior, violence and conflict with the police and educational authorities from early years is characteristic[86] of the syndrome.

Although these males usually do not exhibit obvious physical abnormalities[12, 24, 40, 42, 52, 91], several cases of hypogonadism[11], some with undescended testes, have been reported. Others have epilepsy, malocclusion and arrested development[87], but these symptoms may be fortuitously associated. One case was associated with trisomy 21[61], another with pseudohermaphrodism[35]. The common feature of an acne-scarred face may be related to altered hormone production. The criminally aggressive group were found to have evidence of an increased androgenic steroid production as reflected by high plasma and urinary testosterone levels[12, 43]. If the high level of plasma testosterone is characteristic of XYY individuals, it suggests a mechanism through which this condition may produce behavioral changes, possibly arising at puberty.

Antisocial and aggressive behavior in XYY individuals may appear early in life, however, as evidenced by a case reported by Cowie and Kahn[22]. A prepubertal boy with normal intelligence, at the age of $4\frac{1}{2}$ years, was unmanageable, destructive, mischievous and defiant, over-adventurous and without fear. His moods alternated; there were sudden periods of overactivity at irregular intervals when he would pursue his particular antisocial activity with grim intent. Between episodes he appeared happy and constructive. The boy was over the 97th percentile in height for his age, a fact that supports the view that increased height in the XYY syndrome is apparent before puberty.

It has been suggested that the ordinary degree of aggressiveness of a normal XY male is derived from his Y chromosome, and that by adding another Y a double dose of those potencies may facilitate the development of aggressive behavior[65] under certain conditions. A triple dose (XYYY) would be present in the case reported by Townes et al.[90].

The first reported case of an XYY constitution[39, 79] was studied because the patient had several abnormal children, although he appeared to be normal himself. Until recently, reports of the XYY constitution have been uncommon, probably because no simple method exists for screening the double-Y condition that is comparable to the buccal smear—sex chromatin body technique for detecting an extra X chromosome. Another possible explanation for the rarity of reports on the XYY karyotype is the absence of a specific phenotype in connection with it. Most syndromes with a chromosome abnormality are ascertained because of some symptom or clinical sign that indicates a need for chromosome analysis. Consequently there have been few studies that place the incidence of this chromosome abnormality in its proper perspective to the population as a whole.

. . . Recently a study of the karotypes of 2159 infants born in one year was made by Sergovich et al.[81]. These investigators detected 0.48 percent of gross chromosome abnormalities. In this sample the XYY condition appeared in the order of 1 in 250 males, which would make it the most common form of aneuploidy known for man. The previous estimate was about 1 in 2000 males. If this figure of 1/250 is valid for the population as a whole, it means that the great majority of cases go undetected and consequently must be phenotypically normal and behave near enough to the norm to go unrecognized.

Several cases of asymptomatic males have been published, including the first one described (Sandberg *et al.*[79] and Hauschka *et al.*[39]), which proved to be fertile. It appears that the sons of XYY men do not inherit their father's extra Y chromosome[59a].

Another fertile XYY male, reported by Leff and Scott[53], had inferiority feelings, was slightly hypochondriacal and obsessional, and not very aggressive. He gave a general impression of emotional immaturity. He was 6 feet, 6 inches tall, healthy, with normal genitalia and electroencephalogram. His IQ was 118. Wiener and Sutherland[93] discovered by chance an XYY male who was normal; he was 5 feet, 9½ inches tall, with normal genitalia and body hair, normal brain waves, and with an IQ of 97. He exhibited a cheerful disposition and mild temperament, had no apparent behavioral disturbance and never required psychiatric advice. This case supports the idea that an XYY male can lead a normal life.

SOCIAL AND LEGAL IMPLICATIONS OF THE YY SYNDROME

The concept that when a human male receives an extra Y chromosome it may have an important and potentially antisocial effect upon his behavior is supported by impressive evidence[15, 21]. Lejuene states that "There are no born criminals but persons with the XYY defect have considerably higher chances." Price and Whatmore[74] describe these males as psychopaths, "unstable and immature, without feeling or remorse, unable to construct adequate personal relationships, showing a tendency to abscond from institutions and committing apparently motiveless crimes, mostly against property." Casey and coworkers[16] examined the chromosome complements in males 6 feet and over in height and found: 12 XYY among 50 mentally subnormal and 4 XYY among mentally ill patients detained because of antisocial behavior; also 2 XYY among 24 criminals of normal intelligence. They concluded that their results indicate that an extra Y chromosome plays a part in antisocial behavior even in the absence of mental subnormality. The idea that criminals are degenerates because of bad heredity has had wide appeal. There is no doubt that genes do influence to some extent the development of behavior. The influence may be strongly manifested in some cases but not in others. Some individuals appear to be driven to aggressive behavior.

Several spectacular crime cases served to publicize this genetic syndrome, and it has been played up in newspapers, news magazines, radio and television. In 1965 Daniel Hugon, a stablehand, was charged with the murder of a prostitute in a cheap Paris hotel. Following his attempted suicide he was found to have an XYY sex chromosome constitution. Hugon surrendered to the police and his lawyers contended that he was unfit to stand trial because of his genetic abnormality. The prosecution asked for five to ten years; the jury decided to give him seven.

Richard Speck, the convicted murderer of eight nurses in Chicago in 1966, was found to have an XYY sex chromosome constitution. He has all the characteristics of this syndrome found in the Scottish survey: he is 6 feet 2 inches tall, mentally dull, being semiliterate with an IQ of 85, the equivalent of a 13-year-old boy. Speck's face is deeply pitted with acne scars. He has a history of violent acts against women. His aggressive behavior is attested by his record of over 40 arrests. Speck was sentenced to death but the execution has been held up pending an appeal of the conviction.[1]

In Melbourne, Australia, Lawrence Edward Hannell, a 21-year-old laborer on trial for the stabbing of a 77-year-old widow, faced a maximum sentence of death. He was found to have an XYY constitution, mental retardation, an aberrant brain wave pattern, and a neurological disorder. Hannell pleaded not guilty by reason of insanity, and a criminal court jury found him not guilty on the ground that he was insane at the time of the crime.

A second Melbourne criminal with an XYY constitution, Robert Peter Tait, bludgeoned to death an 81-year-old woman in a vicarage where he had gone seeking a handout. He was convicted of murder and sentenced to hang, but his sentence was commuted to life imprisonment.

Another case is that of Raymond Tanner, a convicted sex offender, who pleaded guilty to the beating and rape of a woman in California. He is 6 feet 3 inches tall, mentally disordered, and has an XYY complement. A superior court judge is attempting to decide whether Tanner's plea of guilty of assault with intent to commit rape will stand, or whether he will be allowed to plead innocent by reason of insanity.[2]

---

[1] Speck's death sentence was among those nullified by a U.S. Supreme Court decision in 1971, setting aside death sentences in cases where persons opposed to capital punishment are systematically excluded from jury duty.

The high court subsequently ruled that the death sentence constitutes "cruel and unusual punishment." In November 1972, Speck was sentenced to prison for a term of 400 to 1200 years. Incidentally, defense attorneys have announced that further research has revealed Speck's chromosomes to be normal. [Ed.]

[2] After having been convicted, Tanner attempted to enter a belated plea of not guilty by reason of XYY-chromosome-related insanity. But the original conviction was upheld, and Tanner is now incarcerated in a state hospital for the criminally insane. [Ed.]

Criminal lawyers in the United States have already begun to request genetic studies of their clients. In October of 1968 a lawyer for Sean Farley, a 26-year-old XYY man in New York who was charged with a rape-slaying, maneuvered to raise the issue of his client's genetic defect in court.

Many questions are raised by the double Y syndrome—basic social, legal and ethical questions—which will become more and more insistent as the implications of chromosome abnormalities take root in the public mind. Is an extra Y chromosome causally related to antisocial behavior? Is there a genetic basis for criminal behavior? If a man has an inborn tendency toward criminal behavior, can we fairly hold him legally accountable for his acts? If a criminal's chromosomes are at fault, how can we rehabilitate him?

The evidence to date is inadequate to prove conclusively the validity of the syndrome and convict all of the world's estimated five million XYY males of innate aggressive or criminal tendencies. But if the concept is proved, what then? The first step would seem to be to identify the XYY infants in the general population. This suggests the need for a nationwide program of automatic chromosome analysis of all newborns.

How should society deal with XYY individuals? If they are genetically abnormal, they should not be treated as normal. If the XYY condition dooms a man to a life of crime, he should be restrained but not punished. Mongolism also is a chromosome abnormality, and afflicted individuals are not held responsible for their behavior. Some valuable suggestions on the legal aspects of the double Y syndrome have been published recently by Kennedy McWhirter[59]. Elsewhere, Kessler and Moos[51] claim that definitive concepts relating to the YY syndrome have been accepted prematurely.

If all infants could be karyotyped at birth or soon after, society could be forearmed with information on chromosome abnormalities and perhaps it could institute the proper preventive and other measures at an early age. Although society can not control the chromosomes (at least at the present time) it can do a great deal to change certain environmental conditions that may encourage XYY individuals to commit criminal acts.

The theory that a genetic abnormality may predispose a man to antisocial behavior, including crimes of violence, is deceptively and attractively simple, but will be difficult to prove. Extensive chromosome screening with prospective follow-up of XYY males will be essential to determine the precise behavioral risk of this group. It is by no means universally accepted yet. Many geneticists urge that we should be cautious in accepting the interpretation that the double Y condition is specifically associated with criminal behavior, and particularly so with reference to the medico-legal validity of these concepts.

LITERATURE CITED

1. Anonymous. The YY syndrome. *Lancet* 1:583–584. 1966.
2. ———. Criminal behavior—XYY criterion doubtful. *Science News* 96: 2. 1969.
3. Baikie, A. G., W. M. Court Brown, K. E. Buckton, D. G. Harnden, P. A. Jacobs, and I. M. Tough. A possible specific chromosome abnormality in human chronic myeloid leukemia. *Nature* 188:1165–1166. 1960.
4. ———, O. Margaret Garson, Sandra M. Weste, and Jean Ferguson. Numerical abnormalities of the X chromosome. *Lancet* 1:389–400. 1966.
5. Balodimos, Marios C., Hermann Lisco, Irene Irwin, Wilma Merrill, and Joseph F. Dingman. XYY karyotype in a case of familial hypogonadism. *J. Clin. Endocr.* 26:443–452. 1966.
6. Barr, M. L. Sex chromatin and phenotype in man. *Science* 130:679. 1959.
7. ——— and D. H. Carr. Sex chromatin, sex chromosomes and sex anomalies. *Canad. Med. Assn. J.* 83:979–986. 1960.
8. ———, D. H. Carr, H. C. Soltan, Ruth G. Wiens, and E. R. Plunkett. The XXYY variant of Klinefelter's syndrome. *Canad. Med. Assn. J.* 90: 575–580. 1964.
9. Belsky, Joseph L. and George H. Mickey. Human cytogenic studies. *Danbury Hospital Bull.* 1:19–20. 1965.
10. Boczkowski, K. and M. D. Casey. Pattern of DNA replication of the sex chromosomes in three males, two with XYY and one with XXYY karyotype. *Nature* 213:928–930. 1967.
11. Buckton, Karin E., Jane A. Bond, and J. A. McBride. An XYY sex chromosome complement in a male with hypogonadism. *Human Chromosome Newsletter* No. 8, p. 11. Dec. 1962.
12. Carakushansky, Gerson, Richard L. Neu, and Lytt I. Gardner. XYY with abnormal genitalia. *Lancet* 2:1144. 1968.
13. Carr, D. H. Chromosome studies in abortuses and stillborn infants. *Lancet* 2:603–606. 1963.
14. ———, M. L. Barr, and E. R. Plunkett. An XXXX sex chromosome complex in two mentally defective females. *Canad. Med. Assn. J.* 84:131–137. 1961.
15. Casey, M. D., C. E. Blank, D. R. K. Street, L. J. Segall, J. H. McDougall, P. J. McGrath, and J. L. Skinner. YY chromosomes and antisocial behavior. *Lancet* 2:859–860. 1966.
16. ———, L. J. Segall, D. R. K. Street, and C. E. Blank. Sex chromosome abnormalities in two state hospitals for patients requiring special security. *Nature* 209:641–642. 1966.
17. ———, D. R. K. Street, L. J. Segall, and C. E. Blank. Patients with sex chromatin abnormality in two state hospitals. *Ann. Human Genet.* 32: 53–63. 1968.
18. Close, H. G., A. S. R. Goonetilleke, Patricia A. Jacobs, and W. H. Price. The incidence of sex chromosomal abnormalities in mentally subnormal males. *Cytogenetics* 7:277–285. 1968.

19. Conan, P. E. and Bayzar Erkman. Frequency and occurence of chromosomal syndromes. I. D-trisomy. *Am. J. Human Genet. 18*:374–386. 1966.

20. ——— and ———. Frequency and occurrence of chromosomal syndromes. II. E-trisomy. *Am. J. Human Genet. 18*:387–398. 1966.

21. Court Brown, W. M. Sex chromosomes and the law. *Lancet 2*:508–509. 1962.

21a. ———. Males with an XYY sex chromosome complement. *J. Med. Genet. 5*:341–359. 1968.

22. Cowie, John and Jacob Kahn. XYY constitution in prepubertal child. *Brit. Med. J. 1*:748–749. 1968.

23. DeCapoa, A., D. Warburton, W. R. Breg, D. A. Miller, and O. J. Miller. Translocation heterozygosis: a cause of five cases of *cri du chat* syndrome and two cases with a duplication of chromosome number five in three families. *Am. J. Human Genet. 19*:586–603. 1967.

24. Dent, T., J. H. Edwards, and J. D. A. Delhanty. A partial mongol. *Lancet 2*:484–487. 1963.

25. Edwards, J. H., D. G. Harnden, A. H. Cameron, V. Mary Crosse, and O. H. Wolff. A new trisomic syndrome. *Lancet 1*:787–790. 1960.

26. Eggen, Robert R. Chromosome Diagnostics in Clinical Medicine. Charles C Thomas, Springfield, Ill. 1965.

27. Ferguson-Smith, M. A., Marie E. Ferguson-Smith, Patricia M. Ellis, and Marion Dickson. The sites and relative frequencies of secondary constrictions in human somatic chromosomes. *Cytogenetics 1*:325–343. 1962.

28. Ford, C. E. and J. L. Hamerton. The chromosomes of man. *Nature 178*: 1020–1023. 1956.

29. ———, K. W. Jones, O. J. Miller, Ursula Mittwoch, L. S. Penrose, M. Ridler, and A. Shapiro. The chromosomes in a patient showing both mongolism and the Klinefelter syndrome. *Lancet 1*:709–710. 1959.

30. ———, K. W. Jones, P. E. Polani, J. C. DeAlmedia, and J. H. Briggs. A sex-chromosome anomaly in a case of gonadal dysgenesis (Turner's syndrome). *Lancet 1*:711–713. 1959.

31. ———, P. E. Polani, J. H. Briggs, and P. M. F. Bishop. A presumptive human XXY/XX mosaic. *Nature 183*:1030–1032. 1959.

32. Forssman, H. and G. Hambert. Incidence of Klinefelter's syndrome among mental patients. *Lancet 1*:1327. 1963.

33. Fraccaro, M., K. Kaijser, and J. Lindsten. Chromosome complement in gonadal dysgenesis (Turner's syndrome). *Lancet 1*:886. 1959.

34. ———, M. Glen Bott, P. Davies, and W. Schutt. Mental deficiency and undescended testia in two males with XYY sex chromosomes. *Folia Hered. Pathol.* (Milan) *11*:211–220. 1962.

35. Franks, Robert C., Kenneth W. Bunting, and Eric Engel. Male pseudohermaphrodism with XYY sex chromosomes. *J. Clin. Endocr. 27*:1623–1627. 1967.

36. Fraser, J. H., J. Campbell, R. C. MacGillivray, E. Boyd, and B. Lennox. The XXX syndrome—frequency among mental defectives and fertility. *Lancet 2*:626–627. 1960.

37. Gates, William H. A case of non-disjunction in the mouse. *Genetics 12*: 295–306. 1927.

38. Hamerton, J. L. Sex chromatin and human chromosomes. *Intern. Rev. Cytol.* 12:1–68. 1961.
39. Hauschka, Theodore S., John E. Hasson, Milton N. Goldstein, George F. Koepf, and Avery A. Sandberg. An XYY man with progeny indicating familial tendency to non-disjunction. *Am. J. Human Genet.* 14:22–30. 1962.
40. Hayward, M. D. and B. D. Bower. Chromosomal trisomy associated with the Sturge-Weber syndrome. *Lancet* 2:844–846. 1960.
41. Hunter, H. Chromatin-positive and XYY boys in approved schools. *Lancet* 1:816. 1968.
42. Hustinx, T. W. J. and A. H. F. van Olphen. An XYY chromosome pattern in a boy with Marfan's syndrome. *Genetica* 34:262. 1963.
43. Ismail, A. A. A., R. A. Harkness, K. E. Kirkham, J. A. Loraine, P. B. Whatmore, and R. P. Brittain. Effect of abnormal sex-chromosomal complements on urinary testosterone levels. *Lancet* 1:220–222. 1968.
44. Jacobs, P. A., A. G. Baikie, W. M. Court Brown, and J. A. Strong. The somatic chromosomes in mongolism. *Lancet* 1:710. 1959.
45. ———— and A. J. Keay. Chromosomes in a child with Bonnevie-Ullrich syndrome. *Lancet* 2:732. 1959.
46. ———— and J. A. Strong. A case of human intersexuality having a possible XXY sex-determining mechanism. *Nature* 182:302–303. 1959.
47. ————, A. G. Baikie, W. M. Court Brown, T. N. MacGregor, N. MacLean, and D. G. Harnden. Evidence for the existence of the human "super female." *Lancet* 2:423–425. 1959.
48. ————, Muriel Brunton, Marie M. Melville, R. P. Brittain, and W. F. McClemont. Aggressive behavior, mental subnormality and the XYY male. *Nature* 208:1351–1352. 1965.
49. ————, W. H. Price, W. M. Court Brown, R. P. Brittain, and P. B. Whatmore. Chromosome studies on men in a maximum security hospital. *Ann. Human Genet.* 31:330–347. 1968.
50. Kesaree, Nirmala and Paul W. Woolley. A phenotypic female with 49 chromosomes, presumably XXXXX. *J. Pediat.* 63:1099–1103. 1963.
51. Kessler, Seymour and Rudolph H. Moos. XYY chromosome: premature conclusions. *Science* 165:442. 1969.
52. Kosenow, W. and R. A. Pfeiffer. YY syndrome with multiple malformations. *Lancet* 1:1375–1376. 1966.
53. Leff, J. P. and P. D. Scott. XYY and intelligence. *Lancet* 1:645. 1968.
54. Lejeune, J. M., M. Gautier, and R. Turpin. Etude des chromosomes somatiques de neuf enfants mongoliens. *Compt. Rend. Acad. Sci.* 248:1721–1722. 1959.
55. ————, J. Lafourcade, R. Berger and M. O. Rethore. Maladie du cri du chat et sa reciproque. *Ann. Genet.* 8:11–15. 1965.
56. Lubs, H. A., Jr., E. V. Koenig, and L. H. Brandt. Trisomy 13–15: A clinical syndrome. *Lancet* 2:1001–1002. 1961.
57. Lyon, M. F. Gene action in the X-chromosome of the Mouse (*Mus musculus* L.). *Nature* 190:372–373. 1961.
58. MacLean, N., D. G. Harnden, W. M. Court Brown, Jane Bond, and D. J. Mantle. Sex-chromosome abnormalities in newborn babies. *Lancet* 1:286–290. 1964.

59. McWhirter, Kennedy. XYY chromosome and criminal acts. *Science 164*: 1117. 1969.

59a. Melnyk, John, Frank Vanasek, Havelock Thompson, and Alfred J. Rucci. Failure of transmission of supernumerary Y chromosomes in man. Abst. Am. Soc. Human Genet. Annual Meeting, Oct. 1–4, 1969.

60. Mickey, George H. Chromosome studies in testicular feminization syndrome in human male pseudohermaphrodites. *Mammalian Chromosome Newsletter* No. 9, p. 60. 1963.

61. Migeon, Barbara R. G trisomy in an XYY male. *Human Chromosome Newsletter* No. 17. Dec. 1965.

62. Milcu, M., I. Nigoescu, C. Maximilian, M. Garoiu, M. Augustin, and Ileana Iliescu. Baiat cu hipospadias si cariotip XYY. *Studio si Cercetari de Endrocrinologie* (Bucharest) 15:347–349. 1964.

63. Miller, Orlando J. The sex chromosome anomalies. *Am. J. Obstet. Gynec.* 90:1078–1139. 1964.

64. Minckler, Leon S. Chromosomes of criminals. *Science 163*:1145. 1969.

65. Montagu, Ashley. Chromosomes and crime. *Psychology Today* 2:43–49. 1968.

66. Mudal, S. and C. H. Ockey. The "double male": a new chromosome constitution in Klinefelter's syndrome. *Lancet* 2:492–493. 1960.

67. Nowell, P. C. and D. A. Hungerford. A minute chromosome in human granulocytic leukemia. *Science 132*:1497. 1960.

68. Painter, T. S. The chromosome constitution of Gates "non-disjunction" (v-o) mice. *Genetics 12*:379–392. 1927.

68a. Palmer, Catherine G. and Sandra Funderburk. Secondary constrictions in human chromosomes. *Cytogenetics 4*:261–276. 1965.

69. Patau, K. The identification of individual chromosomes, especially in man. *Am. J. Human Genet. 12*:250–276. 1960.

70. ————, D. W. Smith, E. Therman, S. L. Inhorn, and H. P. Wagner. Multiple congenital anomalies caused by an extra chromosome. *Lancet 1*: 790–793. 1960.

71. Penrose, L. S. The Biology of Mental Defect. Grune and Stratton, New York. 1949.

72. Pergament, Eugene, Hideo Sato, Stanley Berlow, and Richard Mintzer. YY syndrome in an American negro. *Lancet* 2:281. 1968.

73. Pfeiffer, R. A. Der Phanotyp der Chromosomenaberration XYY. *Wochenschrift* 91:1255–1256. 1966.

74. Price, W. H. and P. B. Whatmore. Behavior disorders and the pattern of crime among XYY males identified at a maximum security hospital. *Brit. Med. J.* 1:533. 1967.

75. ————, J. A. Strong, P. B. Whatmore, and W. F. McClement. Criminal patients with XYY sex-chromosome complement. *Lancet 1*:565–566. 1966.

76. Puck, T. T., A. Robinson, and J. H. Tjio. A familial primary amenorrhea due to testicular feminization. A human gene affecting sex differentiation. *Proc. Exper. Biol. Med. 103*:192–196. 1960.

77. Reitalu, Juhan. Chromosome studies in connection with sex chromosomal deviations in man. *Hereditas 59*:1–48. 1968.

78. Ricci, N. and P. Malacarne. An XYY human male. *Lancet 1*:721. 1964.

79. Sandberg, A. A., G. F. Koepf, T. Ishihara, and T. S. Hauschka. An XYY human male. *Lancet* 2:488–489. 1961.

80. ———, Takaaki Ishihara, Lois H. Crosswhite, and George F. Koepf. XYY genotype. *New England J. Med.* 268:585–589. 1963.

81. Sergovich, F., G. H. Valentine, A. T. L. Chem, R. A. H. Kinch, and M. S. Smout. Chromosome aberrations in 2159 consecutive newborn babies. *New England J. Med.* 280:851–855. 1969.

82. Smith, D. W., K. Patau, and E. Therman. The 18 trisomy syndrome and the D_1 trisomy syndrome. *Am. J. Dis. Child.* 102:587. 1961.

83. ———, K. Patau, E. Therman, S. L. Inhorn, and R. I. Demars. The D_1 trisomy syndrome. *J. Pediat.* 62:326–341. 1963.

84. Sohval, Arthur R. Sex chromatin, chromosomes and male infertility. *Fertility and Sterility* 14:180–207. 1963.

85. ———. Chromosomes and sex chromatin in normal and anomalous sexual development. *Physiol. Rev.* 43:306–356. 1963.

86. Telfer, Mary A., David Baker, Gerald R. Clark, and Claude E. Richardson. Incidence of gross chromosomal errors among tall criminal American males. *Science* 159:1249–1250. 1968.

87. Thorburn, Marigold J., Winston Chutkan, Rolf Richards, and Ruth Bell. XYY sex chromosomes in a Jamaican with orthopaedic abnormalities. *J. Med. Genet.* 5:215–219. 1968.

88. Tjio, J. H. and A. Levan. The chromosome number of man. *Hereditas* 42:1–6. 1956.

89. ———, T. T. Puck, and A. Robinson. The somatic chromosomal constitution of some human subjects with genetic defects. *Proc. Natl. Acad. Sci.* 45:1008–1016. 1959.

90. Townes, Philip L., Nancy A. Ziegler, and Linda W. Lenhard. A patient with 48 chromosomes (XYYY). *Lancet* 1:1041–1043. 1965.

91. Vignetti, P., L. Capotorti, and E. Ferrante. XYY chromosomal constitution with genital abnormality. *Lancet* 2:588–589. 1964.

92. Welch, J. P., D. S. Borgaonkar, and H. M. Herr. Psychopathy, mental deficiency, aggressiveness and the XYY syndrome. *Nature* 214:500–501. 1967.

93. Wiener, Saul, and Grant Sutherland. A normal XYY man. *Lancet* 2: 1352. 1968.

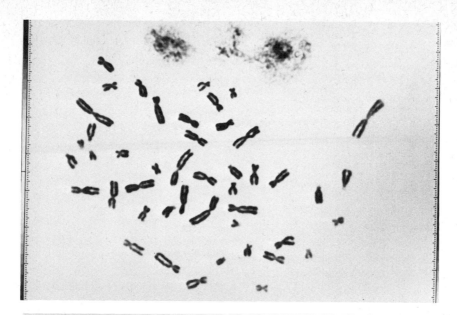

Karyotyping of human chromosomes is a process of isolating, photographing, and studying the structure of the set of 23 pairs of tiny DNA aggregates that embody the entire hereditary potentialities of the individual. To study the chromosomes more efficiently, the faint edges of the chromosomes are "enhanced" by the use of a computer. The missing areas of each chromosome are reconstructed by the computer, resulting in sharper chromosome outlines. *Top*, a photomicrograph of a human karyotype in a blood cell. *Bottom*, the chromosome set after computer enhancement. The X chromosome is number 25, and the Y number 29. (Photographs courtesy Dr. Kenneth R. Castleman, and Jet Propulsion Laboratories, NASA.)

PART TWO

THE EVOLUTIONARY SIGNIFICANCE OF PRIMATE BEHAVIOR

During the last two decades, studies of primates have been undertaken in the context of almost all of the sciences of behavior. Although such studies were raised to a high level of technique a generation ago, notably by the pioneer student of macaques and gibbons, C. R. Carpenter, the great period of primate research has been the 1960s and 1970s.

Psychologists such as Harry and Margaret Harlow have conducted significant and sophisticated laboratory experiments on infant monkeys, and on the ways in which emotional development is mediated by different relationships to play peers and maternal deprivation. Specialists in comparative psychology have specialized in studying the cognitive characteristics of different species of monkeys and apes. Ethologists have chosen to study free-ranging primate groups to document hypotheses about species-specific behavior patterns and their relation to facets of natural environment and population density and distribution.

Because apes and monkeys most closely resemble man in structure and physiological capacity, many studies have been undertaken with anthropoids as subjects prior to subjecting human astronauts to the rigors of a space environment.

Anthropologists have also conducted studies of primates, in the time-honored "observer-participant" mode of ethnography. Today field studies of infrahuman primates have become an integral part of the discipline of physical anthropology. As one might expect, the orientation of anthropological primate study has been evolutionary: using the understanding gleaned from infrahuman primates to supplement fossil evidence and archeological data in the reconstruction of human origins and the evolution of man's capacity for culture.

In reconstructing human origins, early anthropologists attempted to use descriptions of primitive people and the behavior of children as well as observations of the insane. They reasoned that commonalities in the behavior of these groups would represent "throwbacks" to the primitive conditions of mankind. A great deal of information about human behavior was gained by these studies, but this very information led to the demise of "throwback" theories. It was discovered that various tribal peoples of the world were no more "primitive" than civilized men. All living groups of men are recognized today as members of the single species *Homo sapiens*, possessing equivalent biological potentialities.

Children and the mentally ill, whose behaviors were thought to be atavistic, are now recognized as exhibiting immature and pathological forms of uniquely human behavior, that has little in common with the behavior of competent nonsapient adults who were our ancestors.

Today there is also the danger of uncritical use of the observations made by students of primate behavior. Such use can lead to simplistic "biological" explanations for highly complex human institutions and human behavior. For example, employing the device of selecting field data to fit the hypothesis, writers of popular books have "explained" human warfare and modern nationalism simply as the acting out of basic animal proclivities and drives.

Making valid use of the rapidly accumulating data on contemporary primate behavior is an endeavor second in difficulty only to the field studies themselves. We must remember that no living primate species is ancestral to us, or for that matter, "lower" than we are. Chimpanzees and macaques are just as highly adapted to their ecological circumstances as we are to ours.

The delicate and important point to determine in each case is *which* kind of primate behavior is significant for reconstructing each aspect of human evolution, on valid phylogenetic grounds. Obviously the characteristics of chimpanzees are more significant for human evolutionary reconstruction than the characteristics of ducks. The line leading to man diverged from the line leading to chimpanzees much more recently than the divergence of birds and mammals. But chimpanzees in the course of their evolution as chimpanzees have developed characteristics to make them into better chimpanzees, to specialize them to cope with conditions of tropical forest life, without significant cultural assistance. In evolving toward more efficient chimpanzee-hood, they certainly lost a number of features in which they were once more similar to our ancestors than they are today. Similarly, we have also lost a number of traits in which our ancestors resembled chimpanzees more than we do today. Pointing to data on contemporary chimpanzee behavior as evidence of a particular trait of early man or proto-man is a questionable technique. To employ infrahuman primates as special pleaders for one or another of the ideological views on the dilemma of modern man, as Lorenz, Ardrey, Tiger and Fox, and others have done, makes good conversation but less effective science.

Today anthropologists approach the question of "origins" by seeking the commonalities pervading the behavior of man's nonhuman relatives. If these contemporary nonhuman primates share certain behavioral traits, reasonable inferences may be made that man's nonhuman ancestors did also. The selections here are useful examples of this method of comparison of man and the infrahuman primates.

J. H. Crook reviews data from a number of primate field studies to develop a hypothesis on the interaction between species-specific behavioral capabilities and environmental factors. Jane Van Lawick-Goodall discusses agonistic behavior between individual chimpanzees, and leans over backwards *not* to draw general conclusions about the relevance of such data in explaining aggressive behavior in man.

Clifford Jolly attempts a reconstruction of the origins of hominid bipedalism and instrumental culture based on a careful analogy with a well-studied contemporary terrestrial primate, the baboon.

And finally, J. H. Napier compares the modes of locomotion in different primate groups in order to describe changes in locomotion in evolutionary terms.

Chimpanzee. Note the quadrupedal stance, and the "knuckle-walk" of the fore-limbs. (Photographs by Ron Garrison, by permission of San Diego Zoo.)

Left: Note the upright posture of this chimpanzee and the way the hallux is at an angle to the other toes. Free-ranging chimpanzees frequently shift to a bipedal stance for display, and in carrying objects. *Right:* The gorilla is the largest of all primates (this specimen is not fully grown). Possessed of a rudimentary heel, gorillas can move efficiently on the ground. (Photographs by Ron Garrison, by permission of San Diego Zoo.)

The quadrupedal posture of this mature male gorilla illustrates the heavy weight balance toward the forelimbs supported on the knuckles. (Photograph by Ron Garrison, by permission of San Diego Zoo.)

Baboon male, infant, and female. Note the sexual dimorphism exemplified by the difference in body size, in the brow ridges, and the hair pattern. (Photograph by Ron Garrison, by permission of San Diego Zoo.)

5

SOCIAL ORGANIZATION AND THE ENVIRONMENT: ASPECTS OF CONTEMPORARY SOCIAL ETHOLOGY

John Hurrell Crook

It is a commonplace in the academic arena that the broadness of a generalization is often in inverse correlation with the amount of hard data available. It is frequently easier to derive statements from a paucity of information than to theorize on the basis of extensive data. Nowhere is this more obvious than in the area relating our knowledge about subhuman social behavior to the understanding of human behavioral evolution.

On the basis of early returns, scientists and nonscientists alike have succumbed to the temptation of erecting theories of the evolution of human social behavior, using as evidence the studies of free-ranging infrahuman primates. As long as only a few terrestrial and arboreal species had been observed, it was possible for Robert Ardrey and Desmond Morris, for example, to construct models for the evolution of human behavior reminiscent of the "doctrine of original sin" model so prevalent before the scientific revolution. Man's "aggressiveness" or "territoriality" or "cooperativeness" or whatever the basic behavioral trait might be, could be traced in pursuasive prose from analogous behavior in nonhuman primates.

Even such reputable students of animal behavior as Lorenz and Tinbergen joined the discussion to provide underpinnings for such ideas about man's intrinsic aggressiveness. The spirit of the time has undoubtedly encouraged the acceptance of the view that in some baffling way man is a genetically evil species, cursed by his genes to reenact in every clime and time a vicious drama of struggle and barbarity.

Reprinted from *Animal Behaviour.* Part 2, May 1970, 18:197–209. Ballière, Tindall and Cassell, Ltd., Publishers. By permission of the author and the publishers.

*But the period of the 1960s saw an explosion of primate studies.
As more data became available, the less tenable became any theories
of social behavior that relied on the concept of "species-specific"
gene complexes as the mechanism of reconstructing the evolution
of human behavioral capabilities.*

*The author of the following selection is, in a sense, a member
of the "second generation" of behavioral scientists who have made
good use of the plentiful data now available on the social behavior
of infrahuman primates. In this article the student can see the
careful way in which data about nonprimate species is evaluated
and related to the problems of behavioral evolution, both human
and infrahuman. He should note, too, how ecological factors seem
to mediate the social behavior of animal populations, and perceive
the kinds of information still needed before theories of behavioral
evolution based on the argument from animal to man can be
constructed.*

*John Hurrell Crook has done behavioral research on Gelada
baboons, as well as nonprimate species. He is a former colleague
of the late K.R.L. Hall, and at present is a member of the Psychology Department of the University of Bristol, England.*

INTRODUCTION

During the rapid development of ethology in the last decade two diverging lines of research have become particularly apparent. In the first
and more voluminous development the classical ethology of Lorenz and
Tinbergen has flowered into a rigorous and lively area of study depending
fundamentally upon physiological research and the approaches of experimental biology. The main fields of investigation continue to be motivation analysis, developmental studies and the evolution of species specific behavior. A number of major textbooks presenting this material have
appeared. Preeminent among them is Hinde's (1966) remarkable cover-
age of the subject including an important attempt to synthesize approaches derived from behaviorist psychology with those of ethology.

The second development comprises a rapidly growing interest in the
relations between ecology, population dynamics and social behavior (e.g.
Klopfer 1962). The social emphasis here is not so much upon the traditional ethographic study of behavior patterns shown between conspecific
individuals usually studied in dyadic interaction, but upon the relations
between individuals and the natural group considered as the social environment within which they live and to which they are adapted. This
second development is the subject of the present paper and the field of
study will be termed Social Ethology. The links between the socially and

the physiologically oriented wings of ethology remain very close, particularly in such areas as the endocrinology of social interaction and developmental studies. These connections illustrate the continuing interdependence of the branches of the subject.

Curiously enough, in spite of an early emphasis by such workers as Espinas (1878), Kropotkin (1914) and Allee (1938), social ethology has not flowered in the clear cut manner of physiological ethology. Indeed, the recent popular accounts of the subject, avidly read both inside and outside the academic world, far from presenting a stable front of established knowledge, have revealed fundamental contrasts in theoretical orientation and interpretation and have given rise to noisy speculation regarding the inferences that may be made from animal to human social behavior.

One of the major reasons for this lack of clarity lies, I think, in the failure of ethologists generally to consider social behavior as a group process. Social behavior has been treated mainly in terms of reciprocal interactions between individuals presenting stereotyped signals to one another, signals moreover that were commonly species specific and which could be broadly considered as innate. Given such material, it is not difficult to interpret such behavior as the relatively straightforward outcome of neo-Darwinian selection. It would then follow that society, treated as a matrix of such behavioral interactions, is likewise a direct product of natural selection and adapted to the particular circumstances that have moulded it. A number of recent studies of social structure, admirable in other respects, have begun with this a priori supposition. If this were indeed true one could compare societies in the same way as the classical ethologists compared the behavior of individuals. Unfortunately, societies, being inadequately characterized by such an approach, are unlikely to be programmed entirely in this way as indeed studies of intraspecific variation make very clear.

Although the emphasis on fixed action patterns in communication has been one focus of mainstream ethology, an alternative and minority viewpoint has taken a very different stand, one moreover of increasing significance today. The sociologists Emile Waxweiler (1906) and Raphael Petrucci (1906) working in Brussels between 1900 and the First World War developed a well defined social ethology of which sociology referring to man was itself considered a part (Crook 1970a). Petrucci studied social structures as such: spatial dispersion, numbers in groups, group composition, etc., and he pointed out that there were few correlations between a taxonomy of social organizations and the classification of species. At each phyletic level, he saw a marked tendency for similar societies to emerge in parallel adaptation to similar conditions. He concluded that spatial dispersion, group composition and relations between individuals were directly responsive to the environment and that the factors programming the system included such features as food supply, predation and the requirements for sexual reproduction in differing habitats. The limitation on the range of social structures was determined only by the limited variability of the determining conditions. Petrucci concluded that since societies were determined directly by extrinsic factors they cannot be compared in the way that biologists compared morphological characteristics. And it followed, of course, that the social evolution of man was not to be explained purely in Darwinian terms.

Petrucci had taken an extreme position, but the almost total historical neglect of his work is quite unjustified. Recent studies of social organization, particularly in ungulates (Estes 1966) and primates (Crook 1970b), have revealed important intra-specific variations in group composition and inter-individual relations which for the most part appear to conform well with contrasts in the ecology of the demes concerned. Such social characteristics, furthermore, are more labile than the patterns of signalling which formerly comprised the main descriptions of social behavior.

It seems therefore that any statement about contemporary social ethology must begin with at least the following propositions.

(i) Social structure as a group characteristic cannot be conceived as a species specific attribute or property in quite the same way as has commonly been done with, for example, wing color or leg length. Instead, social structure is a dynamic system expressing the interactions of a number of factors within both the ecological and the social milieux that influence the spatial dispersion and grouping tendencies of populations within a range of lability allowed by the behavioral tolerance of the species.

(ii) Historical change in a social structure consists of several laminated and interacting processes with different rates of operation. Thus while the direct effect of environment may mould a social structure quickly, the indirect effects of this on learned traditions of social interaction come about more slowly and genetic selection within the society even more slowly still.

(iii) Because a major requirement for biological success is for the individual to adapt to the social norms of the group in which it will survive and reproduce it follows that a major source of genetic selection will be social, individuals maladapted to the prevailing group structure being rapidly eliminated. Social selection is thus a major source of biological modification. In advanced mammals it is perhaps of as great an importance as natural selection by the physical environment.

(iv) Lastly a methodological point. It seems desirable in considering group characteristics to shift the research emphasis away from questions concerning informational sources (i.e. relative significance of genetics, traditions, environmental programming, etc.) to direct analysis at the level of the social process itself. This shift would do much to bypass the never ending sterility and unreality of the nature–nurture controversy when applied to social life. An understanding of a social process will in itself help to define the nature of the factors involved in its programming.

From this basis we may attempt a brief conspectus of contemporary social ethology focusing upon three interdependent perspectives.

(a) *Socio-ecology*: the comparative study of social structure in relation to ecology. The main focus here is upon correlations between social organizations and contrasts in ecology.

(b) *Socio-demography*: the relations between social organization and population dynamics, including the role of social behavior as a mortality factor and hence as an important source of genetic selection.

(c) *Social systems research*: the study of the actual behavioral processes that maintain group structure, bring about social change and which may cause the social elimination of some individuals rather than others.

SOCIO-ECOLOGY

Within the last eight years a number of ornithological studies have shown quite clearly that the social structures of species populations correlate closely with ecology. Social structure in fact is one aspect of a whole syndrome of characteristics that are in the broadest sense adaptive. Huxley (1959) used the term "grade" to refer to the characteristic life styles of species adapted to a particular biotype. The species belonging to the same grade commonly show similar social organization. Such correlations have been well demonstrated by research on gulls and terns, penguins, Ploceine weaver birds, Estrildidae, Icterids, Sulidae, and a number of other groups that have been intensively studied comparatively in the field. The work of Tinbergen (1964, 1967), E. Cullen (1957), J. M. Cullen (1960), Stonehouse (1960), Crook (1964, 1965), Immelmann (1962, 1967), Orians (1961), Nelson (e.g. 1967), and Pitelka (1942 and unpublished) comes particularly to mind and Lack (1968) has recently published a major review.

The comparative approach of the ornithologists has now been applied to mammalian groups and has shown especial utility in studies of primates and ungulates. Both these groups are also relatively easy to study in the field, at least when compared with such cryptic beasts as insectivores, bats, rodents and many small carnivores. Furthermore, certain common themes occur in both primates and ungulates suggesting that we may soon have some important integrative principles by the tail.

The primate story due to Altmann, Carpenter, DeVore, Gartlan, Hall, Imanishi, Itani, Kummer, Rowell, Struhsaker, Washburn and other recent workers is perhaps most clearly illustrated by reference to the old-world Cercopithecoidea now extensively studied across a wide range of habitats.

Both interspecific and intraspecific population comparisons have been made (reviewed by Crook 1970b).

A survey of current studies suggests a tentative model of socio-ecological relations in these animals. Generally speaking, population demes consisting of one-male groups or "harems," together with peripheral males or all-male groups, occur in a variety of spatial arrangements, in the least ecologically stable savannah and Sahael areas having long harsh dry seasons and relatively low predation frequency. In ecologically more stable woodland savannah and light forest with less extreme annual climatic fluctuation and presumed to have higher potential predation frequency multi-male troops are common. Finally, in tropical and semitropical forests an increasing number of species have been found once more to live in small one-male groups often confined to small territories. Peripheral non-social males also occur and recent studies suggest a high rate of male interchange between groups, the female membership of which remains relatively constant.

It has been argued that this range of social contrasts is adapted to differences in the food resources and their seasonal availability in the differing habitats. In the more arid areas food resources are limited in the long dry season at which period dispersion in small groups, the reduction of males to a singleton in reproductive groups and the separate foraging of all-male groups allow optimum food availability to females which are commonly pregnant or lactating. The most effective evidence, although not conclusive, comes at present from the Gelada baboon in Ethiopia. These data are furthermore unconfounded by grouping tendencies influenced by the shortages of safe sleeping sites (Crook 1966; Crook & Aldrich-Blake 1968).

Geladas live near canyons and sleep on the gorge cliffs. During the day they wander either in harems or in congregations of independent harems and all-male groups that may together number several hundred. In the rains the animals move slowly in large herds over rich food resources. As the dry season progresses and food shortage becomes visibly apparent, Geladas travel rapidly in small groups often of single harem or all-male group size and the greater part of each day (e.g. up to 70 per cent of all afternoon activity) is spent in feeding. At this season all-male groups tend to disperse away from the canyon more markedly than do the harems. This appears to reduce the food competition between the two types of group. As we shall see such behavior also occurs in certain ungulates and leads to a differential mortality between the sexes, females being far more fortunate than the non-reproductive males.

In richer savannah the large troops of baboons or macaques live in areas where food resources normally appear to be sufficient to withstand sustained exploitation by sizeable groups and furthermore such groups offer increased protection against the many potential predators.

The discovery of one-male groups in forests poses difficult theoretical problems but we may perhaps employ in this primate context Ashmole's (1961, 1963) argument applying his explanation of sea-bird breeding biology to the reduced clutch sizes of forest bird species when compared with those of open country relatives (see Lack 1966, pp. 266–270). From Ashmole's viewpoint we may argue that the relatively unchanging forest environment, together with the high density of individuals present due to the great productivity of the biotope, produces conditions in which the population is very close to food shortage throughout the year. Under such circumstances it would be advantageous for individuals to live in small spatially circumscribed groups with low male representation. This would once more permit an optimum allocation of resources to reproductive females.

As Aldrich-Blake (1970) has remarked this view is certainly a simplification. While productivity in tropical forests may vary relatively little, this is not true within the home ranges of individual groups. In any one range there are relatively few food resources, say fruit trees, and these are in fruit only periodically. Feeding conditions, even given a wide variety of plant foods, are likely to vary greatly from week to week. Nevertheless he considers that feeding ecology seems very likely to be involved in an explanation of the contrasting spatial dispersions of forest monkey species. Current studies of the autecology of certain species and the synecology of the monkey populations of given forests will go far to explain these differences.

It remains plausible, however, that social forces alone may play a greater role in the determination of primate social behavior than is at present known. The Gelada and the Hamadryas baboons (Kummer 1968) live under quite comparable conditions and have social structures that at first sight resemble one another closely. Yet the fine details of group dynamics and the individual qualities that make for reproductive success differ greatly. Adaptation to contrasting group processes within groups of comparable structure may lead to the selection of animals of very different behavioral character. Contrasts in social processes within similar social structures may thus be based on differences in behavior that are the results of marked and long-term social selection.

Recent work on a wide range of African ungulates by Jarman (1968) reveals a fascinating range of social structures that partially parallels those of primates. In forest and forest fringes most species such as duikers and dik-dik are small in size, show little sex dimorphism, show no sexual segregation, have territories and live in pairs or alone. These animals are browsers. Other forest and savannah forms, oribi and bushbuck, for example, and also a number of montane animals not treated by Jarman live in small groups of up to about twelve. Males are a little larger than females. Some show territorialism and all are grazer–browsers. In savannah and open savannah the social units are various, usually consisting of one-male breeding groups, all-male groups and sometimes lone males. The one-male breeding groups may be territorial or move as parts of herds. In some species such as the wildebeeste (Estes 1966) some populations are migratory and gregarious, others nearby in a more stable ecology being stationary and territorial. These animals are browser–grazers and grazer–browsers. Out on open grassland plains occur vast herds of mixed sexes and ages (eland, buffalo) in which several varieties of one male group or multimale reproductive units occur. These are mostly of nomadic grazers.

Jarman interprets the survival value of these various grouping tendencies mainly in terms of food availability, particularly in relation to seasonal differences in resources. For example, he argues that the dispersed condition of many forest browsing ungulates is related to the spatial and temporal scattering of suitable food plants. The large nomadic herds of grass plains by contrast are a function of the need for protection from predators in an environment offering little cover and which allows large congregations to form. Nomadism is related to the danger of overexploitation of resources in a given locality, the large expanse of available habitat and the seasonality of the richness of the grass cover.

Jarman's (1968) study of species with one-male reproductive units living along the Kariba river is especially important, for he shows that in the dry season the one-male groups occupy the food-rich riverine areas while the excess males are dispersed further inland where they get less food and are subject to higher predation. He demonstrates that such a dispersal also means that a predator has a higher probability of encountering a male than a female prey animal. Clearly the one-male group structure is highly advantageous to the individual females and single males living in the breeding unit and is a direct cause of much non-breeding male mortality.

We cannot yet tell whether all mating systems of one-male group living ungulates necessarily have these effects. Jarman has nevertheless contributed important evidence for the involvement of a social factor in population dynamics, which goes a long way to vindicate arguments used in explaining the similar social structures in primates.

SOCIO-DEMOGRAPHY

The dispute in socio-demography has centered upon two alternative viewpoints. Lack (1954, 1966) originally interpreted population dynamics in birds as largely the outcome of interacting density-dependent environmental factors. He argues that spatial dispersion is a consequence of the natural selection of individuals and that it allows maximum recruitment from breeding units. By contrast Wynne-Edwards (1962) has argued that social behavior influences dispersion to maintain optimum numbers in relation to resources, that dispersion patterns are a consequence of group selection and that socially mediated mortality is the prime factor in population dynamics.

In the hands of the main protagonists these approaches have perhaps both tended to acquire somewhat scholastic attributes. There is in fact a serious dearth of critical studies and Chitty (1967) has called for a more open-ended theoretical and a more experimental approach to the whole problem.

Wynne-Edwards' work has focussed attention upon the question whether social attributes promoting dispersion do in fact play a major role in population dynamics as mortality factors and therefore also as important social selection pressures in evolution. A demonstration that this may often be the case would not however necessarily support Wynne-Edwards' theory as a whole. Indeed it would allow (see below) the construction of "open adaptive" models (Buckley 1967) of the socio-demographic process, perhaps representing quite closely Lack's more recent position (1968).

TERRITORY AS A SOCIAL MORTALITY FACTOR

Some six studies on a variety of birds appear to demonstrate the significance of territory in this respect. Jenkins, Watson & Miller (1963) working on grouse and ptarmigan, Delius (1965) on skylarks, Tompa (1964) on song sparrows and also Patterson (1965) and Coulson (1968) on certain colonial seabirds have all published material strongly supporting Kluyver & Tinbergen's (1953) original views concerning the great tit.

The studies of grouse, song sparrows, skylarks and Dutch tits all indicate that either the autumnal or spring occupation of territories causes a dispersion of the population in excess of the carrying capacity of the local habitat. The excluded birds may either leave the location and suffer increased mortality through various causes in the peripheral and suboptimal environments or they may remain as a nonbreeding population overlapping with the breeding population and to some extent perhaps competing for food and ready to occupy vacated territory as soon as it appears.

The work on gulls has shown that there is survival value in terms of reproductive success in occupying the preferred territories in the center of a colony rather than peripheral sites. Coulson in particular showed that intense competition for central sites in a kittiwake colony resulted in their occupation by heavy-weight birds. Both sexes surviving at least 5 years from the time of first breeding were heavier there than were neighbors that died within this time. In the center of the colony there was also a larger mean clutch size, higher hatching success and more fledged young than at the periphery.

It appears that spatial dispersion whether in conventionally territorial or colonial-territorial mating systems involves higher individual survival and better reproductive success for the occupiers of prime sites than for those animals forced to a periphery.

However, while the breeding stock of these species was commonly determined by dispersal, this was not always invariably the case. For example, on food-poor grouse moors, numbers varied more annually than on food-rich ones, suggesting a more direct control there by extrinsic environmental factors involving food shortage. Similarly with the skylarks, in one spring following a harsh winter the population was well below the carrying capacity of the local preferred habitat and no dispersal effects were operative that year.[1]

These findings recall Kluyver & Tinbergen's (1953) report that the numbers of great tits vary annually more in less preferred woods than in the most preferred areas nearby. Working at Oxford, Krebs (personal communication) reports that there is now evidence that the territorial factor may be more significant in the control of tit numbers than had been earlier supposed.

SOCIAL MORTALITY IN FLOCKS

Some years ago Lockie (1956), working then with Corvids, argued that individual distance and dominance-subordination phenomena in bird flocks had survival value in that in an encounter over a food item the loser could avoid its opponent without damage and the winner would win without a fight. Either way energy valuable to the individual is conserved. Under conditions of food shortage, however, the loser will progressively starve. The accumulative loss to a population mounts gradually, the relatively subordinate dying first. Were all birds equal presumably the cutback would be sharp and sudden with little mortality differential.

The work of Murton, Isaacson & Westwood (1964) has shown very clearly the importance of such socially mediated mortality in British wood pigeon populations. In flocks the feeding rates of these birds are greater in the middle and to the rear than in the van. Only a proportion of the pigeons can gain entry to the preferred flock center. Others are pushed to the periphery. Those in front, harried by those behind them, eat less and commonly flee from flock to flock, usually again landing in front and being hustled. Under limiting conditions the effect of such behavior leads to differential mortality.

Murton et al. (1964) note that the social effects allocating individuals to starving or nonstarving sections of a population do in fact adjust flock

---

[1] Conceivably this may always be the case in more northerly populations, subject to hard winters, annually.

size within limits to food availability and do maintain the highest survival rate relative to supplies. However the system sometimes breaks down. One year persistent heavy snowfall forced the pigeons into sudden conflict for limited Brassica plants. Then almost the whole population suffered a severe weight reduction at the same time. The effectiveness of social factors in limiting mortality thus depends on food type, item size and dispersion.

Comparable findings emerge from studies of the African savannah weaver bird, *Quelea quelea*, studied in the field by Ward (1965) and in the laboratory by Crook & Butterfield (1970). Ward, an ecologist, had shown that during the dry season period of reduced food availability these weavers show a rapid drop in weight, particularly marked in females. At the same time the proportion of males in the population increased. Ward concluded that competition for food led to an elimination of females. Laboratory work shows that males are dominant over females in mixed sex groups and experimental studies suggest that estrogen inhibition of the LH effects otherwise maintaining a low threshold for aggression in males is responsible for this sex difference. An effect of male dominance appears to be a major reduction in the number of females ready for reproduction at the onset of the breeding season. Unlike many species of weaver, the *Quelea* is obligatorily monogamous yet, even so, many nests in colonies are never occupied. It seems that in the rather harsh Sahael environment where *Quelea* lives the practice of monogamy allows a male to assist the female more in rearing the brood than would be the case had he several mates. Both his and the female's reproductive success may indeed be maximized by the procedure. The natural selection of the behavioral features determining this monogamy are likely to have arisen within the context of the shortage of available females for breeding. And this, as we have seen, is a consequence of differential socially-mediated mortality between the sexes in the dry season.

In general, the available studies show that the control of numbers is brought about by numerous factors, some environmental and some social. At any one period the key factor involved may be extrinsic, for example, food shortage, while under other conditions social factors such as territorial behavior may be the prime regulator through their effects on dispersal. Sometimes, one may suppose, that several factors interact to produce a given outcome and that no clear-cut key factor is operating. In addition the effects of an extrinsic factor may be mediated or buffered by a social process such as the intra-flock spacing mechanism to produce gradual and selective mortality rather than an abrupt fall.

Such a process of population control may be relatively easily modelled using a computer and allocating arithmetic values to hypothetical "Availability to Demand ratios" for each of the several commodities likely to act as controlling factors. Such a model is open and adaptive. A relative steady state is maintained by virtue of the limited variance of each factor. However, should one or more factor move by stepwise change to another range of variance, such a population may be conceived as adjusting to the new levels or new ecological "legislation," as Solomon (1964) calls it.

Such a control model, inherent in the views, for example, of Chitty and his associates differs markedly from the closed homeostatic model used by Wynne-Edwards. In the latter, the animal's perpetual calculating of the relation between its own numbers and its resources, even including those it has yet to exploit, functions as a Sollwert giving the fixed point about which the homeostat functions. There is in fact little evidence for such a model and the one proposed here appears to concord best with reality.

THE SOCIAL PROCESS

The traditional approach of ethology to social interaction has consisted in the study of reciprocal behavior usually between members of dyads and the signal patterns used in such behavior. The dyadic relations are classi- fied in terms of context, courtship, mating, etc., and the sum of such features (an ethogram) often treated as an adequate account of the social process. At least with advanced mammals, this now appears most unlikely to be the case.

Imagine attempting to understand a game of football by means of a study of the dyadic interactions between individual players. As Ray Bird- whistell has repeatedly emphasized, such an approach is sterile. To under- stand the game, the social location of each individual as a role player in relation to each and every other player needs description. Then, with the ball in motion, the relations between these relationships become apparent and the rules may be determined. To gain comparable information for behavior within mammal societies is an exacting task but one in which considerable progress has been made recently, using both wild and captive groups of primates, mostly macaques. To conclude this paper a brief account of the current perspectives in social primatology will be pre- sented, drawing mainly on the recent work of Japanese, American and European workers.

For many years the structure of primate groups was analyzed primarily in terms of dominance. The existence of a status hierarchy was generally thought to stabilize relationships through the reduction of social tension, each animal knowing its place. Often animals are found to cooperate either in the enforcement of existing rank relations or, by contrast, in upsetting them. Cooperation in social control emerges as an important problem area in primate social research. Recently it has become apparent that the simple dominance concept was not only inadequately defined and carrying many unwarranted motivational overtones but that the description of group structure in dominance terminology was in many species not only a difficult task but also an inappropriate procedure (Gartlan 1968).

In an important discussion of dominance in a captive baboon group Rowell (1966) infers that relative rank depends upon a continuous learning process in relation to rewards in interindividual competition for environmental or social goals. This occurs, moreover, against a background of differential kinship status and the observational learning of the behavioral styles of companions. Relative rank is much affected by health. Dominance ranking is based upon the approach–retreat ratio in encounters between two individuals in a group. Measures of rank by differing criteria do not however necessarily correlate and Rowell found no single criterion for high rank.

Rowell (1966) shows that apparent rank is a function of the behavior of the relatively subordinate. Higher rankers, at least in Cercopithecoid primates, evidently feel free to initiate interactions. These initiations commonly lead to some suppression in an ongoing activity by a subordinate or to an outright conflict. Subordinates learn to avoid such situations. Avoidance learning leads to behavioral restraint that leaves higher rankers even greater freedom of movement, easy access to commodities and freedom to initiate behavior with others. In competition for commodities in short supply, low rankers are likely to suffer deprivation and in social relations repeated constraint may involve physiological "stress" and concomitant behavioral abnormality.

Hall & DeVore (1965) describe the "dynamics of threat behavior" in wild baboon troops. A male's dominance status relative to others is a function not only of his fighting ability but also his ability to enlist the support of other males. In one group studied two adult males formed a central hierarchy, the pivot around which the social behavior of the group was organized. When one of these two died the remaining one was unable to prevent the third ranking male in cooperation with a newcomer (a subordinate male that had left another troop) from establishing themselves as central. The third male and the newcomer had evidently become affiliated when both had been relatively peripheral in the group structure. Common mutually supportive behavior seems to have been the pre-condition for their "success" in assuming high rank later. By so doing they gained the freedom to express behavior in the absence of previous constraints and to initiate behavior as and when they wished.

Wilson (1968) provides further information on mutual support in a study of the rhesus troops on Cayo Santiago Island. Young males tend to leave the smaller groups and move into the all-male peripheral areas of larger ones. When they do this they are commonly attacked unless they gain the protection of another male already established there. It so happens that males that give support are usually relatives, even brothers, who originated from the same natal group as the "protegé."

The inadequacies of the dominance terminology have led Bernstein & Sharpe (1966), Rowell (1966) and Gartlan (1968) to describe the social positions of individuals in a group in terms of roles. Roles are defined in terms of the relative frequencies (e.g. per cent of group occurrence) with which individuals perform certain behavioral sequences. When the behavior set of an individual or class of individual is distinct the animal is said to show a "role."

Bernstein (1966) emphasizes particularly the importance of the role of "control animal" in primate groups and shows that such a role may occur in a group of capuchins, for example, in which no clear status hierarchies can be established. The prime responses of a control animal are assuming a position between the group and a source of external disturbance or danger, attacking, and thereby stopping, the behavior of a group member that is distressing another, and generally approaching and terminating cases of intragroup disturbance. Whether or not a control animal is also recognizably the "dominant" or a "leader" (in the sense of determining direction of march) depends upon the social structure in which he or she is situated.

Social position in primate groups may be well described in terms of roles but little attention has been given as yet to an appropriate set of descriptive terms. It is one thing to say an animal shows a "role," another to say precisely what is meant. Using concepts derived from writers such as Nadel (1957) and Sarbin (1959) we may describe a primate's social behavior in terms of the individual's age and sex status, social position and group type affiliation. In any given group each individual shows characteristic patterns of response in relation to others in the group, to older animals, to dominant animals, to subordinates, to peers of comparable kinship, rank, etc. The range of an animal's behavior patterns in relation to others in the group comprise its social behavior repertoire, as shown with its companions in the group. The sum of the proportions of group scores of such characteristics shown by an individual defines its social position in the group. An observer may prepare a social position matrix by allocating the relative frequency of interaction patterns shown by each member to a particular cell. In a more general statement each individual may be defined by its proportion of the total of the various interaction patterns shown within the group.

It so happens that macaque social positions may be categorized into consistent types that recur repeatedly in new groups formed from the division of the old or in separately analyzed independent groups. Such categories are termed "roles." There is, for example, the control male role, the central sub-group secondary male role (competitive with another in the sub-group, dominant to peripheral males but subordinate to the central animal and commonly supportive of him), the peripheral male role, the isolate male, the central and peripheral female roles. We may also bracket together certain types of behavior to describe roles of animals of high status kinship and low status kinship respectively. Now, these styles of behavior may be called roles because they are not fixed and immutable aspects of the life of a given individual. On the disappearance of a control animal another male typically adopts the same role. A male rising or falling in a dominance hierarchy changes his behavior. Fallen males may, however, march out of the group with affiliated females and establish a new group in which they may adopt the role of a central animal. Isolate males may enter a group and form a central sub-group, one of them perhaps being the control.

While not every conceivable role is necessarily present in any given group there is an overall consistency that makes this approach meaningful. Certain roles are usually characteristic of some kinds of groups and not of others—the harem "overlord" occurs in Hamadryas, Patas and Gelada groups but not in other baboons or among macaques.

The behavior characteristic of a particular role is not the "property" of the individual playing it. Such behavior is not fixed by conditioning so that the individual remains forever the same. Physiological and social changes impel behavioral shifts so that in a lifetime individuals may play many roles in their social structures. Social mobility in the sense of role changing is an important attribute of primate groups. It seems more characteristic of males than of females. Young and subadult males go solitarily, live peripherally or in all-male groups or shift from one reproductive unit to another. Females are more loyal to their natal group providing the more fixed social positions around which males revolve. This mobility is an expression of a set of tensions or forces characteristic of group life. At least three sets of factors interact in this social sorting process (Crook 1970c). These are: (i) the maturing and aging of individual group members, (ii) the growth and splitting of groups, and (iii) intragroup competition for environmental and social commodities together with the affiliative cooperation and subterfuge that this entails.

The third set is the most critical at any one time, the other factors acting over longer periodicities. There appear then at any one time to be two opposed social processes operating within a baboon or macaque troop. One consists of the assertive freely expressed utilization of available physical and social commodities by high status animals which has the effect of constraining the behavior of juveniles and subordinates. The other consists of the adoption of behavioral subterfuge by certain subordinates whereby such behavioral constraint can be avoided (e.g. by temporary solitary living or by splitting up the group into branch groups permitting a greater freedom of expression to the new leaders).

This "subterfuge" is presumably not conscious or deliberate. It appears to arise from the need of certain subordinates to free themselves from the behavioral constraints to which they have been exposed. There are in fact relatively few known "routes" along which such an aspirant may move. These are: (a) temporary solitarization, followed by take-over of a small branch group; (b) the use of an affiliated companion to ensure the rise of a dominant animal into a higher or even a centrally located position; and (c) the care given by "aunts" and "uncles" to infants and babies as a means of entering high-status sub-groups and so to affiliate with them.

Examples of the last case are particularly interesting. Itani (1959) showed that adult male macaques become interested in juvenile animals when the latter are 1 year old. At this time females are in the birth season and cease protecting the young of a previous year. The males appear to show interest in young animals only in this period. Their behavior consists of hugging and sitting with the infant, accompanying it on the move, protecting it from other monkeys and dangerous situations and grooming it. Although Itani found no relationship between dominance as such and the frequency of child care he did find the behavior to be especially developed in middle-classed animals of the caste structured troops. The behavior indeed seemed pronounced in animals that were sociable, not aggressive and oriented towards high caste animals. Itani suggests that certain monkeys by showing care to high caste children succeed in being tolerated by leaders and their affiliated females. Such animals may therefore rise in rank. Protectors usually behave in a mild manner which may facilitate both their association with infants and their rise in rank. It is not clear, however, whether the effect on rank lasts for more than the birth season. Itani gives no details of permanent changes in social structure so produced. The adult males appear to protect 1-year-old males and females equally but more second-year females than males are protected owing to the social periphalization of young males. Protected 2-year-olds are moreover often poorly grown individuals that were protected by the same male in the previous year. Protection has a further interesting effect in that adults more readily learn to investigate the new foods a progressive infant has discovered than would otherwise be the case.

Itani found cases of male care common only in three of eighteen troops investigated. It occurred very rarely in seven others and was not observed in the remaining eight. The behavior thus appears to be a local cultural phenomenon. It undoubtedly provides additional chances of survival for young animals likely to be relatives, increases the rate of spread of new patterns of behavior in the group and, since its frequency differs between troops, may increase the reproductive success of some monkey groups over others.

Recent observations on the Barbary Macaque (*Macaca sylvana*) in the Moyen Atlas of Morocco show that adult males in wild groups show an extraordinary amount of interest in young babies of the year. An interest in babies so young was relatively uncommon in Itani's study. Furthermore, the male Barbary Macaque does not seem, on present evidence, to limit his interest to a particular infant. He appears to appropriate babies from females in the group and to groom and care for them for short periods usually under 15 minutes duration. Babies may, moreover, move away from their mothers to accompany males, commonly riding off on their backs. Babies also move from one male to another. Males carrying babies frequently approach males without them in such a way as to encourage the approached animal to engage in a mutual grooming session with the baby as target. In some of these sequences babies are presented while on the back of the approacher, the animal turning its rear towards the other male (as it would do in sexual presenting) as it does so. In one case the approached male was seen to mount and thrust an animal that had done this. It appears too, that most males approaching with babies are relatively juvenile animals. It looks as if relatively subordinate animals are using the babies in some way to improve their relations with higher ranking males (Deag & Crook, 1970).

So far too little is known about the social structure of the Barbary Macaque to show whether there is a caste system comparable to that described from Japan. Whether males are using the babies as a means to increase their social standing and, hence, their freedom to behave without the constraints imposed by low rank, remains unknown. It does seem, however, that in both species not only does the male's behavior increase the chances of survival for young animals but it appears to be closely related to the structure of the monkey group and the patterns of constraints regulating social mobility of individuals within it. We do not know yet whether the behavior of the Barbary Macaque male is restricted, as it is with the Japanese Macaques, to certain troops and localities or whether it is a general phenomenon found throughout the species.

In all these cases individuals closely affiliated to others can combine cooperatively to bring about the liberating effect, an effect which, furthermore, in times of shortage (food, females) would ensure an increased probability of survival and/or reproductive success in addition to the psychological freedom from the effects of "stress" that undoubtedly accompany in some degree a continuing social restraint on individual behavioral expression. Similarly, by enlisting the cooperation of affiliated animals, often relatives, in their "control" behavior, animals that are already established in social positions of high rank will be able to maintain such positions, and the access to commodities it provides.

It seems, then, that cooperative behavior of the high order found in macaque and baboon groups has arisen within the context of competition for access to both physical and social commodities (Crook 1970c discusses). Both direct affiliative behavior and indirect affiliation, for example through a common interest in child-care, provide the basis for common action. As we have seen such cooperation provides numerous advantages for the participants, and not only for the most dominant animals among them. It also provides the behavioral basis for the complex class-structured society of these animals with its tolerance of individual mobility between roles.

Finally, the stable social structure maintained by a powerful clique around a control animal seems to provide the optimum circumstances for maternal security and child rearing. Females form the more cohesive elements of primate groups and, as a consequence of their affiliative relations and kinships links, may play a much greater role in determining who emerges as "control" than is at present known. Males, by contrast, subject to the full force of social competition, are the more mobile animals transferring themselves, as recent research shows, quite frequently from one group to another.

CONCLUSIONS: SIGNIFICANCE FOR HUMAN ETHOLOGY

In this paper I have tried to present a conspectus of the current state of social studies in ethology. Much more could of course have been added and not all would adopt the particular orientation I have used. It is my belief that an effective ethology of man must be based primarily on social considerations. Such a basis may possibly be supplied by a viewpoint such as I have adopted here. Often inherent in this approach is the use of social psychological ideas as a useful adjunct to ethological analysis. This mutuality between two disciplines is likely to develop quickly probably to the benefit of both.

There seems to be a problem in differentiating "human ethology" from the already existing behavioral sciences of man although Waxweiler had in 1906 already made clear that ethology as a broad biopsychological discipline included sociology and its derivatives as a special branch subject focussing upon the problems of man. Today it seems that by human ethology we designate not so much a subject as a way of approaching human behavior viewing the species as a member of the animal kingdom rather than as a separate and peculiar phenomenon. This I believe to be most healthy so long as one remembers that man is indeed a peculiar phenomenon and that mere biological reductionism, a "nothing butism" as Julian Huxley used to say, will get us nowhere.

An attempt to apply classical ethology to man would suggest that an ethological version of kinesics and proxemics was in the offing together with an elaboration of evolutionary theories regarding the adaptive significance of breasts and beards. Although interesting this would seem too curtailed an approach.

My feeling is that human ethology should comprise the whole behavioral biology of man. Such a science, it seems to me, must focus its attention on at least the following problems:

(1) The evolution and cultural history of the basic human grouping structures, families, all-male congregations, etc., and their origins in non-human primate grouping patterns (Reynolds 1966, 1968).

(2) The relations between human individual behavior, group composition and population density. How far do the problems of rodent stress-physiology apply also to man?

(3) Non-verbal communication through facial and postural expression. The evolution of emotional expression in man.

(4) The functioning of non-verbal communication in the control of affect in small groups, the rules of interpersonal ethology in companionship and courtship, the uses of ethology in sensitivity training, control of interpersonal aggression, etc. (Goffman 1963; Argyle 1967).

A programme as broad as this needs a basis within the framework of a biological social ethology such as I have tried to outline in this paper and which must include at least the three perspectives discussed. The scope is clearly very wide with links to anthropology, demography, stress physiology, ecology and particularly social psychology and psychiatry. To obtain workers equipped for research in this area requires an emphasis on training and a focus on such problems within the university syllabus. The day of routine study patterned by the ethological traditions of the past is over. We need fresh orientations and to find them social ethologists must go looking in new places along unexplored paths.

SUMMARY

1. A major development within the ethology of the last decade focusses attention upon the relations between social behavior, ecology and population dynamics. This field may be termed "social ethology" following an early but neglected usage of the term by Waxweiler at the start of the century.

2. Contemporary social ethology comprises three interdependent perspectives: socio-ecology, socio-demography and the study of social processes within natural and experimental groups.

3. In socio-ecology recent studies reveal that close correlations exist between the forms of avian and mammalian social organizations and their respective ecological niches. In particular the adaptive significance of certain mammalian societies comprising, on the one hand, multimale reproductive units and, on the other, those made up of one-male and all-male units is discussed and explanatory hypotheses derived from primate and ungulate data briefly considered.

4. In socio-demography research suggests that socially mediated mortality is of greater significance in the density-dependent control of bird and mammal numbers than had formerly been thought. The relations between

ecological factors extrinsic to and social factors intrinsic to a social organization may be modelled in the form of open adaptive cybernetic systems rather than expressed in terms of analogies to closed mechanical or physiological systems.

5. Studies of social processes in non-human primate groups suggest that some form of role analysis may prove heuristic. The relations between dominance status, affiliative and kinship relations, social subterfuge and competition-contingent cooperation are discussed in an attempt to outline the dynamics of social change within relatively stable group structures.

6. "Human ethology" seems at present to lack adequate definition as an academic discipline. Social ethology may provide the essential biological basis for future research in this area.

REFERENCES

Aldrich-Blake, F. P. G. 1970 Problems of social structure in forest monkeys. In: *Social Behaviour in Birds and Mammals* (Ed. by J. H. Crook). London: Academic Press.

Allee, W. C. 1938 *Cooperation Among Animals.* New York: H. Schuman.

Argyle, M. 1967 *The Psychology of Interpersonal Behaviour.* London: Pelican.

Ashmole, N. P. 1961 The biology of certain terns. D. Phil. Thesis, Oxford.

Ashmole, N. P. 1963 The regulation of numbers of tropical oceanic birds. *Ibis, 103b,* 458–473.

Bernstein, I. S. 1966 Analysis of a key role in a capuchin (*Cebus albifrons*) group. *Tulane Stud. Zool., 13,* 49–54.

Bernstein, I. S. & Sharpe, L. G. 1966 Social roles in rhesus monkey group. *Behaviour, 26,* 91–104.

Buckley, W. 1967 *Sociology and Modern Systems Theory.* New York: Prentice-Hall.

Chitty, D. 1967 What regulates bird populations? *Ecology, 48,* 698–701.

Coulson, J. C. 1968 Differences in the quality of birds nesting in centre and on the edges of a colony. *Nature, Lond., 217,* 478–479.

Crook, J. H. 1964 The evolution of social organisation and visual communication in the weaver bird (Ploceinae). *Behaviour,* Suppl. 10.

Crook, J. H. 1965 The adaptive significance of avian social organisations. *Symp. zool. Soc. Lond., 14,* 181–218.

Crook, J. H. 1966 Gelada baboon herd structure and movement, a comparative report. *Symp. zool. Soc. Lond., 18,* 237–258.

Crook, J. H. 1970a Social behavior and ethology. In: *Social Behaviour in Birds and Mammals* (Ed. by J. H. Crook). London: Academic Press.

Crook, J. H. 1970b The socio-ecology of primates. In: *Social Behaviour in Birds and Mammals* (Ed. by J. H. Crook). London: Academic Press.

Crook, J. H. 1970c Sources of cooperation in animals and man. In: *Man and Beast, Comparative Social Behaviour.* IIIrd International Symposium. Smithsonian Institution, 1969.

Crook, J. H. & Aldrich-Blake, P. 1968 Ecological and behavioural contrasts between sympatric ground dwelling primates in Ethiopia. *Folia primat.*, 8, 192–227.

Crook, J. H. & Butterfield, P. A. 1970 Gender role in the social system of Quelea. In: *Social Behaviour in Birds and Mammals* (Ed. by J. H. Crook). London: Academic Press.

Cullen, E. 1957 Adaptations in the Kittiwake to cliff-nesting. *Ibis*, 99, 275–302.

Cullen, J. M. 1960 Some adaptations in the nesting behaviour of terns. *Proc. XII Int. Orn. Congr. Helsinki*, 1958, pp. 153–157.

Deag, J. & Crook, J. H. 1970 Social behaviour and ecology of the wild Barbary Macaque (*Macaca sylvana* (L.)). (In preparation).

Delius, J. D. 1965 A population study of skylarks, *Alauda arvensis. Ibis*, 107, 465–492.

Espinas, A. 1878 *Des Societés Animales*. Paris: Baillière.

Estes, R. D. 1966 Behaviour and life history of the Wildebeest (*Connochaetes taurinus* Burchell). *Nature, Lond.*, 212, 999–1000.

Gartlan, J. S. 1968 Structure and function in primate society. *Folia primat*, 8, 89–120.

Goffman, E. 1963 *Behaviour in Public Places*. New York: Free Press.

Hall, K. R. L. & DeVore, I. 1965 Baboon social behavior. In: *Primate Behavior* (Ed. by I. DeVore). pp. 53–100. New York: Holt, Rinehart & Winston.

Hinde, R. A. 1966 Animal behaviour. In: *A Synthesis of Ethology and Comparative Psychology*. New York: McGraw-Hill.

Huxley, J. S. 1959 Grades and grades. Systematics Ass. Publ. No. 3. *Function and Taxonomic Importance*, pp. 21–22.

Immelmann, K. 1962 Beiträge zur Biologie und Ethologie australischer prachtfinken (Spermestidae). *Zool. Jb.* (*Syst.*), 90, 1–196.

Immelmann, K. 1967 Verhaltensökalogische Studien an afrikanischen und australischen Estildidae. *Zool. Jb.* (*Syst.*), 94, 1–67.

Itani, J. 1959 Paternal care in the Wild Japanese Monkey, *Macaca fuscata. J. Primat.*, 2, 61–93.

Jarman, P. 1968 The effect of the creation of Lake Kariba upon the terrestrial ecology of the Middle Zambesi valley, with particular reference to the large mammals. Ph.D. Thesis. Manchester University.

Jenkins, D., Watson, A. & Miller, G. R. 1963 Population studies on Red Grouse, *Lagopus I. scoticus. J. anim. Ecol.*, 36, 97–122.

Klopfer, P. H. 1962 *Behavioral Aspects of Ecology*. New Jersey: Prentice Hall, Englewood Cliffs.

Kluyver, H. N. & Tinbergen L. 1953 Territory and the regulation of density in titmice. *Arch. neerl. Zool.*, 10, 266–287.

Kropotkin, P. 1914 *Mutual Aid, a Factor in Evolution*. New York: A. A. Knopf, Inc.

Kummer, H. 1968 Social organisation of Hamadryas baboons. *Bibliotheca Primatologica*, No. 6.

Lack, D. 1954 *The Natural Regulation of Animal Numbers*. Oxford: Clarendon Press.

Lack, D. 1966 *Population Studies of Birds*. Oxford: Clarendon Press.

Lack, D. 1968 *Ecological Adaptations for Breeding in Birds*. London: Methuen.

Lockie, J. 1956 Winter fighting in feeding flocks of Rooks, Jackdaws and Carrion Crows. *Bird Study, 3,* 180–190.

Murton, R. K., Isaacson, A. T. & Westwood, K. T. 1964 The relationships between woodpigeons and their clover food supply and the mechanism of population control. *J. appl. Ecol., 3,* 55–96.

Nadel, S. F. 1957 *The Theory of Social Structure*. Glencoe, Illinois: Free Press.

Nelson, J. B. 1967 Etho-ecological adaptations in the Great Frigate-bird. *Nature, Lond., 214,* 318.

Orians, G. H. 1961 The ecology of blackbird (*Agelaius*) social systems. *Ecol. Monogr., 31,* 285–312.

Patterson, I. J. 1965 Timing and spacing of broods in the Black Headed Gull (*Larus ridibundus*). *Ibis. 107,* 433–459.

Petrucci, R. 1906 Origine polyphyletique, homotypie et non-comparabilité direct des societiés animales. *Trauvaux de l'Institut de Sociologie. Notes et Memoires, 7,* Bruxelles, Instituts Solvay.

Pitelka, T. A. 1942 Territoriality and related problems in North American humming birds. *Condor, 44,* 189–204.

Reynolds, V. 1966 Open groups in hominid evolution. *Man, 1,* 441–452.

Reynolds, V. 1968 Kinship and the family in monkeys, apes and man. *Man, 2,* 209–223.

Rowell, T. E. 1966 Hierarchy in the organization of a captive baboon group. *Anim. Behav., 14,* 420–443.

Sarbin, T. R. 1959 Role theory. In: *Handbook of Social Psychology* (Ed. by G. Lindzey). Cambridge, Massachusetts: Addison-Wesley.

Solomon, M. E. 1964 Analysis of procedures involved in the natural control of insects. In: *Advances in Ecological Research* (Ed. by J. B. Crag), Vol. 2. London and New York: Academic Press.

Stonehouse, B. 1960 The King Penguin *Aptenodytes pategonica* of South Georgia. I. Breeding behaviour and development. *Sci. Rep. Falkland Is. Depend. Surv., 23,* 1–181.

Tinbergen, N. 1964 On adaptive radiation in Gulls (Tribe Larini). *Zool. Meded., 39,* 209–223.

Tinbergen, N. 1967 Adaptive features of the Black Headed Gull (*Larus ridibundus* L.). *Proc. XIV Int. Orn. Congr.,* Oxford, 1966, pp. 43–59. Oxford: Blackwell.

Tompa, F. S. 1964 Factors determining the numbers of Song Sparrows *Melospiza melodia* (Wilson) on Mandarte Island, B. C. Canada. *Acta Zool. fenn., 109,* 1–68.

Ward, P. 1965 Seasonal changes in the sex ratio of *Quelea quelea* (Ploceinae). *Ibis, 107,* 397–399.

Waxweiler, E. 1906 Esquisse d'une Sociologie *Travaux l'Institut de Sociologie. Notes et Memoires, 3,* Bruxelles: Instituts Solvay.

Wilson, A. P. 1968 Social Behaviour of free-ranging rhesus monkeys with an emphasis on aggression. Ph.D. Thesis, University of California, Berkeley.

Wynne-Edwards, V. C. 1962 *Animal Dispersion in Relation to Social Behaviour*. Edinburgh: Oliver & Boyd.

6

SOME ASPECTS OF AGGRESSIVE BEHAVIOR IN A GROUP OF FREE-LIVING CHIMPANZEES

Jane van Lawick-Goodall

For a long time, the chimpanzee (Pan) has been accepted as man's closest relative. Almost every known fact about this interesting animal has been gathered from observations in the laboratory or in the zoos. More recently, field studies of chimpanzees in their natural surroundings in central Africa have contributed fresh insight into the primate substrate of human behavior, particularly the behavior of the earliest hominids who were not far removed from ape status themselves.

However, we should not interpret chimpanzee behavior as directly reflecting the behavior of the earliest men. Chimpanzees are themselves the product of evolutionary processes involving adaptations to special environments, and we, of course, are not descended from any living ape. Nevertheless, field studies of free-ranging groups, such as the following paper by Jane van Lawick-Goodall, enable us not only to appreciate the potentialities of the infrahuman primate behavioral inventory but also to see more clearly the actual ways in which the hominids differ from the apes.

Dr. Jane van Lawick-Goodall is Scientific Director of the Gombe Stream Research Center in Tanzania.

INTRODUCTION

In 1960 I began a field study of the social behavior of wild chimpanzees (*Pan troglodytes* [or *satyrus*] *schweinfurthi*) in the Gombe National Park, Tanzania (East Africa). The park consists of a narrow stretch of rugged mountainous country running for some ten miles along the eastern shores of Lake Tanganyika and running inland three miles or so to the

Reprinted from *International Social Science Journal*, 23:1, 1971. Reproduced by permission of Unesco and the author.

tops of the peaks of the rift escarpment. The area supports between 100 and 150 chimpanzees.

Chimpanzees are nomadic within a fairly large home range (which may be thirty square miles or more for an adult male) and they follow no regular routes in their daily wanderings in search of food. Moreover, unlike many primate species, chimpanzees do not move about in stable, or fairly stable groups, but in small temporary associations, the membership of which constantly changes. The only stable association, over a period of years, is a mother and her younger offspring: such a sub-group freely joins up with and leaves other temporary associations.

In 1963 I established a feeding area where chimpanzees could get bananas and, for the first time, it became possible to make fairly regular observations on a number of individuals. By 1964 some forty chimpanzees were visiting the feeding area, some regularly, others infrequently. I took on research assistants trained in my observation methods and recording techniques. Since 1964 observations on the chimpanzees have been kept up on a daily basis.

It is sometimes possible to follow a group of chimpanzees for hours as it wanders through the forest without observing a single aggressive incident. At the artificial feeding area, however, where the chimpanzees compete for a favored food which is in comparatively short supply, disputes, including fighting, occur more often, giving an opportunity to study the mechanics of aggressive behavior.

THREAT AND ATTACK

Before I discuss the sorts of situation in which chimpanzees most usually show aggressive behavior, let me briefly outline the more obvious sounds, gestures and postures which make up the threat and attack repertoire of the species.

Those aggressive behaviors which do not involve physical conflict, but which merely elicit submissive behavior, avoidance or flight in the individual to whom they are directed may be termed threat; and chimpanzees, like most animals, solve more disputes by means of threat than by actual fighting. A dominant chimpanzee may fix his subordinate with an intent and prolonged stare, he may slightly jerk his chin upward whilst uttering a soft bark, he may raise his arm rapidly, or he may run towards an opponent in an upright position, waving his arms in the air whilst uttering loud yells. These patterns are of interest since they closely resemble some of the aggressive repertoire of man himself. The chimpanzee has other threat patterns too, of course, such as hitting out towards another with the back of the hand, swaggering and swaying from foot to foot in an upright position, and running towards another chimp stamping and slapping on the ground with feet and hands.

When actually attacking an opponent a male chimpanzee often tries to jump onto his victim's back and stamp hard with his feet. A small chimpanzee may be actually raised from the ground and slammed down repeatedly, or it may be dragged along by one limb. Other attack patterns include biting, hitting, grappling, pulling out hair and scratching. Female chimpanzees are more likely to clinch, rolling over and over on the ground together, or to pull out each others' hair or to scratch.

Most attacks last no more than half a minute and even those which appear, to a human observer, to be extremely vicious, seldom result in injury to either the aggressor or his victim—other than the loss of a handful of hair perhaps, or a slight scratch.

Most adult male chimpanzees hurl rocks or branches, usually at random, during their "charging displays." (These displays are usually performed on arrival at a food source, when two groups meet up, at the onset of heavy rain and so on.) However, in addition to this random throwing, many male chimpanzees and some females deliberately use objects as weapons in aggressive contexts. Most of our adult males, for instance, throw rocks or other objects at baboons during competition for bananas at the feeding area. Chimpanzees sometimes hurl things at each other and sometimes at human observers. . . . However, whilst some individuals carefully select large and potentially harmful missiles, others throw anything that happens to be close by, including such inappropriate objects as handfuls of bananas! Moreover, whilst most chimpanzees aim quite well, they seldom score hits unless the targets are less than five or six feet away. These chimpanzees also occasionally use sticks as clubs to hit each other, baboons or other objects. . . . Usually, however, whilst a large stick may be wielded forcefully, the chimpanzee then throws or drops his weapon prior to contact with his target.

RESPONSE OF CHIMPANZEE WHO IS THREATENED OR ATTACKED

The response of a chimpanzee who is threatened or attacked varies with respect to the relative social status of the two concerned, the relationship between them, the cause of the dispute, and the violence of the aggressive act. Thus if one chimpanzee mildly threatens another of only slightly lower rank than himself who, for instance, approaches his food too closely, the recipient of the gesture may seem to ignore the threat or, at most, slightly withdraw. If the threatened chimpanzee is much lower in status, the same gesture may elicit rapid withdrawal, loud screaming, submissive behavior—or all of these behaviors. A chimpanzee is likely to direct a more violent form of threat at another who tries to take away some of his share of bananas than at one who merely wants to sneak off with a discarded peel, and the more vigorous the threat, the more frightened or submissive the recipient is likely to be.

Once a subordinate has been attacked it may either crouch to the ground screaming until the attack is over, or it may struggle to escape and rush away. The closer the social status of the victim to the aggressor, the more likely it is that the former will turn and attack in self-defence.

SOME CONTEXTS IN WHICH AGGRESSION IS LIKELY TO OCCUR

Some major causes of aggression amongst our chimpanzees are as follows:

At the artificial feeding area the most frequently observed cause of aggressive incidents was, of course, competition for bananas. . . . Under more normal conditions, however, aggression over food is comparatively rare since most chimpanzee foods are present in abundance.

If a chimpanzee is frustrated in achieving some goal—if, for instance, he cannot open a box to get at the bananas inside, or if he is attacked or threatened by a higher-ranking individual and dare not reciprocate, he frequently redirects his aggression at a lower-ranking animal. Alternatively he may rush off in a charging display, which includes many seemingly aggressive acts such as stamping, shaking branches, hurling rocks and so forth. Such a display often seems to calm a frustrated chimpanzee: afterwards he appears relaxed and at ease.

Sometimes a chimpanzee threatens or attacks a subordinate who has failed to respond correctly to some social signal. For instance, sometimes a male "forces" a female to follow him about: if she does not hurry to his side when he shakes branches at her, he may then actually attack her. One male threatened another who went to sleep instead of reciprocating during a social grooming session.

At times adults appear to be "irritated" by noisy inferiors. One submissive pattern involves the lower-ranking individual bobbing up and down and uttering loud grunts in front of his superior, often actually getting in the way, and sometimes this resulted in an attack from the dominant individual. Pain sometimes seems to make a chimpanzee irritable: thus one male with a newly broken toe charged several times at a group of youngsters playing nearby when the game became noisy.

Strong bonds of affection or friendships may develop between pairs of chimpanzees, particularly between mothers and their offspring (including adult offspring), and siblings. As would be expected, a mother will rush to the defence of her infant, often threatening or attacking the chimpanzee responsible for hurting or frightening her child. She may do the same for her offspring when he is a fully mature male, and he will hurry to her defence in the same way.

Sometimes female chimpanzees join together to threaten, attack or chase away a female from a different area if she arrives at the feeding area. Males have not been observed to show this sort of deliberate aggression towards "strangers" of either sex.

It sometimes happens that when one individual is attacked, other chimpanzees who were not apparently involved in the original dispute hurry over to join in, attacking the victim when it escapes from the original aggressor.

Fear sometimes seems to spark off aggression. When a paralytic disease (probably poliomyelitis) swept through the area, some chimpanzees became paralyzed in one or more limbs. If this resulted in abnormal locomotion, when other chimpanzees saw the stricken chimpanzee for the first time, they initially showed fearful responses (screaming, embracing one another and so forth) and subsequently displayed aggressive behavior towards or actually attacked their injured companions.

A good deal of aggression occurred during interactions between two individuals which I have termed "dominance fights" although, in fact, such encounters only rarely involved actual attack. Two young males, for instance, each holding an approximately equal position in the dominance hierarchy, may commence to show off in a vigorous manner, swaggering about in an upright position and violently swaying branches at each other. Such incidents are sometimes sparked off by causes which may not be at all clear-cut to the human observer. It is in connexion with the dominance status of a male that his charging display appears to play a vital role: in principle, the more frequently and the more vigorously he displays, the higher in the social ladder he is liable to climb. I shall return to the importance of this display below.

Chimpanzees of the Gombe Stream area are, at times, active hunters and prey upon young baboons, monkeys, young bushbucks and so on. However, the behavior of a predator towards its prey cannot necessarily be considered as aggressive. After all, a man may catch a fish for his supper without feeling any more aggressive towards the fish than to the apples he may pick off a tree to eat or the meat he buys in a butcher's shop. On the other hand, if he shoots a buffalo for meat and the buffalo charges him, then aggression, in the usual sense of the word, will certainly be roused. When chimpanzees hunt young baboons, the adult baboons often threaten or attack the hunters: the chimpanzees then become aggressive towards the baboons. Finally, it may well be that, during a hunting episode, particularly the actual killing, the physiological processes may be similar to those involved in other types of aggressive behavior.

SUBMISSIVE AND REASSURANCE BEHAVIOR

After being threatened or attacked, a subordinate then often approaches the aggressor and directs towards him submissive, or appeasing behavior. This includes such gestures and postures as turning the rump towards the aggressor, crouching on the ground in front of him, holding out a hand towards him, touching or kissing him. The dominant chimpanzee, in many cases, responds by gestures such as reaching out to touch the subordinate, holding its hand, patting it gently on the head, back or other part of the body . . . kissing it, briefly grooming or embracing it. Such behavior, on the part of the aggressor, serves to reassure the subordinate: a youngster who crouches screaming and tense on the ground gradually relaxes and quietens under the soft patting of a male who, a few moments before, was pounding him up and down on the ground.

Some individuals, particularly juvenile and young adolescent males, show a definite need for such reassurance: if the aggressor ignores their approach and submission they may maintain their appeasing gestures and postures, whilst screaming and whimpering, until he does respond. One youngster used to throw temper tantrums, screaming and hurling himself about on the ground, if his submissive behavior was ignored. The need for such contact sometimes results in an obvious conflict situation: thus after a violent attack an adolescent male showed a tendency to approach the aggressor which was counteracted by a tendency to flee, so that he approached in a series of zig-zags as he alternately approached and turned from the big male.

Reassurance behavior is undoubtedly of great importance in maintaining the relaxed relationships which are normally apparent between most of the different individuals of a wild chimpanzee group. It is of interest that so many of the gestures and postures involved in chimpanzee submissive and reassurance behavior so closely resemble our own, not only in appearance but, more importantly, the contexts in which they occur.

AGGRESSION AND THE RISE TO DOMINANCE

As I have already mentioned, the charging display of the male chimpanzee plays an important role in bettering, or at least maintaining his social status. Whilst most displays are not deliberately directed towards another individual, they do, nevertheless, serve to impress other chimpanzees present. Even much older individuals sometimes hurry out of the way of an adolescent male when he rushes close by during a vigorous charging display.

The best example of the apparent significance of the male charging display relates to the sudden rise in status of the mature male, Mike. Mike was fully mature when I first got to know him in 1962. He was then among the lowest ranking of all the socially mature males, and at times was threatened and even attacked by nearly all the others.

Early in 1964 Mike, during one of his charging displays, seized hold of an empty four-gallon kerosene can that was lying in his way and dragged it behind him as he ran. The can made a loud noise as it bumped along the ground and the other chimps present at the feeding area rushed out of the way. After this, Mike began to use cans more and more frequently during his displays. . . . Other males had also dragged or hit [kerosene] cans during their charging displays, but only Mike, it seemed, was able to take advantage of such artificial props. After displaying several times with them, presumably because they simply happened to be to hand, Mike began to hunt deliberately for the cans prior to displaying. After a few weeks he learned to keep up to three of them ahead of him, hitting or kicking them, whilst rushing at top speed across the ground. He made a tremendous noise in this way, and the other males became increasingly fearful.

An example will serve to illustrate Mike's tactics. One morning Mike was sitting by himself, staring at a nearby group of males. The other males were engaged in a social grooming session. After a while Mike got up and walked very calmly over to my tent, selected two kerosene cans, and returned to his place. He continued to stare at the group and then began to rock to and fro, almost imperceptibly (a sign of frustration or unease in a chimpanzee). As his rocking became more vigorous, so his hair gradually began to stand on end (autonomic behavior associated with almost any violent emotion in a chimpanzee). Finally he began the series of hooting calls which typically precede and accompany a charging display, and then he charged straight towards the other males, each of whom fled. Mike and the cans vanished down a track and, after a moment, the group reassembled and recommenced grooming. A couple of minutes later, however, Mike's hooting began again and he reappeared, charging towards the group, hitting and kicking his cans. Again the group fled. After displaying thus for a third time Mike stopped and sat, breathing hard, his hair still erect, in the spot where the males had been grooming earlier. One by one the others approached, grunting nervously, reaching to touch or kiss him in submission. Eventually Mike was the center of the grooming group.

Four months after his first display with a kerosene can Mike had achieved a top-ranking position in the group, usurping the position of the former dominant male, Goliath. Shortly after this we hid all the cans, for not only did we dislike the noise but we were sometimes hit by one of them during a display. By then, however, Mike's position was assured. For almost a year Mike himself nevertheless seemed uneasy in his top-ranking position. He displayed very frequently and very vigorously, and he was constantly attacking his subordinates, particularly the females. Often it seemed he attacked another over the merest trifle, or for no obvious reason at all.

Several times during the months subsequent to Mike's new position we observed spectacular "dominance fights" between him and the previous top-ranking male, Goliath. At no time did we see either male actually touch the other during such an encounter, save occasionally with the ends of the branches which they swayed vigorously at each other. Each time it seemed that after a while Goliath's nerve suddenly broke, and he thereupon hurried towards Mike with submissive gestures. The two then engaged in long social grooming sessions which served to calm the tensions that had built up between them.

During those first uneasy months of Mike's supremacy, we twice observed groups of five adult males "gang up" on Mike; but though he rushed away screaming at first, when he was finally cornered up a tree he turned and displayed: his aggressors broke ranks and fled.

As Mike became more secure in his position he became increasingly less aggressive and more and more tolerant of his subordinates. He is still top-ranking male today, although two young males, both of whom often ignore Mike's charging displays instead of running out of the way, are obviously giving the old male cause for concern.

DISCUSSION

The chimpanzee is very close to man in many ways. Recent biochemical research suggests that, in some respects, the chimpanzee is biochemically as close, or closer, to man than he is to the gorilla. Again the communication patterns of chimp and man, on the emotional non-verbal level, together with the contexts in which they may occur, are often remarkably similar. . . . Thus an understanding of the aggressive behavior of chimpanzees may be of help in understanding some of our own.

In the chimpanzee, actual fighting, compared with the use of threat and bluff is, under normal circumstances, relatively infrequent. When we introduced an artificial element into the life of the Gombe chimpanzees, namely a feeding area (where many chimps gathered together more regularly than they would have done otherwise and where they competed for a food in relatively short supply), this resulted in an increase of all types of aggression, including physical attack. Even so, whilst some flights looked vicious, they seldom resulted in serious injuries to either aggressor or victim, and no chimpanzee has been observed to kill another.[1]

Humans, for the most part, also make use of threat and bluff, including

---

[1] But see J. D. Bygott 1972 Cannibalism among wild chimpanzees. *Nature*, Lond., 238:410–411. [Ed.]

verbal aggression, far more than physical fighting. However, when people do fight they are far more likely to harm one another than are chimps. This, of course, is principally because men so often use weapons; it is ironical, in a way, that the forms of attack most likely to kill involve the least strenuous physical exertion—such as shooting. It would almost seem that, in the evolutionary sense, man has not yet adjusted behaviorally to the comparatively recent acquisition of weapon use in intra-specific fighting.

Sometimes the causes of aggression seem quite similar in man and chimp. A mother, whether chimp or human, is likely to feel aggressive towards anyone or anything which threatens to harm or does harm her baby. Two individuals, whether chimp or human, may become aggressive when competing for a particular object which each wants, whether this be a bunch of bananas in the case of chimps, or an attractive girl in the case of humans. Also many chimpanzees, like many humans, are constantly alert for a chance to better their social position, and sometimes this leads to aggression in both species.

However, whilst the biological roots of aggression may not be too dissimilar in man and chimp, the expression and causes of ill-feeling and anger in man have been immeasurably complicated by his development of self-esteem and pride, the acquisition of moral values, the hunger for material possessions and the development of a spoken language.

Chimpanzees usually settle a dispute immediately. The subordinate simply approaches his superior with appeasing gestures, irrespective of whether or not he was attacked for some misbehavior. The aggressor responds with reassurance gestures and the victim is calmed. This does not mean that, even when a dispute appears to be over, the subordinate may not try and retaliate subsequently, if he sees a good chance. But it does mean that, for the most part, high- and low-ranking individuals are able to coexist peacefully and enjoy relaxed relationships with each other for much of the time. There are no individuals, so far as we know, who live in constant terror of retribution at every turn.

Human relationships, however, are necessarily far more complex. Often pride prevents a person from making an apology after a dispute, even when he knows he was in the wrong. Thus two individuals, normally friends, may avoid each other for days: indeed, some quarrels are never made up. Sometimes differences of opinion over a moral or religious matter can only be solved by the persons concerned shaking hands and "agreeing to differ," but only too often they do no such thing and part with feelings of animosity, each feeling certain that *he* is "right."

The development of man's superior intellect has complicated his patterns of inter-individual communication in all respects, and this is, of course, particularly true of the development of a verbal language. In so many human interactions words have largely taken over from gestures.

Words are a fairly recent acquisition in the evolution of our species, and they can so easily be misinterpreted. When one chimpanzee gestures to another, and if the individuals belong to the same community, there is no misunderstanding: the recipient of the signal interprets the message correctly. In most cases, no doubt, even chimpanzees from different areas would have no difficulty in understanding each other's signals. Indeed, some chimpanzee gestures, particularly those relating to threat and submission, can be correctly interpreted even by naïve human observers and by baboons. When we turn to humans, however, there may be misunderstandings even at the family level as to the meaning of verbal signals, let alone when people of different nations and different cultures are trying to communicate. How much human aggression, I wonder, is the result of simple misinterpretation: how often do words serve to rouse aggression in man when the speaker means to do no such thing? Conversely, a verbal threat is not always recognized as such so that a correct response is not made.

One final point: through verbal language a human being can communicate with a very large number of people; words can be used successfully to rouse people to love or to hate. Whilst a chimpanzee, by means of simple gestures and calls, can sometimes rouse a member of his immediate family to come to his defense, man, by communicating his ideas and ideals, his hopes and his fears, can awaken sympathy in people far beyond his family circle. If he is skilful in using words he can unite an entire group, sometimes almost an entire nation, so that each individual feels himself part of a super-family, and then, if one member of this family is threatened, aggression may be roused in the others.

Is it, ironically, man's very ability for loving and self-sacrifice which, so often, leads him to kill others of his species?

7

EVOLUTIONARY ASPECTS OF PRIMATE LOCOMOTION

J. R. Napier

*A crucial element in understanding the behavior of the earliest
ancestors of man is reconstructing early locomotor behavior—the
ways in which animals moved. A variety of locomotor adaptations
are seen in the modern primates: brachiation, knuckle walking,
bipedalism, quadrupedalism, arboreal leaping, scrambling, hanging,
crawling, to name but a few. Which modes of locomotion are
most germane to an understanding of the evolution of hominid
bipedalism? Again, the study of the fossil record, combined with
an ecological approach to the analysis of the locomotor behavior
of modern monkeys and apes, can yield valuable clues about the
probable directions of the evolutionary sequence leading to man's
upright posture.*

*Dr. Napier provides just such a discussion in the following paper.
In preparation for this publication, he has corrected and brought
the article up to date, as of 1971.*

*Dr. Napier is with the Unit of Primate Biology, Smithsonian
Institution, and the Royal Free Hospital School of Medicine,
London, England.*

ABSTRACT   Both neontological and phylogenetic studies are necessary to
interpret primate locomotion. Reference to palaeoprimatology and palaeocology,
for instance, will lead to a fuller understanding of the origins of such gaits as
the vertical clinging and leaping of *Tarsius*, *Indri* and *Propithecus*.

Evolutionary trends in posture and locomotion are discussed. The postural
trend has been towards maintenance of trunk verticality and the locomotor
trend towards an increasing dependence on the forelimbs among arboreal primates. Three stages are recognized in the phylogenetic course of arboreal locomotor adaptation: Stage A. Vertical clinging and leaping; Stage B. Quadrupedalism; Stage C. Brachiation.

The role of prehensility of the hand in the evolution of locomotor types is

By permission of the author. An earlier version of this article appeared in the
American Journal of Physical Anthropology, Vol. 27:3 No. 3, 1967.

discussed in relation to forest morphology and, in particular, to stratification. Finally a scheme of evolution, set in the framework of ecology, for Old World Monkey groups is presented.

The study of locomotion is T-shaped. It has two components, a vertical or phylogenetic one and a horizontal or neontological element. In order to understand the adaptive significance of the locomotion of living primates and of the part locomotion has played in the dispersal and taxonomic differentiation of the Order, reference must be made to the evolutionary history of the group, particularly in the context of past ecology. Phylogenetically, as Washburn ('50) was the first to point out, locomotor adaptations have provided the principal milestones along the evolutionary pathway of primates from the Eocene to the Pleistocene.

There are many puzzling gaits and strange locomotor behavioral habits amongst living primates, such as the leaping of langurs, the slow-climbing quadrupedalism of the African and Asian lorises, the vertical clinging and leaping of certain Madagascan lemurs such as *Indri* and *Propithecus*, or the knuckle-walking gait of *Pan* and *Gorilla* that are not easy to understand if viewed wholly in terms of the present. Only by considering these primates in the light of past ecological, geological, climatological, and zoogeographical conditions can one hope to rationalize what seem to be inexplicable adaptations in terms of present ecology.

The ecology of certain vertical clinging and leaping prosimians, such as *Tarsius*, *Indri*, and *Propithecus*, provides an example of the importance of the evolutionary approach in interpreting the behavior of living primates. In very broad terms the habitats of the tarsier on the one hand and the Madagascan lemurs on the other are much the same. The tarsier is a dweller in southeast Asian tropical rain forests and the lemur in various types of tropical forest in Madagascar; only in a restricted region of southwest Madagascar does *Propithecus* live in a specialized habitat, the Euphorbia scrub forests. In terms of diet, activity rhythm, and predator interactions the two groups of prosimians however show marked dissimilarities. *Tarsius* is insectivorous and carnivorous; the Madagascan lemurs are frugivorous and leaf-eating. *Tarsius* is nocturnal and *Indri* and *Propithecus* are diurnal. There are few animal predators that constitute any danger for the lemurs, but the situation is more hazardous for *Tarsius* which is exposed to attacks from tree-dwelling carnivores, rodents, and owls. It is simple enough to understand why *Tarsius* occupies its particular niche on trunks of trees and saplings where insects, food and small reptiles are abundant, and why therefore a vertical clinging posture is of survival value. It is also easy to understand, in view of the predator situation, that the specialized leaping component of the tarsier's locomotion provides a rapid and effective avoidance mechanism (Walker, personal communication). It is not so apparent why a vegetarian, diurnal lemur living in a relatively predator-free environment should possess this highly specialized locomotion pattern. Is the inclusion of these two groups of prosimians within the same locmotor group an example of convergence, or is it the result of parallelism of common inheritance? Convergence would seem unlikely in view of the differing aspects of their present ecology; parallelism would seem unlikely for the same reasons unless the origin of this locomotor specialization can be shown to lie in the distant past when ancestors of tarsiers and lemurs occupied the same sort of habitat. In fact it is becoming apparent that vertical clinging and leaping as a way of life is a very ancient primate capacity indeed (Napier and Walker, '67; Simons, '67).

Osteological characters that have been correlated with this gait have been recognized in the following families of Eocene prosimians: Adapidae (*Smilodectes, Notharctus*), Microsyopidae (*Microsyops*), Omomyidae (*Hemiacodon* etc.), Anaptomorphidae (*Tetonius*), and Tarsiidae (*Necrolemur*). These characters have also been recognized in some of the few postcranials of certain non-hominoid primates of the Fayum fauna. All these forms were *small*, not much bigger than *Tarsius*, all, presumably, to judge by what is known of their dentition, were largely insectivorous, and all lived in an ecological setting by all evidence not far removed from that existing today in tropical forest biota. The evidence so far is suggestive that vertical clinging and leaping was the earliest arboreal locomotor adaptation of the primates and that from this type have stemmed all other locomotor variants of the Order (Napier and Walker, '67; Napier, '67) including the bipedalism of man.

Tarsiers are behaviorally and morphologically the most specialized of this group; the indris, the sportive lemurs and the sifakas are morphologically more generalized vertical clingers, lacking for instance the calcaneal elongation seen in tarsiers. Other members of the locomotor group such as the galagos, while as specialized morphologically as the tarsier (Hall-Craggs, '65), are rather more generalized behaviorally. *Hapalemur* is modified in the direction of quadrupedalism. Outside the group as defined are certain members of the genus *Lemur* (e.g. *L. catta*), which behaviorally and morphologically retain a number of vertical clinging characters but are more usually classified among the quadrupeds.

Tarsiers can be regarded as survivors of a fully specialized Eocene stock that has remained in the forests in which it evolved. Indriids, on the other hand, as survivors of a less specialized stock of Eocene prosimians, the Adapidae, ultimately reached the utopian forests of Madagascar and retained their primitive locomotor habit in the absence of any evolutionary pressures to the contrary. Indriids increased in size and, as a corollary, altered their dietary requirements and therefore their activity rhythm.

TRENDS IN PRIMATE EVOLUTION

The ecological trend in primate evolution has been towards life in trees. The morphological trend has been in the direction of adaptations that ensure survival in this habitat, i.e. binocular vision, a high degree of neuromuscular coordination, acquisition of prehensile extremities, lengthening of the forelimb, and an increasing mobility of the joints of the elbow and shoulder. Principal among the behavioral trends (Washburn, '50) are those concerned with posture and locomotion. The postural trend has been towards the maintenance of trunk verticality, and the locomotor trend towards an increasing dependence on the forelimbs.

Postural Trend

The majority of primates are capable of sitting, clinging, or hanging in the vertical position and many of standing or even walking upright. Certain prosimian families include genera in which a vertical resting posture is habitual. These animals comprise the vertical clinging group already discussed. There are, however, other Lemuriformes, for instance *Lemur catta,* that are somewhat intermediate in their locomotor adaptations, and in which trunk verticality while sitting on the ground or on branches of trees is a striking feature of their behavioral repertoire. Jolly ('67) describes this posture of *L. catta* and *Propithecus verreauxi verreauxi* as "sunning" behavior.

Among catarrhine primates long periods of sitting, during the day as well as the night, are facilitated by the presence of specialized pads, the ischial callosities (Washburn, '57). The pads are lacking in the New World monkeys which usually adopt a horizontal sleeping posture. The Pongidae, Hylobatidae, and Hominidae employ the sitting posture frequently. Hominidae are habitually bipedal and have evolved a specialized striding gait (Napier, '63, '64, '67); Pongidae (Elftman, '44) and Hylobatidae (Straus, '41; Prost, '67) are only facultatively bipedal. Bipedalism is occasionally observed in certain Cebidae such as *Ateles, Lagothrix,* and *Cebus* and in many Cercopithecidae (*see* Hewes, '61).

Truncal uprightness is an ancient primate possession, as Wood Jones ('16) believed when he asserted "arboreal uprightness preceded terrestrial uprightness"; as Straus ('62) emphasized, and as recent reviews of Eocene prosimian postcranial material have confirmed (Simons, '67). Uprightness constitutes a fundamental trend of primate evolution. From the vertical clinging adaptation, which is believed to be the primary arboreal adaptation of primates, to brachiation, which is the most recent,

trunk verticality has been the predominating characteristic of primate posture. It is not really surprising (Napier, '67) that one primate, man, has
developed and refined it to such advantage.

ARBOREAL LOCOMOTOR TREND

The trend to convert an essentially hindlimb-dominated gait into a
forelimb-dominated one is discussed here only briefly. This trend, shown
diagrammatically in Figure 7-1, comprises the elements of three separate
stages. These stages are named in terms of the specialized gaits which
have arisen out of them:
Stage A. Vertical clinging and leaping.
Stage B. Quadrupedalism.
Stage C. Brachiation.
In order to visualize the actual evolutionary stages comprising this
trend it is necessary to develop the following hypothesis. A primate with

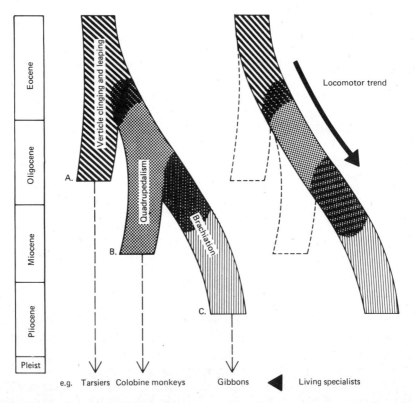

FIGURE 7–1—Trends in primate locomotion. A, vertical clinging and leaping;
B, quadrupedalism; and C, brachiation. *Left*, the three locomotor stages; *right*,
the overall trend.

the morphology and behavior of a vertical clinging form but without the extreme specializations of a committed vertical clinger, such as the tarsier, underwent an adaptive shift in the direction of quadrupedalism by means of a relative increase in the length of the forelimbs and by acquisition of a prehensile hand with a divergent and mobile, though not a strictly opposable, thumb. These adaptations, taken to the logical conclusion, could have led to commitment to a wholly quadrupedal way of life (B). Alternatively, as a result of a further adaptive shift, selection could have operated more and more strongly in favor of forelimbs functioning as suspensory rather than as supporting structures, leading finally to the evolution of a wholly arm-swinging type of morphology and behavior (C). Straus ('62) was one of the first to point out the existence of a forelimb-dominant trend in primate evolution. He asserted that it provided a substrate for the development of both true brachiation and the erect bipedal mode of locomotion.

This hypothesis clearly suggests that certain "types" of locomotor behavior of primates are of greater antiquity than others. Vertical clinging made its first appearance in the fossil record in the Eocene with *Necrolemur, Smilodectes, Notharctus*, etc. (Simons, '67). Quadrupedalism among hominoids is not recognized until the Miocene with *Proconsul africanus* (Napier and Davis, '59), *Pliopithecus vindobonensis* (Zapfe, '60) and among cercopithecoids with *Mesopithecus pentelici* (Gaudry, 1862) of the Miocene-Pliocene boundary. Brachiation makes its debut in the fossil record with *Oreopithecus bambolii* (Strauss, '63). Furthermore, in the broadest terms, these three principal patterns of primate locomotion, characterizing prosimians, cercopithecoids, (or ceboids), and hominoids respectively, reflect three main grades of organization through which primates have passed during their evolution.

No mention is made in this paper of the secondary or ground-living locomotor adaptations of primates such as digitigrade quadrupedalism, knuckle-walking, or bipedalism; these secondary adaptations and their relation to the primary arboreal ones will be discussed elsewhere.

EVOLUTION OF PREHENSILITY

Increase in size during primate phylogeny appears to be one of the major factors contributing to the evolution of prehensility of the hands. Prehensility is much enhanced by the presence of a divergent and independently mobile thumb, although a certain degree of prehensility is possible with a non-opposable thumb as in *Callithrix* and *Saguinus* and *Tarsius*. In *Tarsius* the non-opposable pollex is associated with a widely divergent "pseudo-opposable" hallux (for definition of these terms *see* Napier, '61; Napier and Napier, '67). These are adaptations for the vertical clinging posture where the weight of the body is supported principally by the feet, although substantial aid is provided by the friction of the tail (Sprankel, '65).

Provided that the ratio of branch size to body size remains high, arboreal animals can maintain their balance in trees without the help of prehensile extremities (Figure 7-2a), particularly if their center of gravity is low. For treeshrews, such as the feather-tailed *Ptilocerus lowii*, one of the smallest of primates, safety in trees is assured by the presence of clawed extremities that pierce and grip the bark. With increase in body size (or decrease in branch size) a critical point is approached when further increase in body size produces a potentially unstable system and any lateral displacements of the center of gravity are liable to lead to overbalancing (Figure 7-2b). Stability in these circumstances is assured by the possession of prehensile extremities by which the body can be *suspended* below the branch (Figure 7-2c). Three types of suspension are seen among primates:

(1) *Quadrupedal suspension*: practiced, for instance, by pottos habitually and by many other quadrupedal forms occasionally.

(2) *Pedal suspension*: practiced occasionally by orangutans, chimpanzees, howlers, spider monkeys, and woolly monkeys, also occasionally by galagos and pottos. Both feet are usually involved.

(3) *Manual suspension*: practiced by gibbons and siamangs habitually, occasionally by great apes, South American prehensile-tailed monkeys, and some colobines.

Manual suspension, employing one hand or two, is associated with structural adaptations in the hand, elbow, and shoulder. The hook-type of hand is the ideal adaptation for this purpose. The thumb in specialized arboreal hands, such as those of brachiators, is more of a liability than an asset; and in most specialized arboreal primates, other than the gibbons, it is short and attenuated. Opposability of the thumb, present in all Recent catarrhine monkeys and apes, is an essential concomitant of the arboreal grip of quadrupedal primates, whether the forelimb is acting in support or suspension. The mechanism for true opposability (Napier, '61) apparent in the carpo-metacarpal joint of the thumb of pongids, for instance, is strong evidence that the ancestral pongids passed through a quadrupedal stage. Indeed, for different reasons, the post-cranial skeleton of *Dryopithecus* (*Proconsul*) *africanus* provides evidence that pongids passed through such a stage. Such a phylogenetic sequence would be in accordance with the arboreal locomotor trend discussed above and shown in Figure 7-1.

Lack of prehensility of the extremities, combined with large body size and body weight, severely restricts the range of any animal moving in the crowns of trees—particularly towards the periphery of the crown where fruits and leaves are found in greatest abundance. With the evolution of

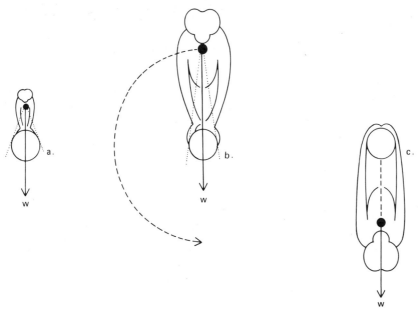

FIGURE 7–2—Relation between branch size and body size. For explanation, see text.

prehensility of the limbs a new ecological niche, the fine tertiary branches and leaf-bearing terminals of the canopy, could be exploited to the full. Primates could then, with perfect security, move into a small-branch setting, as Avis ('62) has called it, where fruits and leaves are at their most abundant.

In terms of the overall primate locomotor trend discussed above, the evolution of prehensility of the hands constituted a critical turning point at which a generalized vertical clinging form shifted into a new locomotor-feeding adaptive zone where the forelimbs were beginning to acquire a more functionally significant role in locomotion. It seems likely that in the Upper Eocene or Lower Oligocene, a dichotomy occurred in the catarrhine stock, which set in motion a major series of morphological and physiological adaptations that led ultimately to the emergence of two separate superfamilies, the Hominoidea and the Cercopithecoidea (Figure 7-3). Such an event is likely to have been primarily locomotor in origin. A most probable explanation at this point in time (Eocene-Oligocene) is that in the two groups, distinct dietary specializations evolved in association with habitat differences; the pre-hylobatids and the pre-pongids becoming largely fruit-eaters and the pre-cercopithecoids largely leaf-eaters (Napier, '70). The Fayum fauna from the middle of the Oligocene provide some fossil evidence for this hypothesis. The overt dental specializations of the two groups were already well established in *Parapithecus* and in *Aeolopithecus* and *Aegyptopithecus* (Simons, '65). There is little post-cranial evidence from the Fayum, but from the limb morphology of such Miocene pongids as D. (*Proconsul*) *africanus* and *Pliopithecus vindobonensis*, which are predominantly monkey-like in their morphology (Clark and Leakey, '51; Napier and Davis, '59; Zapfe, '60), it is safe to assume that in Fayum in the Middle Oligocene, the marked post-cranial adaptations, which now separate quadrupeds and brachiators, had not yet evolved. The early catarrhines then were prehensile-limbed, arboreal quadrupeds (for want of a more precise word) with an intermembral index of the order of 70 to 80, indicating that the hindlimbs were still considerably longer than the forelimbs—possibly a reflection of their evolution from a "generalized" vertical clinger.

Subsequently, as we know, the three groups of catarrhine primates (the monkeys, the great apes, and the gibbons) underwent a striking locomotor divergence. It is possible that the basis of this divergence was in some way related to the stratification of the forest canopy. As we know from the late A. H. Booth's papers ('56, '57) and from Haddow ('52) and Sanderson ('48), primates show considerable interspecific variation in the level at which they eat, sleep, and feed in the canopy. We know from studies of forest morphology, such as those of Richards ('57, '63) and Keay ('53) in West Africa, that the environment, vis-à-vis crown shape, branch form and size, distribution of fruit and leaves, temperature and humidity, and distribution of Diptera, such as mosquitoes, is also most variable. Three strata are usually recognized in climax rain forests (Richards, '57), and they are variously described as A, B, and C or upper, middle, and lower strata; other strata exist, the shrub layer, herb layer, and so on, which are probably important as primate habitats, but they will not be considered here. It has been suggested elsewhere (Napier and Napier, '67) that for usage in primate biology it is sufficient to divide the canopy into "open" and "closed" regions (Figure 7-4). The open canopy is the A or upper stratum where crowns of trees arise out of a candelabra-shaped arrangement of branches and are broad and discontinuous. Here the fruit and foliage are largely restricted to the smaller, more peripheral branches. The closed canopy usually comprises the B and C layers where contiguous crowns are taller than they are wide and where the leaves and fruit are more homogeneously distributed.

The principal implications for primate locomotion are as follows:

(1) Open canopy with its slender unsupported branches puts an intense demand on prehensility of extremities. In the closed canopy however the branches are more rigid, more interlocked, and demand less acrobatic activity.

(2) The gaps between the crowns of the trees of the open canopy demand that movement from tree to tree must have a big leaping or arm-

swinging element; in the closed canopy, the contiguity of the crowns means that primates can move through the forest by established aerial pathways with less need to indulge in much leaping and arm-swinging.

Deploying this sort of theme in an evolutionary setting, one might suppose that the dichotomy of the major catarrhine groups was related to the morphology of Eocene forests and that the proto-cercopithecoids were closed-canopy dwellers and the proto-pongids were inhabitants of the open canopy. It is clear that this hypothesis must remain a hypothesis until we know a great deal more about the nature of early Tertiary forests, the stratification and primate distribution in modern forests, the distribution, in terms of open and closed canopy, of the varieties of fruits and leaves that primates feed on.

The flowering of grasslands from the end of the Oligocene to present times is a pivotal concept in the locomotor phylogeny of primates. We visualize that the diminishing forests forced the ancestral hominids and the ancestral ground-living cercopithecines into adopting a new ecological niche which, as we have recently suggested (Napier, '67), has been in the woodland savannahs rather than in the traditional "open grassland."

Arising from the cercopithecoid stem, the group that entered this new niche were the ancestral baboons and their allies (Figure 7-3). The more conservative cercopithecoids remained arboreal and evolved into modern Colobinae, the leaf-eating monkeys. From the hominoid stem, similarly, a progressive group adopted a woodland savannah habitat and ultimately evolved as hominids, leaving behind in the forests the conservative (in an

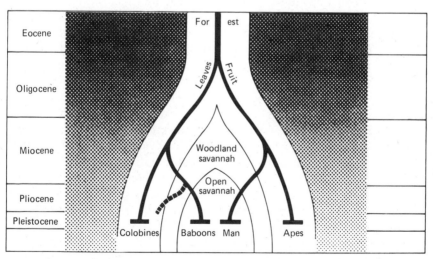

FIGURE 7–3—Diagram showing the possible deployment of catarrhine primates in relation to major vegetational zones and in relation to geological time. Broken line indicates the return pathway of *Cercopithecus* and *cercocebus* groups from woodland savannah to high forest.

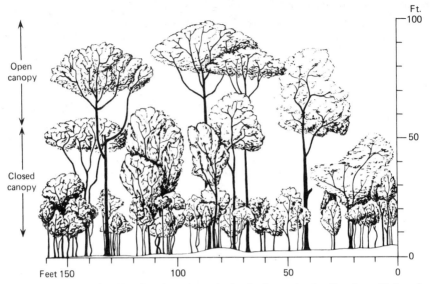

FIGURE 7–4—Profile of a plot of tropical rain forest in the Southern Bakunda Forest Reserve, West Africa, showing stratification of the canopy into open and closed layers. (Redrawn from Richards, 1963.)

ecological sense) ancestral apes, which forms became progressively more adapted to an arboreal habitat in accordance with the locomotor trend discussed above.

Thus, presumably during the Miocene, the four major catarrhine stocks sorted themselves out. This scheme omits any mention of such genera as the guenons and the mangabeys; their phylogeny is far from clear (Tappen, '60). It seems possible that both the *Cercopithecus* and *Cercocebus* groups are at present *secondarily* arboreal, having returned to the trees after a brief (*Cercopithecus*) and a not-so-brief (*Cercocebus*) flirtation with ground-living life in the late Miocene or Early Pliocene. The patas monkey with its clear-cut *Cercopithecus*-like affinities (Verheyen, '62) may be a segment of the *Cercopithecus* stock that stayed on as a ground-dweller. Certainly the patas monkey's highly evolved terrestrial locomotor adaptations do not suggest a very recent adaptive shift. The savannah monkey (*C. aethiops* species-group) on the other hand appears to be a relative newcomer to ground-living life, and in its present deployment in more open habitats we may be observing an adaptive shift in the process of evolution. Properly studied this species-group should provide important clues as to how the Miocene trend towards ground living took place in the ancestral cercopithecinal stock.

This paper has been concerned principally with the evolutionary roots of the locomotion of living primates; but the ecological approach to phylogeny which it invokes is equally applicable to other aspects of behavior. Andrew ('63) has described the humanoid grunts given in greeting situations by *Papio, Theropithecus, Macaca sylvanus,* and *Homo;* these vocalizations are clearly expressions of convergence under similar environmental conditions. That they seem largely unconnected with closeness of phyletic affinity is suggested by the rarity of this behavior in the forest-living baboon, *Mandrillus,* which indulges in greeting grunts to about the same extent as do some of the forest-living guenons, the De Brazza monkey, *Cercopithecus neglectus,* for example.

LITERATURE CITED

Andrew, R. J. 1963 Evolution of vocalization in monkeys and apes. Symp. Zool. Soc. Lond., *10*:89–101.

Avis, V. 1962 Brachiation: The crucial issue for man's ancestry. Southwestern J. Anthrop., *18*:119–148.

Booth, A. H. 1956 The distribution of primates in the Gold Coast. J. W. Afr. Sci. Assoc., *2*:122–133.

——— 1957 Observations on the natural history of the Olive Colobus monkey, *Procolobus verus* (van Beneden). Proc. Zool. Soc. Lond., *129*: 421–430.

Clark, W. E. Le Gros, and L. S. B. Leakey 1951 The Miocene Hominoidea of East Africa. Fossil Mammals of Africa, No. 1. Brit. Mus. (N. H.), London.

Elftman, H. 1944 The bipedal walking of the chimpanzee. J. Mammal., *25*:67–70.

Gaudry, A. 1862 Animaux Fossiles et Géologie de l'Afrique. Paris.

Haddow, A. J. 1952 Field and laboratory studies on an African monkey, *Cercopithecus ascanius schmidti* Matschie. Proc. Zool. Soc. Lond., *122*: 297–394.

Hall-Craggs, E. C. B. 1965 An osteometric study of the hind-limb of the Galagidae. J. Anat. Lond., *99*:119–126.

Hewes, G. W. 1961 Food transport and the origin of human bipedalism. Amer. Anthrop., *63*:687–710.

Jolly, A. 1967 Lemur Behavior. Univ. of Chicago Press, Chicago.

Jolly, C. J. 1965 The origins and specialisations of the long-faced Cercopithecoidea. University of London. Ph.D. Thesis.

Keay, R. W. J. 1953 An outline of Nigerian vegetation. Government Printer, Lagos, Nigeria.

Napier, J. R. 1961 Prehensility and opposability in the hands of primates. Symp. Zool. Soc. Lond., *5*:115–132.

——— 1963 Brachiation and brachiators. Symp. Zool. Soc. Lond., *10*:183–195.

——— 1964 Evolution of bipedal walking in the hominids. Arch. Biol. *75*:673–708.

——— 1967 The antiquity of human walking. Sci. Amer., *216*(4):56–66.

——— 1970 Paleoecology and Catarrhine Evolution. In: *Old World Monkeys,* J. R. and P. H. Napier (Eds.). Academic Press, New York.

Napier, J. R., and P. R. Davis 1959 The forelimb skeleton and associated remains of *Proconsul africanus*. Fossil Mammals of Africa, No. 16, Brit. Mus. (N. H.), London.

Napier, J. R., and P. H. Napier 1967 A Handbook of Living Primates. Academic Press, London.

Napier, J. R., and A. C. Walker 1967 Vertical clinging and leaping—a newly recognized category of locomotor behaviour of primates. Folia Primat. 6:180–203.

Prost, J. H. 1967 Bipedalism of man and gibbon compared using estimates of joint motion. Am. J. Phys. Anthrop., 26:135–148.

Richards, P. W. 1957 The Tropical Rain Forest. Cambridge Univ. Press, Cambridge.

Richards, P. W. 1963 Ecological notes on West African vegetation. II. Lowland forest of Southern Bakundu Forest Reserve. J. Ecol., 51:123–149.

Sanderson, I. T. 1948 Mammals of the North Cameroons forest area. Trans. Zool. Soc. Lond., 24:623–725.

Simons, E. L. 1965 New fossil apes from Egypt and the initial differentiation of the Hominoidea. Nature, 205:135–139.

———— 1967 Fossil primates and the evolution of some primate locomotor systems. Am. J. Phys. Anthrop., 26:241–253.

Sprankel, H. 1965 Utersuchungen an *Tarsius*. 1. Morphologie des schwanzes. Folia Primat., 3:153–188.

Straus, W. L., Jr. 1941 Locomotion of gibbons. Am. J. Phys. Anthrop., 27:199–207.

———— 1962 Fossil evidence for the evolution of the erect bipedal posture. Clin. Orthopaed., 25:9–19.

———— 1963 The classification of *Oreopithecus*. In: Classification and Human Evolution, S. L. Washburn (ed.), Aldine, Chicago.

Tappen, N. C. 1960 African monkey distribution. Current Anthrop., 1:91–120.

Verheyen, W. N. 1962 Contribution à la craniologie comparée des Primates. Ann Musée Roy. Afrique Centrale-Tervuren, Belgique, Ser. 8, Sci. Zool., 105:1–256.

Walker, A. C. 1967 Patterns of extinction among the subfossil Madagascan lemuroids. In: *Pleistocene Extinctions*, P. S. Martin (ed.), Yale Univ. Press, New Haven.

Washburn, S. L. 1950 The analysis of primate evolution with particular reference to the origin of man. Cold Spring Harbor Symp., 15:67–78.

———— 1957 Ischial callosities as sleeping adaptations. Amer. J. Phys. Anthrop., 15:269–280.

Wood Jones, F. 1916 Arboreal Man. Ed. Arnold, London.

Zapfe, H. 1960 Die Primatenfunde aus der miozänen Spaltenfüllung von Neudorf an der March, Tschechoslowakei. Schweiz. Palaeontol. Abh., 78:4–285.

8

THE SEED-EATERS:
A NEW MODEL OF HOMINID
DIFFERENTIATION BASED ON
A BABOON ANALOGY

Clifford J. Jolly

*How did the earliest hominids become differentiated from the
populations that were to evolve into the pongids? What ecological
conditions favored the divergence of the early forms that led to man?
What factors made it advantageous to depart from an "ape" way
of life to take up a mode that was to include habitual bipedalism
and the use and fabrication of implements and weapons that came
to characterize the australopithecines of the early Pleistocene?*

*We simply do not know. Sherwood Washburn and others have
elaborated upon the role that tool use itself must have played in the
evolution of bipedalism. Other researchers have tried to reconstruct
the environments of Africa and Asia during the Miocene-Pliocene
epoch, during which the earliest stages of the divergence of pongidae
and hominidae probably occurred.*

*In the following paper, Professor Clifford Jolly reviews some of
the theories of hominid differentiation developed by other writers,
and then provides one of his own. He reconstructs the first
faltering steps of the emergence of the early hominids by the
use of cautious inference from the behavior and ecology of a
contemporary terrestrial primate, the baboon.*

*Dr. Jolly is a member of the Anthropology Faculty at New York
University.*

Despite years of theorizing, and a rapidly accumulating body of fossil
evidence, physical anthropology still lacks a convincing causal model of
hominid origins. Diverse lines of evidence point to a later common an-

Reprinted from *Man*, 5:1, March 1970, pages 5–26. By permission of the
author and the publisher.

cestry with the African pongids than with any other living primate, and studies of hominid fossils of the Basal and Early Lower Pleistocene (Howell 1967) have elucidated the complex of characters which at that time distinguished the family from African and other Pongidae (Le Gros Clark 1964). It is also possible to argue that the elements of the complex form a mutually reinforcing positive feedback system. Bipedalism frees the forelimb to make and use artefacts; regular use of tools and weapons permits (or causes) reduction of the anterior teeth by taking over their functions; the elaboration of material culture and associated learning is correlated with a cerebral reorganization of which increase in relative cranial capacity is one aspect. Bipedalism is needed to permit handling of the relatively helpless young through the long period of cultural conditioning, and so on.

Preoccupied with the apparent elegance of the feedback model, we tend to forget that to demonstrate the mutual relationship between the elements is not to account for their origin, and hence does not explain *why* the hominids became differentiated from the pongids, or why this was achieved in the hominid way. From their very circularity, feedback models cannot explain their own beginnings, except by tautology, which is no explanation at all. In fact, the more closely the elements of the hominid complex are shown to interlock, the more difficult it becomes to say what was responsible for setting the feedback spiral in motion, and for accumulating the elements of the cycle in the first place. Most authors seem either to avoid the problem of origins and causes altogether (beyond vague references to "open country" life), or to fall back upon reasoning that tends to be teleological and often also illogical. This article is an attempt to reopen the problem of origins by examining critically some of the existing models of hominid differentiation, and to suggest a new one based on a fresh approach.

PREVIOUS MODELS OF HOMINID DIFFERENTATION

Direct fossil evidence for the use of "raw" tools or weapons is necessarily tenuous, and that for the use of fabricated stone artefacts appears relatively late (Howell 1967). Nevertheless, as Holloway has pointed out (1967), the currently orthodox theory regards these elements as pivotal in the evolution of the hominid adaptive complex, probably antedating and determining the evolution of upright posture, and certainly in some way determining the reduction of the anterior teeth, the loss of sexual dimorphism in the canines, and the expansion of the cerebral cortex (Bartholomew & Birdsell 1953; Washburn 1963; DeVore 1964). A variant of this theory, proposed by Robinson (1962), sees bipedalism as the primary adaptation (of unknown origin), from which tool-using developed and hence anterior dental reduction.

Holloway (1967) rejects the orthodox, tool-and-weapon-determinant model, partly on the grounds that it postulates no genetic or selectional mechanism for anterior dental reduction, and thus implies Darwin's "Lamarckian" notion of the gradual loss of structures through the inherited effects of disuse. It seems a little carping to accuse Washburn and his colleagues of Lamarckism because they omit to make explicit their view of the selective factors involved. These are in fact stated by Washburn in his reply to Holloway (Washburn 1968): natural selection favors the reduction of canines after their function has been subsumed by artificial weapons since this reduces the chance of accidental injury in intraspecific altercations. Since, as we shall see, orthodox natural selection is adequate to explain the reduction of teeth to an appropriate size following a change of function, it is hard to see why Washburn should avoid the Scylla of Lamarckism only to fall into the Charybdis of altruistic selection. Why should natural selection favour the evolution of a structure that is of no benefit to its bearer, for the benefit of other, unrelated, conspecifics? Even if we swallow altruistic selection, a basic illogicality remains. If the males use artificial weapons to fight other species, why should they bite one another in intra-specific combat? If they do not, then the size of their canines is irrelevant to the infliction of any injury, accidental or otherwise. In any case, the best way to avoid accidental and unnecessary intra-specific injury, of any kind, is to evolve unambiguous signals expressing threat and appeasement without resort to violence. The ability to make and recognize such signals is of advantage to *both* parties to the dispute, and therefore can be favored by orthodox natural selection, is independent of the nature of the weapons used, and is found in the majority of social species, including both artefact-using and non-artefact-using higher primates.

We may now consider the underlying proposition that it was artefact use, which, by making the canine redundant as a weapon, and the incisors as tools, led to their reduction. It is known that hominoids with front teeth smaller than those of living or fossil Pongidae were widespread at the close of the Miocene period: *Oreopithecus* in southern Europe and Africa (Hürzeler 1954, etc.; Leakey 1967), and *Ramapithecus* (probably including *Kenyapithecus*) in India, Africa, and perhaps southern Europe and China (Simons & Pilbeam 1965). If the theory of artefactual determinism is to be applied consistently, regular tool- and weapon-making has to be extended back into the Miocene, and also attributed to Hominoidea other than the direct ancestor of the Hominidae, whether one considers this to be *Ramapithecus*, *Oreopithecus*, or neither. Simons (1965) regards *Ramapithecus* as too early to be a tool-*maker*, but he and Pilbeam (1965) suggest that it was a regular tool-*user*, like the savannah chimpanzee (Goodall 1964; Kortlandt 1967). This is eminently likely, but is no explanation for anterior dental reduction since the chimpanzee has relatively the largest canines and incisors of any pongid, much larger than those of the gorilla, which has never been observed to use artefacts in the wild. To explain hominid dental reduction on these grounds, therefore, we presumably have to postulate that the basal hominids were much more dependent upon artefacts than the chimpanzee, without any obvious explanation of why this should be so. One would also expect signs of regular toolmaking to appear in the fossil record at least as early as the first signs of dental reduction, rather than twelve million years later. The more artefactually sophisticated the wild chimpanzee is shown to be, of course, the weaker the logic of the tool/weapon determinant theory becomes, rather than the other way about, as its proponents seem to feel.

Clearly, some other explanation is needed for anterior tooth reduction, at least at its inception. Recognizing this, Pilbeam and Simons (1965) and Simons (1965) regard tool-use by *Ramapithecus* as compensation rather than cause for anterior tooth reduction, adopting as a causal factor Mills's (1963) suggestion that upright posture leads to facial shortening, and that canine reduction would then follow to avoid "locking" when the jaw is rotated in chewing. The main objection to this scheme (Holloway 1967) is that there is no logical reason why facial shortening should follow upright posture. Indeed, if brachiation is counted as upright posture, then it clearly does not. (Among extinct Madagascan lemurs, for instance, the long-faced *Palaeopropithecus* was a brachiator, the very short-faced *Hadropithecus* a terrestrial quadruped (Walker 1967).) Nor does a reduced canine accompany a short face in, for instance, *Hylobates* or *Presbytis*. Furthermore, the explanation extends only to the canines, and does not account for the fact that incisal rather than canine reduction distinguishes the known specimens of *Ramapithecus* from small female Pongidae.

The same criticism applies to the model proposed by Holloway (1967), who finds an explanation of canine reduction in hormonal factors associated with the adoption of a hominid way of life:

> . . . Natural selection favoured an intragroup organisation based on social cooperation, a higher threshold to intragroup aggression, and a reduction of dominance displays . . . a shift in endocrine function took place so that natural selection for reduced secondary sexual characters (such as the canines) meant a concomitant selection for reduced aggressiveness within the group (1967:65).

Thus, reduced canine dimorphism is apparently attributed to a pleiotropic effect of genetically-controlled reduction in hormonal dimorphism, itself favored by the "co-operative life" of hunting.

This argument is vulnerable on several counts. First, there is no obvious reason why even *Homo sapiens* should be thought less hormonally dimorphic than other catarrhines; in structural dimorphism the "feminized" canine of the male is a human peculiarity, but humans are rather more dimorphic in body-mass than chimpanzees, and much more dimorphic than any other hominoid in the development of epigamic characters, especially on the breast and about the head and neck, which can only be paralleled, in Primates, in some baboons. Equally, there seems little to suggest that human males are any less competitive and aggressive among themselves than those of other species; the difference rather lies in the fact that these attributes are expressed in culturally-determined channels (such as vituperative correspondence in the *American Anthropologist*) rather than by species-specific threat gestures or physical assault, so that expression of rage is postponed and channelled, not abolished at source. It seems unlikely that the basal hominids had departed further than modern man from the catarrhine norm. In fact, an elaboration of dominance/subordination behavior, and thus an intensification of the social bond between males, is often attributed to a shift to "open-country" life (Chance 1955; 1967).

Second, the hypothesis that the canines which are disclosed when a male primate yawns are functioning as "organs of threat" is not unchallenged; Hall (1962) found that in Chacma baboons yawning appeared in ambivalent situations where it could more plausibly be interpreted as displacement. The size of the canines "displayed" by a male in a displacement yawn would be of no consequence to his social relations or his Darwinian fitness.

Third, and most trenchant, we must critically examine the assumptions, accepted by "orthodox" opinions as well as by Holloway, that an increase in meat-eating beyond that usual in primates would follow "open-country" adaptation, and that the peculiarities of the hominids ultimately represent adaptations to hunting. The first of these assumptions is perhaps supported by the fact that chimpanzees living in savannah woodland have been seen catching and eating mammals (Goodall 1965), while those living in rain-forest have not. The flaw lies in the second part of the argument, and is like that in the artefact-determinant theory; the more proficient a hunter the non-bipedal, large-canined, large-incisored chimpanzee is found to be, the less plausible it becomes to attribute the origin of converse hominid traits to hunting. Moreover, the hunting and meat-eating behavior of the chimpanzee does not, to the unbiased eye, suggest the selective forces that could lead to the evolution of hominid characters. Neither weapon-use nor bipedalism is prominent. Prey is captured and killed with the bare hands, and is dismembered, like other fleshy foods, with the incisors. Thus, if a population of chimpanzee-like apes becomes adapted to a hunting life in savannah, there is absolutely no reason to predict incisal reduction, weapon-use, or bipedalism. On the contrary, it is most difficult to interpret the hominid characters of the australopithecines functionally as adaptations to life as a carnivorous chimpanzee. Incisal reduction would make for less efficient processing of all fleshy foods, including meat. A change from knuckle-walking, which can be a speedy and efficient form of terrestrial locomotion, to a mechanically imperfect bipedalism (Washburn 1950; Napier 1964) would scarcely improve hunting ability, especially since a knuckle-walking animal can, if it wishes, carry an artefact in its fist while running (cf. illustration in Reynolds & Reynolds 1965:382). Once these characters existed as preadaptations in the basal hominids, they may well have determined that when hunting was adopted as a regular activity, it was hunting of the type that we now recognize as distinctively human, but to use this as an explanation of their first appearance is inadmissibly teleological.

This view is supported by the absence of fossil evidence for efficient hunting before the latter part of the Lower Pleistocene. It seems most unlikely that the hominid line would become partially and inefficiently adapted to hunting in the Miocene, only to persist in this transitional phase until the Lower Pleistocene (becoming, meanwhile, very specialized dentally, but no better at hunting or tool-making!), when a period of rapid adaptation to hunting efficiency took place. Perhaps recognizing this, adherents of the "predatory chimpanzee" model tend to situate the hominid-pongid divergence in the late Pliocene, and regard all known fossils of basal Pleistocene Hominidae as representative of a short-lived transitional phase of imperfect hunting adaptation (Washburn 1963). This is a view that is intrinsically unlikely, and difficult to reconcile with the fossil evidence of Tertiary hominids. The obvious way out of the dilemma is to set aside the current obsession with hunting and carnivorousness, and to look for an alternative activity which is associated with "open-country" life but which is functionally consistent with the anatomy of basal hominids.

Impressed by the bipedal charge of the mountain gorilla, and his tendency to toss foliage around when excited, two authors (Livingstone 1962; Wescott 1967) have suggested that therein might lie the origin of human bipedalism and the other elements of the hominid complex. The objection to this notion is again that it is illogical to invoke the behavior of living apes to explain the origin of something that they themselves have not developed; if upright display leads to habitual bipedalism, why are gorillas still walking on their knuckles? Conversely, if hominid bipedalism were initially used solely in display, why should they have taken to standing erect between episodes? Even if we grant that the savannah is more predator-ridden than the forest (a view often stated but seldom substantiated, even for the Recent, let alone the Tertiary), it is difficult to believe that attacks were so frequent as to make defensive display a way of life.

The occasional bipedalism, tool- and weapon-use, and meat-eating of the pongids are useful indicators of the elements that were probably part of the hominid repertoire, ready for elaboration under particular circumstances. To explain this elaboration, however, we must look *outside* the normal behavior of apes for a factor which agrees functionally with the known attributes of early hominids. As we have seen, "hunting" is singularly implausible as such a factor. The object of this article is to suggest an alternative, based initially on the observation that many of the characters distinctive of basal hominids, as opposed to pongids, also distinguish the grassland baboon *Theropithecus* from its woodland-savannah and forest relatives *Papio* and *Mandrillus*, and are functionally correlated with different, but no less vegetarian, dietary habits.

THEROPITHECUS-HOMINID PARALLELISMS

The assumption is made here that both hominids and living African pongids are descended from Dryopithecinae, a group intermediate between the two in most of its known characters (most of which are dental), though rather closer to Pongidae than to Hominidae. The chimpanzee can then be seen as manifesting evolutionary trends away from the ancestral condition more or less opposite to those of the Hominidae, while the gorillas retain a more conservative condition, at least dentally. This model would work as well on the less likely assumption that the chimpanzee represents the primitive condition. It is also assumed that the Cercopithecine genera *Theropithecus* and *Papio* either diverged from an intermediate common ancestor, or, more probably, that *Theropithecus* and *Mandrillus* have become differentiated in opposite directions from a *Papio*-like form (Jolly 1969). This process can be documented for *Theropithecus* during the course of the Pleistocene (Jolly 1965, 1972).

Table 8-1 summarizes characters by which *either* early Pleistocene Hominidae differ from *Pan*, or *Theropithecus* from *Papio* and *Mandrillus*, listed without regard to their functional interrelationships or significance. Those which distinguish early Hominidae from Pongidae constitute the "Hominid adaptive complex," and are indicated in column A, while those which form part of the "*Theropithecus* adaptive complex" are indicated in column B. Rectangles show those features common to the two complexes, of which there are twenty-two out of forty-eight, reasonable *prima facie* evidence for parallelism between them. This hypothesis can be tested by checking the elements of the complexes for cross-occurrence in *Papio* and *Pan*. If the high number of common characters were simply due to chance, rather than to parallelism, we should not expect significantly fewer of the Hominid characters to appear in *Papio* (as opposed to *Theropithecus*), or significantly fewer of the *Theropithecus* complex characters to occur in *Pan*. In fact, none of these cross-correspondences occurs. There are some grounds, therefore, for assuming the existence of evolutionary parallelism, and perhaps some degree of functional equivalence between the differentiation of *Theropithecus* and that of the basal hominids, and the common features may be used to construct a model of hominid divergence from pongids. To do this, we must examine the functional implications of the "AB" characters.

Of these, only one certain one appears in the "behavior" category, largely because of the impossibility of observing the behavior of fossil forms. Inferences of behavior from structure are, of course, not permissible at this stage of analysis. The single common character is the basic one of true "open-country" habitat, inferred largely from the death-assemblages in which early *Theropithecus* and Hominidae are found, as well as the habitat of *T. gelada*.

Three skeletal "AB" characters are postcranial. The abbreviated fingers and unreduced thumb makes a pollex-index grip possible for the terrestrial

TABLE 8-1

Adaptive characters of the Villafranchian Hominidae and
Theropithecus.
A. *Characters distinguishing early Hominidae from* Pan *and other Pongidae.*
B. *Characters distinguishing* Theropithecus *from* Papio *and Mandrillus.*
C. *Features of the Hominid complex not seen in* Theropithecus.
D. *Features of the* Theropithecus *complex not seen in Hominidae.*

	A	B	C	D	NOTE NO.
1. *Behaviour*					
a. Open-country habitat, not forest or woodland	X	X	—	—	
b. Trees rarely or never climbed when feeding	(X)	(X)	—	—	1
c. One-male breeding unit	(X)	(X)	—	—	1, 2
d. Foraging mainly in sitting position	?	(X)	—	—	1
e. Small daily range	?	(X)	—	—	1
f. More regular use of artefacts in agonistic situations	X	—	X	—	3
g. Regular use of stone cutting-tools	X	—	X	—	4
h. Most food collected by index-pollex precision grip	?	(X)	—	—	1
2. *Postcranial structure*					
a. Hand more adept, Opposability Index higher	X	X	—	—	5, 6
b. Index finger abbreviated	?	X	—	—	7
c. Hallux short and weak	—	X	—	X	7
d. Hallux relatively non-abductible	X	X	—	—	7, 8
e. Foot double-arched	X	—	X	—	8
f. Phalanges of pedal digits 2–5 shorter	(X)	X	—	—	7

g. Ilium short and reflexed	X	—	X	—	9
h. Sacroiliac articulation extensive	X	—	X	—	9
i. Anterior-inferior iliac spine strong	X	—	X	—	9
j. Ischium without flaring tuberosities	X	—	X	—	9
k. Accessory sitting pads (fat deposits on buttocks) present	(X)	(X)	—	—	7
l. Femur short compared with humerus	?	X	—	—	7
m. Distal end femur indicates straight-knee 'locking'	X	—	X	—	9
n. Epigamic hair about face and neck strongly dimorphic	(X)	(X)	—	—	1, 7
o. Female epigamic features pectoral as well as perineal	(X)	(X)	—	—	1, 10

3. *Cranium and mandible*

a. Foramen magnum basally displaced	X	—	X	—	11
b. Articular fossa deep, articular eminence present	X	—	X	—	9
c. Fossa narrow, post-glenoid process appressed to tympanic	X	X	—	—	9, 7
d. Post-glenoid process often absent, superseded by tympanic	X	—	X	—	9
e. Post-glenoid process long and stout	—	X	—	X	7, 12
f. Basi-occipital short and broad	X	X	—	—	9, 7
g. Mastoid process regularly present	X	X	—	—	9, 7, 13
h. Temporal origins set forward on cranium	X	X	—	—	9, 7
i. Ascending ramus vertical, even in largest forms	X	X	—	—	9, 7, 12
j. Mandibular corpus very robust in molar region	X	X	—	—	9, 7, 12
k. Premaxilla reduced	X	X	—	—	9, 7
l. Dental arcade narrows anteriorly	X	X	—	—	9, 7
m. Dental arcade of mandible parabolic, 'simian' shelf absent	X	—	X	—	9
n. Dental arcade (especially in larger forms) V-shaped; shelf massive	—	X	—	X	7

4. *Teeth*

a. Incisors relatively small and allometrically reducing	X	X	—	—	9, 7
b. Canine relatively small, especially in larger forms	X	X	—	—	9, 7
c. Canine incisiform	X	—	X	—	9
d. Male canine 'feminised', little sexual dimorphism in canines	X	—	X	—	9
c. Third lower premolar bicuspid	X	—	X	—	9

TABLE 8-1 *continued*

	A	B	C	D	NOTE NO.
f. Sectorial face of male P$_3$ relatively small and allometrically decreasing	—	X	—	X	7
g. Molar crowns more parallel-sided, cusps set towards edge	X	X	—	—	14, 7
h. Cheek-teeth markedly crowded mesiodistally	X	X	—	—	14, 7
i. Cheek-teeth with deep and complex enamel invagination	—	X	—	X	7
j. Cheek-teeth with thick enamel	X	—	X	—	
k. Canine eruption early relative to that of molars	X	X	—	—	7, 9
l. Wear-plane on cheek-teeth flat, not inclined bucco-lingually	X	—	X	—	9
m. Wear on cheek-teeth rapid, producing steep M1–M3 'wear-gradient'	X	X	—	—	14, 7

[1] Crook 1966; 1967; Crook & Aldrich-Blake 1968. Parentheses indicate behavioral and soft-part features which are present in the living representative of the group (*Theropithecus gelada* or *Homo sapiens*), but which cannot be demonstrated on fossil material.

[2] In all but a very few human societies, where polyandry sometimes occurs (Murdock 1949).

[3] The 'bashed baboons' of the South African cave sites (Barbour 1949; Dart 1949, etc.) are the most direct evidence for this.

[4] Not, apparently, in the earliest Pleistocene hominid sites (Howell 1967).

[5] Napier 1962.

[6] J. R. Napier, personal communication.

[7] Pocock 1925; Jolly 1965; 1969a; 1969b.

[8] Day & Napier 1964.

[9] Clark 1964.

[10] Matthews 1956.

[11] The functional interpretation of this character is disputed (cf. Clark 1964; Biegert 1963).

[12] Leakey & Whitworth 1958.

[13] Not in *T. gelada*, but regular in larger Pleistocene forms.

[14] Simons & Plibeam 1965.

monkeys (Napier & Napier 1967). Bishop (1963) showed that *Erythrocebus patas* made more consistent use of such a grip than its more arboreal relatives, the guenons. Recent work by Crook (e.g. 1966), including filmed close-ups of hand-use in the wild, has made it clear that the gelada (in contrast to, for instance, *Papio*) uses a precision-grip for most of its food-collecting. Food consists mainly of grass-blades, seeds and rhizomes which are picked up singly between thumb and index, and collected in the fist until a mouthful is accumulated. The index is thus continually used independently of the other digits. This feeding method is facilitated by the well-developed pollex and the very short index finger (Pocock 1925; Jolly

1965), a combination giving the gelada the highest "opposability index" (Napier & Napier 1967) of any catarrhine, not excluding *Homo sapiens* (J. R. Napier, personal communication). It is significant that the precision-grip of the gelada, which like other Cercopithecinae has not been seen making or using artefacts in the wild, should far outclass that of the tool- and weapon-using chimpanzee (Napier 1960).

The two common features of the foot are attributable to terrestrial adaptation which requires pedal compactness rather than hallucal gripping-power (Pocock 1925; Jolly 1965); the rest of the foot structure is different in the two forms and reflects the fact that their move to terrestrialism was quite independent and analogous. Most of the postcranial elements of the hominid complex are absent in *Theropithecus*, being related to upright bipedalism (Clark 1964). The post-cranial features of the *Theropithecus* complex are much fewer, expressing the fact that apart from the absence of tree-climbing its locomotor repertoire scarcely differs qualitatively from that seen in its woodland and forest relatives, although the frequency of the elements differs considerably (Crook & Aldrich-Blake 1968).

The mastoid process of the large Pleistocene *Theropithecus* is perhaps unexpected, since in Hominidae it can be related to erect posture (Krantz 1963). However, unlike the hindlimb characters which *Theropithecus* does not share, the mastoid is related to poising the head on the erect trunk, not the trunk upon a hyperextended hindlimb. The gelada spends most of the day in an upright *sitting* position, as, probably, did its Pleistocene relatives, and, when foraging, even moves in the truncally erect position, shuffling slowly on its haunches, hindlimbs flexed under it. Thus, *truncal* erectness is more habitual than in any non-bipedal catarrhine, and the mastoid process becomes explicable. Also, the forelimb is more "liberated" from locomotor function in *Theropithecus* than in any other non-biped, simply because the animal rarely locomotes. Sitting upright allows both hands to be used simultaneously for rapid gathering of small food-objects, a pattern seen more rarely in *Papio* where a tri-pedal stance leaving one hand free is associated with a diet mainly of larger items (Crook & Aldrich-Blake 1968).

The majority of "AB" characters are in the jaws and teeth. The temporal muscles (which are large in *Theropithecus* and some, at least, of the early Hominidae) are set well forward, so that their line of action lies almost parallel to that of the masseters, and their moment-arm about the temporo-mandibular joint is relatively long, as compared to that of the resistance of food-objects between the teeth, thus exerting a high grinding or crushing force per unit of muscular exertion. On the other hand, the gap between the opposing occlusal surfaces is small per unit of muscular extension, limiting the size of objects that can be tackled. Also, the horizontal component of temporal action is reduced, lessening its effectiveness in bilateral retraction of the mandible against resistance ("nibbling"), and in resisting forces tending to displace the mandible forwards, as when objects are held in the hand and stripped through the front teeth. Thus efficiency of incisal action, which is used by catarrhines in fruit-peeling, nibbling flesh of fruits from rinds, stripping cortex from esculent vines and tubers, and, occasionally but very significantly, tearing mammal-meat from bones or skin (DeVore & Washburn 1963; Goodall 1965), is sacrificed to adding the power of the temporals to that of the masseters and pterygoids, used mainly for cheek-tooth chewing.

In the larger forms (of both taxa), with their allometrically longer faces, the forward position of the temporals is preserved by making the ascending ramus of the mandible high but vertical, deepening the posterior maxilla. The tooth row scarcely lengthens, although the face is long from prosthion to nasion, and the cheek-teeth become mesio-distally crowded. In the fruit-eaters the horizontal component of temporal action is maintained by keeping the ascending ramus low as the face lengthens; the corpus becomes elongated and marked diastemata tend to appear in the tooth row. The short basi-occiput, and anterio-posteriorly narrowed articular fossa of both *Theropithecus* and Hominidae can be seen as part of the same functionally-determined developmental pattern. The proportions of the molars and incisors fit the same functional complex. Both *Theropithecus* and the hominids have narrower, smaller incisors than their woodland and forest relatives. Molar area is greater per unit of body-mass, and incisal width less, in *Theropithecus* than *Papio* or *Mandrillus*, and absolute incisal breadth is no greater in *T. oswaldi mariae*, as big as a female gorilla, than it is in *T. gelada*. In *Papio* both incisors and molars increase proportionately to body-mass in larger forms, and in the forest-dwelling *Mandrillus* it is the molars which in males are scarcely larger than those of females half their weight (Jolly 1969, 1972). Among the Hominoidea, the Villafranchian Hominidae are *Theropithecus*-like in their dental proportions, and perhaps in their allometric ratio; the large form "Zinjanthropus" has the most extreme elative incisal reduction, while *Pan* is *Mandrillus*-like in proportions and ratios (Jolly & Chimene in preparation).

In the monkeys, the evidence for molar dominance in *Theropithecus* agrees well with data on diet in the natural habitat. In the few areas where the Pleistocene sympatry of *Theropithecus* and *Papio* still exists in the Ethiopian highlands, *Theropithecus* eats small food objects requiring little incisal preparation, but prolonged chewing, while *Papio* (which elsewhere in its wide range is a most catholic feeder) here concentrates on fleshy fruits and other tree products, most of which require peeling or nibbling with the incisors (Crook & Aldrich-Blake 1968). There seems no good reason against attributing the *Theropithecus*-like incisal proportions and jaw characters of the early hominids to a similar adaptation to a diet of small, tough objects. There is no need to postulate a compensatory use of cutting-tools for food preparation, until it can be shown archaeologically that such tools were being made (cf. Pilbeam & Simons 1965).

To avoid the charge of Lamarckism, I should perhaps suggest some selectional mechanisms leading to incisal reduction in molar-dominant forms. One is a general explanation of the reduction of structures to a size related to their function. Every structure is at once a liability, in that it can become the site of an injury or infection and requires energy and raw materials for its formation and maintenance, and an asset in so far as it performs a homeostatic function. Natural selection will favor the genotype producing a structure of such size and complexity as to confer the greatest *net* advantage. In a monkey or hominoid adapting to a gelada-like diet, each unit of tooth-material allotted genetically to a molar will bring a greater return in food processed than a unit allotted to an incisor. Thus selection should favor the genotype which determines the incisors at the smallest size consistent with their residual function. This "somatic budget effect" differs from Brace's (1963) "random mutation effect" (criticized by Holloway (1966) among others) chiefly in that it proposes a positive advantage in reduction.

A second mechanism is specific to teeth. While dental size is genetically (or at least antenatally) determined, the development of the alveolus depends partly upon the stresses placed upon it during its working life (Oppenheimer 1964). An under-exercised jaw may thus be too small to accommodate its dental series, which tends to become disadvantageously crowded and maloccluded. Natural selection will then favor the genotype which reduces the teeth to a size fitting the reduced alveolus. The "Oppenheimer effect," originally proposed to explain the reduction of complete dentitions (as in the case of *Homo sapiens* after the introduction of cooking and food-preparation), could equally operate on particular dental regions, as in the case of *Theropithecus* and the early hominids, where the incisors were reduced but the molars were, if anything, larger than those ~~gigantopithecus~~ of their forest- and woodland-dwelling relatives.

One of the more surprising findings is that canine reduction is one of the shared characters. This is contrary to the weapon-determinant hypothesis which states that "open-country," terrestrial primates, being more exposed to predation, should have *larger* canine teeth, unless they use artificial weapons. The situation in *Theropithecus* is somewhat complicated by the existence of both allometric and evolutionary trends towards canine reduction (Jolly 1972). The canines decrease in relative size from the Basal to the Upper Pleistocene, when compared between forms of approximately equal body-size. And within each palaeospecies, there is evidence that the canine is smallest, relative to the molars and to the general size of the animal, in the largest forms, where the males were about the same size as a female gorilla. The allometric trend is exactly opposite to that seen in present-day *Papio*, in which males of the largest forms have relatively and absolutely the largest canines. (*Theropithecus gelada*, a very small form of highly-evolved *Theropithecus*, has a male canine size that is relatively large for the genus, but entirely predictable from the allometric ratios between molars and canines characteristic of the whole genus.)

Both the allometric and the evolutionary trends are incompatible with the theory that canine size in males is positively correlated with terrestrial life in non-artefactual primates, but at least two alternative explanations for canine reduction are possible. It may be favored as an adaptation to increased efficiency in rotary chewing, by avoiding canine "locking" and producing more even molar wear; this would be consistent with the evidence for "molar dominance." An early stage of adaptation might involve the canines' being worn flat as they erupted. This would obviously be a wasteful situation which might be expected to be corrected by the "somatic budget" effect, and natural selection.

The other possible explanation relates canine reduction to reduction of the incisors. The fact that the two trends parallel each other so closely in both *Theropithecus* and the Hominidae, both evolutionarily and allometrically, suggests, *a priori*, a relationship between these processes. Since incisal reduction can be plausibly interpreted as an adaptation to small-object feeding, it seems reasonable to treat canine reduction as the secondary, dependent, character. The dependence can be attributed to either, or both, of two mechanisms. First, the direct effects of incisal disuse upon the anterior alveolar region might produce an extended "Oppenheimer effect," acting primarily on the incisors, and, secondarily and less intensely, on the neighboring canines. This might explain the fact that while incisal reduction seems fully evolved already in the rather primitive, probably early, Makapan *Theropithecus*, canine reduction proceeds through the Pleistocene. Alternatively, the dependence might be at the genetic level, with canine reduction being a simple pleiotropic effect of a genotype which primarily determined incisal reduction. There is some evidence for a canine-incisal genetic "field" in both Cercopithecoidea (Swindler *et al.* 1967) and Hominoidea (Jolly & Chimene, unpublished data). It may well be that both selective factors are operative in canine reduction: adaptation to rotary chewing favoring crown height reduction, and effects stemming from incisal reduction acting upon crown-area dimensions. Since the genetic factors determining these two parameters of canine size are most unlikely to be independent of each other, the two processes would be mutually reinforcing.

This scheme for canine reduction in *Theropithecus* is distinct from that of Simons and Pilbeam (1965) who also attempted to explain both incisal and canine reduction (in *Ramapithecus*) in terms of diet. They proposed that canine size was related to its own function in food preparation, thus exposing themselves to Washburn's objection that if this were so canine sexual dimorphism would imply a sexual dietary difference in non-human catarrhines which is not borne out by field observations. The scheme proposed here recognizes the essentially agonistic function of the canine but suggests that its reduction in *Theropithecus* is unrelated to this function, and is a secondary effect of dietary influences on incisors and molars. If as a consequence of a dietary aspect of terrestrial adaptation, and in the complete absence of either bipedalism or use of artificial weapons, a trend towards canine reduction can be initiated in *Theropithecus* (a highly terrestrial catarrhine), then there is no need to postulate that these characters either preceded or accompanied the earliest stages (at least) of canine reduction in Hominidae, which could similarly be attributed to dietary factors.

Having, I hope, established that adaptation to terrestrial life and small-object feeding constitutes at least a reasonable working model for the initial hominid divergence from Pongidae, I should now like to speculate about the characters of soft tissues and social organization, although admittedly these can never be tested against the fossil record.

Unlike the savannah-woodland species, all three truly "open-country" Cercopithecinae (*Theropithecus, Erythrocebus* and *Papio hamadryas*) have a social organization involving exclusive mating-groups with only one adult male. Of the three, the patas is peculiar in that a female-holding male maintains his exclusive sexual rights by vigilant and agonistic behavior directed against other adult males, whose presence he will not tolerate (Hall *et al.* 1965). Patas "harems" do not, therefore, co-exist as parts of a higher-order organization. In the hamadryas and gelada, however, the male maintains cohesion of his group by threatening and chastising his own females when necessary, and many one-male groups co-exist within semi-permanent troops or bands (Kummer 1967; Crook 1967). In both species, cooperation between males is not excluded, and is frequently seen in situations of extra-troop threat. It would not be unreasonable to expect a similar social organization, with permanent, monogamous or polygamous one-male groups set within the matrix of a larger society, to be developed by a hominoid adapting to a gelada-like way of life. This pattern is also, one might add, still distinctive of the vast majority of *Homo sapiens* (Murdock 1949). ~~Therefore~~ - *polygamy is primitive*

In both hamadryas and gelada the attention-binding quality of the adult male is enhanced by the conspicuous cape of fur about his shoulders, which is groomed by his harem. It is likely that this feature has been favored by Darwinian sexual selection (Jolly 1963). Only one other living catarrhine has such striking sexual dimorphism in epigamic hair about the face and neck: *Homo sapiens.* Similarly, bearing the female epigamic features pectorally and ventrally, rather than perineally, is a feature which can be correlated, in geladas, with a way of life in which the majority of the foraging time is spent sitting down (Crook & Gartlan 1966). It is unique to *Theropithecus* among non-human primates, but also occurs in *Homo sapiens.*

Fatty pads on the buttocks, adjacent to the true ischial callosities, are another *Theropithecus* peculiarity (Pocock 1925) which can be plausibly related to the habit of sitting while feeding, and also occur uniquely in *Homo sapiens* among the Hominoidea.

DIVERGENCES BETWEEN THE *THEROPITHECEUS* AND HOMINID ADAPTATIONS

So far I have concentrated on the parallelisms between the *Theropithecus* and hominid adaptive complexes. Their divergences are, however, equally instructive. Columns C and D of Table 8-1 have been added to extract those characters which occur as part of one of the complexes, but not of the other. Since characters were initially selected because they fitted either A or B (that is, they discriminate *within* superfamilies), I omit the large number of attributes which simply indicate that *Theropithecus* is a monkey but the Honinidae are Hominoidea, a point that is not at issue. The last two columns therefore indicate features adaptive to different aspects of the "open-country" habitat, and analogous adaptations to the same aspect. Again, we must examine their functional implications.

The "AB" characters of the jaws and teeth seem to comprise a functional complex related to a diet of small, tough objects. In *Theropithecus,* these are known to be mostly grass blades and rhizomes, accounting for the tendency towards cheek-tooth crown complexity which is convergent upon similar structures in grass-eating animals as diverse as voles, warthogs and elephants. This character is not part of the hominid complex, which instead includes cheek-teeth with relatively low cusps but thick enamel which wear to an even, flat, and uniform surface. Also, the temporo-mandibular joint of the hominids is unique among catarrhines in the possession of an articular eminence upon which the mandibular condyle rides in rotary chewing. Thus, the molar surfaces do not simply grind across one another in chewing, they also swing toward and away from one another. Such molars and jaw action are clearly not adapted to mincing grass blades, but rather to breaking up small, hard, solid objects of more or less spherical shape, by a combination of crushing and rolling such as is employed in milling machines. The efficiency of the combined action lies in its seeking out, by continuous internal deformation, the weaknesses in the structure of the object to be crushed.

The possession of a parabolic dental arcade (rather than the V-shape seen in large *Theropithecus*), and the absence of the "simian" symphyseal shelf, are also explicable on the basis of such a diet. The features cannot be related to tooth-function in any obvious way, nor can they be attributed simply to a Hominoid rather than Cercopithecoid inheritance, since the Hominoid *Gigantopithecus bilaspurensis* has a *Theropithecus*-like arcade and symphysis. A possible functional explanation involves the tongue. Experiments with grain-chewing in *Homo sapiens* suggest that objects not crushed by one masticatory stroke tend to be pushed by the rolling action into the oral cavity, whence they are guided back to the teeth by the tongue. This demands much more constant, agile, lingual motion than is needed in masticating a fibrous bolus of fruit (or turf). Thus, a chewing apparatus of the hominid type might be expected to include a thick, muscular, mobile tongue, accommodated in a large oral cavity. The highly-arched palate, capacious interramal space, and absence of symphyseal shelf may all be interpreted as elements of the large oral cavity, also, incidentally, providing preadaptations to articulate speech.

We can thus distinguish a sub-complex of unique hominid characters which suggests that the "small objects" of the basal hominid diet were solid, spherical, and hard. Many potential foods fit the description, but only one is widespread enough in open country to be a likely staple. This is the seeds of grasses and annual herbs, which still provide the bulk of the calories of most hominids. This is not to say that other resources were not exploited when available, but that the diet of basal hominids was probably centered upon cereal grains as that of the chimpanzee is upon fruit and that of the gorilla on herbage.

Although canine reduction is one of the AB features, the incisiform shape of the canine, and its "feminization" in the male are unique, among catarrhines, to hominids of australopithecine and later grades. Presumably these features represent an extreme stage of reduction in adaptation to rotary chewing, reflected in the flat wear-plane of the cheek-teeth. One might therefore ask why a similar degree of reduction has not appeared in the *Theropithecus* line. Several explanations are possible. First, the chewing motion involved in grass-mincing might not be as demanding as that used for seed-milling. Second, a relatively high canine/incisor ratio and high sexual dimorphism are probably Cercopithecoid heritage characters, which would tend to blunt the effect of "genetic fall-out" from incisal reduction. A reduction in anterior tooth size which leaves the cercopithecoid *Theropithecus* with a canine reduced in size, but still a useful weapon, might reduce the canine of a hominoid beyond the point of usefulness. Perhaps because the canines of all Pongidae are shorter and blunter than those of Cercopithecoidea, a tendency to use artefacts as weapons was probably another hominoid heritage character, thus permitting the Hominidae to compensate for the loss of biting canines favored by rotary chewing, and allowing the extreme degree of reduction represented by incisification and feminization. Alternatively, it may be that *Theropithecus*, a much more recent lineage than the hominids, has not yet had time to achieve full canine reduction.

A second group of C and D characters is postcranial and reflects the fact that while *Theropithecus* is a quadruped, at least when moving more than a few paces, the Villafranchian hominids were evidently bipeds of a sort. Again, we can evoke a combination of adaptational and heritage features in explanation. A gelada-like foraging pattern leads to constant truncal erectness in the sitting position, with the trunk "balanced" on the pelvis, and the forelimbs free. In *Theropithecus*, this behavioral trait (and its associated adaptive features) are superimposed upon a thoroughgoing, cercopithecoid quadrupedalism, producing a locomotor repertoire in which the animal abandons "bipedal" bottom-shuffling for quadrupedal locomotion when it moves fast, or for more than a few paces. The hominoid ancestor of the Hominidae, on the other hand, is most unlikely to have been postcranially baboonlike. The smallest, and best known, dryopithecine, *D. (Proconsul) africanus*, described as a "semi-brachiator" (Napier & Davis 1959), shows limb proportions and other features recalling generalized arboreal climbers like some of the Cebidae, but no distinctively cercopithecoid features. Its larger relatives had probably moved in the direction of truncal erectness, forelimb independence and abductibility, and the other characters distinctive of the Pongidae as a group, which are generally attributed to brachiation and can be seen as the consequences of large size in an arboreal habitat (Napier 1968).

Recently Walker and Rose (1968) and Walker (personal communication) have detected signs of locomotor adaptations very like those of the living African pongids in the fragmentary (and largely undescribed) postcranial remains of the larger African Dryopithecinae. In the basal hominid, therefore, the "gelada" specializations would be superimposed upon a behavioral repertoire and post-cranial structure already attuned to some degree of truncal erectness. This combination of heritage and adaptation may have been the elusive determinant of terrestrial bipedalism, a gait that is inherently "unlikely," and which would thus have begun as a gelada-like shuffle. Locomotion of any kind is infrequent during gelada-like foraging, so that (unlike hunting!) it is an ideal apprenticeship for an adapting biped. Furthermore, as a final bonus, if the hominids were derived from a "brachiator" or knuckle-walker stock, they would have carried, in preadaptation, the high intermembral index which *Theropithecus* has had to acquire as part of his adaptation.

A NEW MODEL
OF HOMINID DIFFERENTIATION:
PHASE 1, THE SEED-EATERS

The anatomical evidence seems to suggest that at some time during the Tertiary, the populations of Dryopithecinae destined to become hominids began to exploit more and more exclusively a habitat in which grass and other seeds constituted most of the available resources, while trees were scarce or absent. However, the great majority of contemporary tropical grasslands and open savannahs (especially those immediately surrounding patches of evergreen rainforest in all-year rainfall areas), are believed to be recent artefacts of burning and clearance by agricultural man (Rattray 1960; Richards 1952; Hopkins 1965; White *et al.* 1954). Under climatic climax conditions the vegetation of the seasonal rainfall tropics would almost always include at least one well-developed tree stratum, ranging from semi-deciduous forests through woodlands to *sahel* where paucity of rainfall inhibits both herb and tree strata (Hopkins 1965).

What, then, would have been the biotope of the grain-eating, basal hominids? The obvious answer is provided by the areas of treeless edaphic grassland which exist, even under natural conditions, within woodland or seasonal forest zones, wherever local drainage conditions cause periodic flooding, and hence lead to perpetual sub-climax conditions by inhibiting the growth of trees and shrubs (Richards 1952; Sillans 1958). These areas range in extent from hundreds of acres, like the bed of the seasonal Lake Amboseli, to a network of strips interlaced with woodlands (*dambos*; Ansell 1960; Michelmore 1939; Sillans 1958). Edaphic grasslands produce no tree-foods, but support a rich, all-year growth of grasses and other herbs, and are the feeding-grounds for many grass-eating animals (Ansell 1960). The remains of Villafranchian hominids are often found in deposits formed in such seasonal waters, as is Pleistocene *Theropithecus* (Jolly 1972), adding some circumstantial evidence that this was their preferred habitat. Seasonality in rainfall, producing a fluctuating water-level, is important to the development of edaphic grasslands, since perennial flooding leads to a swamp-forest climax. While there is little evidence for catastrophic desiccation in the tropics of the kind demanded by some models of hominid differentiation, there are indications that a trend towards seasonality persisted through the Tertiary, especially in Africa (Moreau 1951).

The first stages of grain-feeding adaptation probably took place in a *dambo*-like environment, later shifting to wider floodplains. The change from a fruit (or herbage)-centered diet to one based upon cereals would lead, by the evolutionary processes discussed, to the complex of small-object-feeding, seed-eating, terrestrial adaptations (see Figure 8-1). Other grassland resources obtainable by individual foraging or simple, *ad hoc* co-operation like that seen in the woodland chimpanzee, would also be utilized. Such items as small animals, vertebrate and invertebrate, leafy parts of herbs and shrubs, and occasional fruits and tubers would be qualitatively vital, if only to supply vitamins (especially ascorbic acid and B_{12}), and minerals, and could easily be accommodated by jaws adapted to grain-milling.

The ability to exploit grass-seeds as a staple is not seen in other mam-

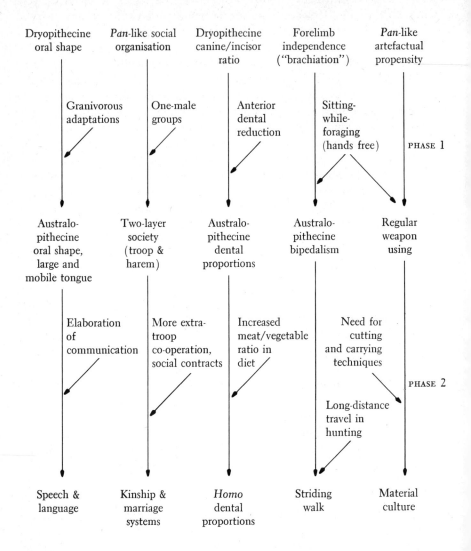

FIGURE 8–1—A model of the development of some of the major hominid characters. During Phase 1 a series of Dryopithecine heritage characters (top line) is modified by the functionally-determined requirements of the sedentary seed-eating complex (second line), producing the characters of evolved Phase 1 hominids (line three). These are the preadaptive heritage characters of Phase 2, which determine the fact that adaptation to the demands of a hunting way of life (fourth line) takes the form of the human traits listed in the bottom line. The illogicalities of previous models tend to arise from omitting the vital second line, inserting the elements of the hunting complex in its place, and invoking feedback.

mals of comparable size, though it is seen in birds and rodents, presumably because the agile hand and hand-eye co-ordination of a higher primate is a necessary preadaptation to picking up such small objects fast enough to support a large animal. With these preadaptations, and the adaptive characters of jaws, teeth and limbs, the basal hominids would have faced little competition in the exploitation of a concentrated, high-energy food (a situation which would hardly have existed had they, as the "hunting" model demands, started to eat the meat of ungulates in direct competition with the Felidae, Canidae, Viverridae, and Hyaenidae). They would thus have attained a stable, adaptive plateau upon which they could have persisted for millions of years, peacefully accumulating the physiological adaptations of a terrestrial, "open-country" species. There is no reason to suppose that they would show radical advances in intellect, social organization, material and non-material culture, or communication, beyond that seen in one or other of the extant higher primates. The "third ape," in Simons's phrase, remained an ape, albeit a hominid ape.

PHASE 2, "HUMAN" HOMINIDS

We do not therefore need to invoke late, "human" characteristics in teleological explanation of initial hominid divergence. However, Phase 1 hominids would be uniquely preadapted to develop such features following a further, comparatively minor, ecological shift. The latter may have involved the increasing assumption by the adult males of the role of providers of mammal meat, with the equally important (but often neglected) corollary that the females and juveniles thereby became responsible for collecting enough vegetable food for themselves *and* the hunters. The adult males would perhaps be behaviorally predisposed to hunting by an existing role as "scouts" (Reynolds 1966). This is as inherently likely in a species adapted to exploiting patches of seeding grasses as it is in the fruit-eating chimpanzee which Reynolds uses as a model of a pre-hominid hominoid. The environmental change prompting the inception of hunting need have been only slight; perhaps an intensification of seasonality in a marginal tropical area which would put a premium on exploiting meat as an additional staple instead of an occasional treat. The dietary change would be small; an increase in the ratio of one high-energy, concentrated food (meat) to another (grain), which would be reflected dentally in a moderate reversal of the Phase 1 back-tooth dominance in favor of the incisal breadth needed to tear meat. The major impact of the change would be upon culture and society rather than upon diet itself. To a female, collecting vegetable food for herself and her mate, there would be great advantage in developing techniques for more rapid harvesting, for carrying the day's booty, and for preparing it by a less laborious means

esp. if males absent

than chewing. On the male side, there would be a premium on the development of cutting-tools for preparing the kill for transport back to the band. The skilful hands, upright posture, and reduced anterior dentition acquired as part of the Phase 1 complex would predispose the hominids to solve these problems of adaptation by the development of their hominoid artefactual propensity into true material culture, a solution which could not be predicted on the basis of the "hunting chimp" model (see Figure 8-1). In both sexes the division of labor would involve constantly postponing feeding for the sake of contributing to the communal bag, and, in males, the impulse to dominate would likewise have to be controlled in the interests of the hunt. The need for co-operation between local bands may have led to the elaboration of truly human kinship systems, in which rights to females are exchanged.

All these factors, and others, were probably related to the evolution of complex forms of symbolic communication (largely, thanks to the seed-eating mouth, in the form of speech rather than gesture), language, ritual, and intellect. Thus the beginning of true hunting and the division of labor would initiate a second period of quantum evolution in hominid history, which we are still experiencing. The effects of this step upon human physical and behavioral evolution have been examined at length by others (e.g. Washburn & Lancaster, 1968), and are beyond the scope of this article. The point to be emphasized here is that this second, distinctively human phase is most comprehensible when it is built upon a firm base of preadaptations which had their initial significance in the seed-eating complex, not upon a chimpanzee-like or semi-human condition. By distinguishing the elements of the Phase 1 adaptive plateau, those of the Phase 2 "hominization process" are thrown into relief.

After some populations had shifted into the Phase 2 cycle, there is no reason why they should not have existed sympatrically with other hominid species which continued to specialize in the Phase 1 niche.

THE FOSSIL RECORD
IN THE LIGHT OF THE NEW MODEL

The new model must now be compared with the fossil record, both to test its compatibility and to relate its events to a timescale.

A medium-sized *Dryopithecus* of the Miocene is a reasonable starting-point for hominid differentiation, and increased seasonality in the Middle-to-Upper Miocene and Lower Pliocene makes it likely that Phase 1 differentiation began at that time. The fragmentary Upper Miocene specimens referred to *Ramapithecus* (Simons 1965), though representing only jaws and teeth, are of precisely the form to be expected in an early Phase 1 hominid: narrow, uprightly-placed and weak incisors, broad, large and mesio-distally crowded cheek-teeth, set in a short but very robust mandibular corpus. Recent work has shown (D. Pilbeam, personal communication) that the molar enamel is thicker than that of contemporary Dryopithecinae, though thinner than that of later hominids, and that the wear-gradient from anterior to posterior molars is steeper. In the Fort Ternan specimen the canine crown is small, but still conical, like that of a Mid-Pleistocene *Theropithecus* female of comparable size. The material is as yet insufficient to show whether or not the male canine was yet "feminized"; this is immaterial to the argument that by Fort Ternan times the Hominidae had entered a granivorous niche in edaphic grasslands. Tattersall's recent (1969) appraisal of the habitat of the Siwalik *Ramapithecus* is consistent with this hypothesis. He describes forested country crossed by watercourses which at the latitude of the Siwaliks must have fluctuated with a seasonal rainfall, resulting in *dambo*-like conditions. The very incompleteness of the *Ramapithecus* material enables predictions to be made to test the seed-eating model. We may predict that the mandibular ascending ramus of *Ramapithecus* will be found to be relatively vertical, its postorbital constriction narrow, supraorbital ridge projecting and face concave in profile; the postcranial skeleton should show a short ilium and short, stout phalanges, but also rather long arms and short, stout legs.

Among the Early Pleistocene hominids, the "robust" australopithecines show exactly the combination of characters to be expected from long-term Phase 1 adaptation. Indeed, a major advantage of the two-phase model is that it makes sense of the apparent paradox of these hominids. Their specializations (such as "superhuman" incisal and canine reduction) are related to the seed-eating complex, while their apparent primitiveness (represented by characters like relatively small cranial capacity and comparatively inefficient bipedalism [Napier 1964; Day 1969; Tobias 1969]) is simply *absence* of Phase 2 specializations.

Robinson (1962; 1963) is one of the few to see the robust australopithecines as representative of a primitive stage of hominid evolution, rather than a late and "aberrant" line, and to recognize that basal hominids are unlikely to have been more carnivorous than pongids. He therefore comes closest to the present scheme, but does not solve the paradox of *robustus* by recognizing the significance of *small-object* vegetarianism to the characters of Phase 1 adaptation. Recent discoveries suggest that robust australopithecines have a time-span in Africa running into millions of years; this would be compatible with our model of Phase 1 differentiation leading to an adaptive plateau, but not with schemes which see all australopithecines as incompetent and transient hunters, nor those that see the robust group as a late "offshoot."

The population represented by the specimens called *Homo habilis* (Leakey *et al.* 1964) fit the model as early, but clearly differentiated

Phase 2 hominids, as their describers contend, with a dentition in which the trend to back-tooth dominance has been partially reversed, a cutting-tool culture, and increased cranial capacity. At several African sites there is evidence of contemporary and sympatric Phase 1 (*robustus*) hominids, as predicted by the model.

If this interpretation is correct, there would seem ample justification for referring *habilis* to the genus *Homo*, on the grounds of its departure along the path of Phase 2 adaptation.

The "gracile" australopithecines (excluding *habilis*) might fit into one of three places on the model (and, conceivably, different populations referred to this species do in fact fit in different places). Possibly they are evolved Phase 1 hominids whose apparently unspecialized dental proportions might be attributed to an allometric effect of smaller size, as in *Theropithecus*. This interpretation, however, is unlikely, mainly because the size-difference between the robust and gracile groups seems insufficient to account for their divergences in shape by allomorphosis alone. Alternatively, they might be a truly primitive (Phase 1) stock from which both robust australopithecines and Phase 2 hominids evolved. This view has been widely espoused, and if dental proportions were all, it would be most plausible. However, the known *africanus* specimens are probably too late and too cerebrally advanced to be primitive. Most likely, they are at an early stage of Phase 2 evolution, with their osteodontokeratic culture, perhaps improved bipedalism, and some cerebral expansion beyond that seen in the *robustus* group (Robinson 1963). In this case their anterior dentition could be seen as secondarily somewhat enlarged from the primitive condition. This interpretation would favor sinking *Australopithecus* in *Homo*, while retaining *Paranthropus* for the evolved Phase 1 forms, as Robinson suggested.

The nature of an evolutionary model, concerned with unique events, is such that it cannot be tested experimentally. Its major test lies in its plausibility, especially in its ability to account for the data of comparative anatomy, behavior, and the fossil record inclusively, comprehensively, and with a minimum of sub-hypotheses. It should also provide predictions which are in theory testable, as with a more complete fossil record, thus enabling discussion to move forward from mere assertion and counter-assertion. An evolutionary model which is designed to account for nothing beyond the data from which it is derived, may be entertaining, but has about as much scientific value as the *Just So Stories*.

While none of the previous models of hominid differentiation is without plausibility, none is very convincing. Too few of the elements of the hominid complex are accounted for, and too often the end-products of hominid evolution have to be invoked in teleological "explanation." On the other hand, the nature of the causal factors invoked, especially behavioral ones, is often such as to make the hypothesis untestable.

The model presented here is based upon the nearest approach to an experimental situation that can be found in evolutionary studies, the parallel adaptation to a closely similar niche by a related organism. While based initially upon diet, and dental characters, it also accounts for hominid features as diverse as manual dexterity, shelfless mandible and epigamic hair, and for features of the fossil record such as the apparent paradox of *Paranthropus*, and the fact that hominid (or pseudohominid) dentitions apparently preceded tools by several million years. On the other hand, there seem to be no major departures from logic or from the data. It is therefore suggested that the two-phase model, with a seed-eating econiche for the first hominids, should at least be considered as an alternative working hypothesis against which to set new facts and fossils.

REFERENCES

Ansell, W. F. H.   1960.   *Mammals of Northern Rhodesia.* Lusaka: Government Printer.

Barbour, G. B.   1949   Ape or man? *Ohio Sci. J.* 49, 4.

Bartholomew, G. A. & J. B. Birdsell   1953   Ecology and the protohominids. *Am. Anthrop.* 55:481–98.

Biegert, J.   1963   The evaluation of characteristics of the skull, hands and feet for primate taxonomy. In: *Classification and human evolution* (ed.) S. L. Washburn (Viking Fd Publ. Anthrop. 37). Chicago Aldine.

Bishop, A.   1963   Use of the hand in lower primates. In *Evolutionary and genetic biology of the primates* (ed.) J. Buettner-Janusch. New York: Academic Press.

Brace, C. L.   1963   Structural reduction in evolution. *Am. Naturalist* 97: 39–49.

Chance, M. R. A.   1955   The sociability of monkeys. *Man* 55:162–5.

———   1967   Attention structure as the basis of primate rank orders. *Man* (N.S.)2:503–18.

Clark, W. E. Le Gros   1964   *The fossil evidence for human evolution* (2nd edn). Chicago: Univ. Press.

Crook, J. H.   1966   Gelada baboon herd structure and movement: a comparative report. *Symp. zool. Soc. Lond.* 18:237–58.

———   1967   Evolutionary change in primate societies. *Sci. J.* 3:(6)66–72.

———   & P. Aldrich-Blake   1968   Ecological and behavioural contrasts between sympatric ground dwelling primates in Ethiopia. *Folia primat.* 8: 192–227.

———   & J. S. Gartlan   1966   Evolution of primate societies. *Nature, Lond.* 210:1200–3.

Dart, R. A.   1949   The predatory implemental technique of *Australopithecus*. *Am. J. phys. Anthrop.* (N.S.) 7:1–38.

Day, M. H.   1969   Femoral fragment of a robust Australopithecine from Olduvai Gorge, Tanzania. *Nature, Lond.* 221:230–3.

———   & J. R. Napier   1964   Hominid fossils from Bed I Olduvai Gorge, Tanganyika: fossil foot bones. *Nature, Lond.* 201:967–70.

DeVore, I. 1964 The evolution of social life. In: *Horizons in anthropology* (ed.) S. Tax. Chicago: Univ. Press.

———— & S. L. Washburn 1963 Baboon ecology and human evolution. In: *African ecology and human evolution* (eds) F. C. Howell & F. Bourlière (Viking Fd Publ. Anthrop. 36). Chicago: Aldine.

Goodall, J. 1964 Tool-using and aimed throwing in a community of free-living chimpanzees. *Nature, Lond. 201*:1264–6.

———— 1965 Chimpanzees of the Gombe Stream Reserve. In: *Primate behavior: field studies of monkeys and apes* (ed.) I. DeVore. New York: Holt, Rinehart & Winston.

Hall, K. R. L. 1962 The sexual, agonistic and derived social behaviour patterns of the wild chacma baboon, *Papio ursinus. Proc. zool. Soc. Lond. 139*: 283–327.

————, R. C. Boelkins & M. J. Goswell 1965 Behaviour of patas monkeys (*Erythrocebus patas*) in captivity, with notes on the natural habitat. *Folia primat. 3*:22–49.

Holloway, R. L. Jr. 1966 Structural reduction through the probable mutation effect: a critique with questions regarding human evolution. *Am. J. phys. Anthrop. 25*:7–11.

———— 1967 Tools and teeth: some speculations regarding canine reduction. *Am. Anthrop. 69*:63–7.

Hopkins, B. 1965 *Forest and savanna*. Ibadan, London: Heinemann.

Howell, F. C. 1967 Recent advances in human evolutionary studies. *Quart. Rev. Biol. 42*:471–513.

Hürzeler, J. 1954 Zur systematischen Stellung von *Orecopithecus. Verh. naturf. Ges. Basel 65*:88–95.

Jolly, C. J. 1963 A suggested case of evolution by sexual selection in primates. *Man 63*:178–9.

———— 1965 The origins and specialisations of the long-faced Cercopithecoidea. Thesis, University of London.

———— 1969 The large African monkeys as an adaptive array. Paper presented at Wenner-Gren Symposium 43, *Systematics of the Old World Monkeys*.

———— 1972 The classification and natural history of *Theropithecus (Simopithecus*) baboons of the African Pleistocene, *Bull. Brit. Mus. Nat. Hist. Geol. Series, 1972*.

Kortlandt, A. 1967 Experimentation with chimpanzees in the wild. In: *Progress in primatology* (eds) D. Starck *et al*. Stuttgart: Gustav Fischer.

Krantz, G. S. 1963 The functional significance of the mastoid process in man. *Am. J. phys. Anthrop. 21*:591–3.

Kummer, H. 1967 *Social organisation of hamadryas baboons*. (Bibliotheca primat 6). Basel: Karger.

Leakey, L. S. B. 1967 Notes on the mammalian faunas from the Miocene and Pleistocene of East Africa. In *Background to evolution in Africa* (eds) W. W. Bishop & J. D. Clark. Chicago: Univ. Press.

———— & T. Whitworth 1958 *Notes on the genus* Simopithecus, *with description of a new species from Olduvai* (Coryndon Mem. Mus. occ. Pap. 6). Nairobi: Coryndon Memorial Museum.

————, P. V. Tobias & J. R. Napier   1964   A new species of the genus *Homo* from Olduvai Gorge. *Nature, Lond.* 202:3–9.

Livingstone, F. B.   1962   Reconstructing man's Pliocene pongid ancestor. *Am. Anthrop.* 64:301–5.

Matthews, L. H.   1956   The sexual skin of the gelada. *Trans. zool. Soc. Lond.* 28:543–8.

Michelmore, A. P. G.   1939   Observations on tropical African grasslands. *J. Ecol.* 27:283–312.

Mills, J. R. E.   1963   Occlusion and malocclusion in primates. In: *Dental anthropology* (ed.) D. R. Brothwell. Oxford: Pergamon Press.

Moreau, R. E.   1951   Africa since the Mesozoic with particular reference to certain biological problems. *Proc. zool. Soc. Lond.* 121:869–913.

Murdock, G. P.   1949   *Social structure.* New York: Macmillan.

Napier, J. R.   1960   Studies of the hands of living primates. *Proc. zool. Soc. Lond.* 134:647–57.

————   1962   Fossil hand bones from Olduvai Gorge. *Nature, Lond.* 196: 409–11.

————   1964   The evolution of bipedal walking in the hominids. *Arch. Biol. (Liège)* 75 Suppt, 673–708.

————   1967   The antiquity of human walking. *Sci. Am.* 216:56–66.

————   1968   Evolutionary aspects of primate locomotion. *Am. J. phys. Anthrop.* 27:333–42.

————   & P. R. Davis   1959   The forelimb skeleton and associated remains of *Proconsul africanus. Brit. Mus. (Nat. Hist.) Foss. Mam. Afr.* 16:1–69.

————   & P. H. Napier   1967   *Handbook of living primates.* London: Methuen.

Oppenheimer, A.   1964   Tool use and crowded teeth in Australopithecinae. *Curr. Anthrop.* 5:419–21.

Pilbeam, D. R. & E. L. Simons   1965   Some problems of Hominid classification. *Am. Sci* 53:237–59.

Pocock, R. I.   1925   External characters of the catarrhine monkeys and apes. *Proc. zool. Soc. Lond.* (2):1479–579.

Rattray, J. M.   1960   *The grass cover of Africa.* New York: FAO.

Reynolds, V.   1966   Open groups in hominid evolution. *Man* (N.S.) 1: 441–52.

————   & F. Reynolds   1965   Chimpanzees in the Budongo forest. In: *Primate behavior: field studies of monkeys and apes* (ed.) I. DeVore. New York: Holt, Rinehart & Winston.

Richards, P. W.   1952   *The tropical rain forest: an ecological study.* Chicago: Univ. Press.

Robinson, J. T.   1962   The origins and adaptive radiation of the Australopithecines. In: *Evolution und Hominisation* (ed.) G. Kurth. Stuttgart: Gustav Fischer.

————   1963   Adaptive radiation in the Australopitecines and the origin of man. In: *African ecology and human evolution* (eds) F. C. Howell & F. Bourlière (Viking Fd Publ. Anthrop. 36). Chicago: Aldine.

Sillans, R.   1958   *Les savanes de l'Afrique centrale.* Paris: Éditions P. Lechevalier.

Simons, E. L.   1965   The hunt for Darwin's third ape. *Med. Opinion Rev.*
1965 (Nov.) 74–81.

——— & D. Pilbeam   1965   Preliminary revision of the Dryopithecinae.
Folia primat. 3:81–152.

Swindler, D. R., H. A. McCoy & P. V. Hornbeck   1967   Dentition of the
baboon (*Papio anubis*). In: *The baboon in medical research*, vol. 2 (ed.)
H. Vagtborg. Austin: Texas Univ. Press.

Tattersall, I.   1969   Ecology of the north Indian *Ramapithecus*. *Nature,*
Lond. 221:451–2.

Tobias, P. V.   1969   Cranial capacity in fossil Hominidae. Lecture, American
Museum of Natural History, May 1969.

Walker, A. C.   1967   Locomotor adaptations in recent and fossil Madagascan
lemurs. Thesis, University of London.

——— & M. Rose   1968   Fossil hominoid vertebra from the Miocene of
Uganda. *Nature, Lond.* 217:980–1.

Washburn, S. L.   1950   Analysis of primate locomotion. *Cold Spr. Harb.*
quart. Symp. Biol. 15:67–77.

——— 1963   Behavior and human evolution. In: *Classification and human*
evolution (ed.) S. L. Washburn (Viking Fd Publ. Anthrop. 37). Chicago:
Aldine.

——— 1968   On Holloway's 'Tools and teeth'. *Am. Anthrop.* 70:97–101.

——— & C. S. Lancaster   1968   The evolution of hunting. In: *Man the*
hunter (eds) R. B. Lee & I. DeVore. Chicago: Aldine.

Wescott, R. W.   1967   Hominid uprightness and primate display. *Am. An-*
throp. 69:78.

White, R. O., S. V. Venkatamanan & P. M. Dabadghao   1954   The grassland of India. *C. R. int. Congr. Bot.* 8:46–53.

PART THREE

THE DYNAMICS OF
HOMINID EVOLUTION

It is tempting to glorify man and his uniqueness among animals, stressing his behavioral flexibility, the variety of his life styles, and the potentialities that are encompassed in his capacity for symbolic transaction, toolmaking, and learning. Certainly, although perhaps without meaning to, cultural anthropologists have provided abundant evidence to lead us to consider the uniqueness of man. Yet this has had one unfortunate result for anthropology as a field, namely, to split off the study of human evolution from the main body of anthropology. Core anthropology has been cultural anthropology, and it has required the tortuous manipulations of anthropologists and observers of anthropology as a unified discipline to relate the activities of the investigators of human evolution to that of the investigators of human culture.

One of the encouraging results of the recent reexamination of human evolution and its relation to culture has been the realization that all anthropologists really are talking about the same animal—an animal that evolved the capacity for culture-building.

If we are ever to understand how human beings evolved from non-humans, if we are ever to grasp the significance of human body form and changes in structure evidenced by the hominid record, we must study human evolution in a larger context than simply the close examination and comparison of fossils. The realization that human evolution is in large part the evolution of man's capacity for culture now influences most research in physical anthropology. It is to the larger comprehension of human evolution that the modern generation of paleoanthropologists are addressing themselves. Results of the reconstruction of the history of plant life and animal life in areas inhabited by human populations have gone into the synthesis, as have studies of the relation between anatomical structure and behavior as well as the study of the evolutionary processes themselves. Precise determination by dating, by careful stratiographic and chemical analysis, has also helped to wring every last bit of data from the fossil record.

In the articles that follow, some of the major problems of hominid evolution are discussed. Simons describes some of the methods commonly employed in dating evolutionary events. Howell discusses the interplay between biological characteristics and protohominid population, and the features of the hominid ecological niche that led to man. Mayr, as a zoologist, offers a classification of human fossils that reflects their probable evolutionary relationships. Newman attempts an evolutionary reconstruction of man's present adaptation to climate. Finally, Lancaster and Washburn evaluate the role of hunting in the evolution both of human structure and human emotional capacity.

New fossils keep turning up, and new ways of making inferences from the fossil record are always being developed. Our view of human evolution alters from one decade to the next, and we should not be surprised or troubled by different interpretations synthesized at different times. If, despite the collection of new data and the development of new theories, the generalizations remain the same, then our endeavors are not really scientific at all.

The views represented in this section remind us forcibly that careful observation of fossils themselves is more important than ever. Despite the proliferation of theories about human evolution, the newer physical anthropology continues to depend on a close reading of the fossil record.

With the discovery of new fossil material in the East Rudolf region of Kenya, in East Africa, the dating of the australopithecines has been extended almost a million years. Fifteen specimens, in sediments from the late Pliocene-Early Pleistocene, have been found. Radioisotope dating makes them about 2,600,000 years old. *Above*, a left lateral view of specimen KNM.ER.406. *Below*, KNM.ER.406, shown here in frontal view, is the skull of an adult. Note the crest, massive face, and large eye orbits. Photographs courtesy R.E.F. Leakey. Reproduced by permission of Natural Museum of Kenya, Nairobi.

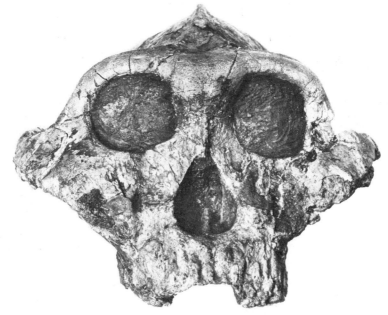

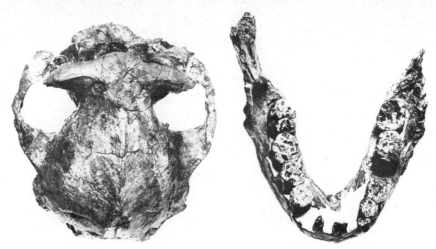

Left, the supraorbital torus of KNM.ER.406 is seen in this vertical view of the fossil. *Right*, shown in this superior view of the massive KNM.ER.729 mandible, broad and rounded, with a slight thickening, or torus, at the base.

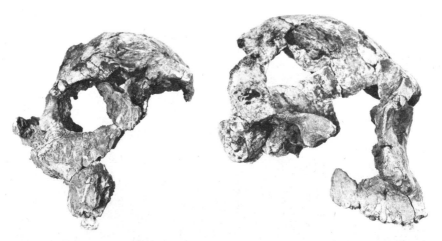

KNM.ER.732, a more fragmentary cranial fossil, is an adult australopithecine, more gracile in features, without a sagittal crest or sharply defined supraorbital torus. The fossil is shown here in frontal and right lateral views. Photographs courtesy R.E.F. Leakey. Reproduced by permission of Natural Museum of Kenya, Nairobi.

9

UNRAVELING THE AGE
OF EARTH AND MAN

E. L. Simons

*The chronological dimensions of evolution are truly staggering.
When we stop to think that the earliest historical cultures, such
as the ancient Egyptians, are only 120 or so generations removed
from us in time, we realize how long it takes for evolutionary
mechanisms to transform populations and evolve new species.*

*The time span for hominid evolution covers as much as 10
million years, perhaps close to a million generations. Of course,
long stretches of that time sequence are not yet known to us. Much
of what we do know, however, of the course of hominid evolution
is derived from the close study of the fragmentary fossil remains
of populations that are ancestral to modern man.*

*To make sense of the fossils, anthropologists must have accurate
information about when a particular extinct population was alive.
Because of this, the sophisticated chemical methods of dating fossil
material that have come into use in the last couple of decades are a
boon to the paleonthropologist who is attempting to work out the
probable relationships among fossil populations. In the following
article, Dr. Elwyn Simons discusses some of the techniques of
dating, and the ways they have been employed to illuminate the
hominid fossil record. Dr. Simons, working out of the Peabody
Museum of Yale University, is accustomed to dealing with vast
spans of geological time; his research has focused on the fossils of
the earliest "apes" (25 million or more years ago), and the shadowy
time zone of the transition of some hominid populations into the
first bipedal, tool-using hominids (perhaps 10 to 15 million years
ago). In this article he justifies our methods of dating, which are
just as important as the conclusions they yield.*

For thousands of years man has asked and tried to answer a dual
question—how old is the earth and how long is it since the appearance of

Reprinted from *Natural History*, Vol. LXXVI, February, 1967, pages 52–29,
by permission of the author and publisher. Copyright 1967 *Natural History*.

primates and the emergence of modern man? The answers have often been as interesting—and far ranging—as the riddle itself. It was not until about thirty years ago, however, that researchers began offering reasonably satisfactory figures. And it is only since the late 1950's that an abundance of precise dates have become available to chronologists who deal with the prehistoric past.

Previously, of course, there had been no lack of estimates. They occur, for example, in the sacred books of many religions. In India the Brahmins considered earth and time as eternal, while some Hindu chronologies indicate that thirteen million years had passed since their "golden age" began. Manetho, the Egyptian historian, listed dynasties of gods and demigods estimated to have reigned on earth for some thirty-six thousand years—from the first dynasty of the gods to the thirteenth historical dynasty. Hebrew priestly writers provided a chronology dating creation at about 4000 B.C., and they seem to have based their creation story, in part, on earlier Babylonian myths, some of which implied an even greater time since creation.

Among many Christians it was thought that the Bible narrative and its chronology could be taken literally; and in 1654 Archbishop Ussher of Armagh, Ireland, seemed to find no uncertainties in the Bible record when he calculated from biblical genealogies that the earth had been created in 4004 B.C. In fact, one of Ussher's contemporaries determined the precise time—the twenty-third day of October at nine o'clock in the morning. Jewish scholars, however, studying the same Old Testament sources, dated the world's creation as October 7, 3761 B.C.; with this date the Jewish calendar begins.

Many modern theologians have challenged, sometimes ridiculed, such literal interpretation of scriptural sources. More than fifteen hundred years ago, St. Augustine proposed that the six days of creation might signify only a logical rather than real succession, and at present even conservative Christian scholars usually agree that one can interpret the six days of creation as standing for six general periods, or eras, rather than for specific 24-hour periods. Meanwhile, with the growth of scientific knowledge in the eighteenth and nineteenth centuries, it became increasingly clear to scholars that the earth must be much more than a few thousand years old. A number of ingenious procedures for estimating the age of the earth, of life, or of the oceans were attempted. All these, however, involved relative methods—no broadly applicable "absolute" or "chronometric" ways of dating rocks were yet known. In 1715, for instance, when the astronomer Halley proposed the first scientific procedure for determining an approximate age for the earth, he assumed that if on initial condensation the primal oceans were fresh water, one could calculate their age by dividing the total amount of sodium salts already present in the oceans by the average amount that is added to the oceans each year from the rivers of the world. Later scientists elaborated this approach, but produced estimates that were too small. We know now that accurate determination of the amount of sodium that land erosion adds each year is uncertain, that the amount added per annum has fluctuated greatly and was probably unusually high during the Cenozoic Era. We also know that salt is removed in deep-sea sediments and as salt deposits produced by evaporation in large inland bays. If the addition of salt had always proceeded at the present rate of about 400 million tons of salt yearly, then one would have to conclude that the seas are only about 100 million years old; the uncertainties are such that this can only be a minimum estimate.

Another early procedure for estimating the earth's age involved measuring the world annual rate of deposition of sediment, which is brought about by the agents of erosion—principally wind and water—and dividing this into the total estimated thickness of sediments. For if geologists can estimate the average rate at which sedimentation occurs, they can then determine how long it has been depositing the measurable thickness of geologic layers.

In 1799 William Smith, an English canal surveyor, announced one of the basic concepts upon which the science of geology depends. He pointed out that each rock formation he had observed contained fossils peculiar to itself, and that the fossils present in each formation "always succeed one another in the same order." Fossils thus became the chief instrument for quickly identifying the relative age and sequence of each rock layer. Since the day of William Smith, geologists have separated the series of rock layers into six main divisions or eras: the Azoic (without life), Archeozoic (beginning life), Proterozoic (early life), Paleozoic (ancient life), Mesozoic (middle life), and Cenozoic (recent life, including development of mammals).

Adding up the total known thickness of rocks deposited during the Cenozoic, Mesozoic, and Paleozoic Eras, Professor James D. Dana of Yale was able, as early as 1876, to calculate that the relative durations of these eras were in the ratio of 1:3:12. But not having an accurate estimate of the average rate of sediment deposition throughout geologic time, he was unable to successfully convert these ratios to numerical values. Moreover, parts of the sedimentary sequence, called the "stratigraphic column," were missing.

The divisions between geologic periods come where early geologists observed great unconformities. An unconformity is a gap between two layers of rock, usually caused by a period of widespread erosion during which all the strata that would have filled the gap have been destroyed. One can easily see that when layers of shale, sandstone, and limestone are laid down on one another, like layers in a cake, those on the lowest levels are older than those above them. But sometimes these relatively flat layers become folded in the processes of mountain-building, and the crests of such folds may eventually be eroded away, leaving a series of worn edges upon which new flat-lying layers may be deposited. Consequently, between the old, folded layer and the newer one immediately above it, the time required to build and erode the mountain is represented only by a surface, seen in cross section as merely a discontinuity. Such a surface is an unconformity, and such gaps make the geologic record seem like a book from which pages have been torn at intervals. So far, it has been impossible to discover all these missing pages in existing deposits. And in the early days of geology there was no way of estimating the lapse of time corresponding to these great unconformities other than by noting the relative amount of difference or the extent of evolution between the fossils above and below the gap.

Within these limitations, by measuring the greatest thickness of each rock layer and adding up the results, one finds that during the Paleozoic, Mesozoic, and Cenozoic Eras, a total thickness of about 68½ miles would have accumulated if deposition had continued in one spot for the whole period of time. This estimate does not include the lapses of time during all unconformities; if one allows for them it would probably increase the total height to nearly 100 miles. One estimate of the average rate of sedimentary deposition is about one foot every 880 years, which would indicate that the last three geologic eras have lasted some 440 million years.

Accurate measurement of the older three eras is impossible because of frequent gaps in our knowledge of sedimentary deposition, but it is thought that they probably lasted over three times as long, or about one and a half billion years.

Despite their ingenuity, the sedimentary methods of dating the age of the earth or the oceans were never too accurate. Their inadequacies were carefully analyzed in 1917 in an important review of "Rhythms and the Measurement of Geologic Time," by Joseph Barrell. Among other things, he pointed out that estimates of time lapse from sediments are uncertain because erosional rates have fluctuated in the past, and the thickest sediments usually accumulate in narrow marine troughs that contain little of the total product of nearby continental erosion.

By the nineteenth century, a number of geologists had taken another tack. They were estimating the length of time since the development of earliest life. In 1867 one of the founders of the science of geology, Sir Charles Lyell, conjectured that 240 million years might account for the successional changes in animal and plant species. But a contemporary estimated only 60 million years for the same span (since the Cambrian Period). The smaller figure led Charles Darwin to write that it could hardly account for the total development and evolution of organisms. At about the same time, however, an even younger age for the earth was proposed by the physicist Lord Kelvin. His estimate came from determining the rate of cooling of the planet from a state of completely molten rock to its present composition, with only a molten core. In a series of papers, he decreased his first estimate of 400 million to only 20 million years. Obviously, this latter figure greatly distressed most contemporary geologists and paleontologists—it did not allow enough time for known changes that had taken place, particularly in earth's fauna. Kelvin, however, had not taken into account an extra source of heat—unknown to him—that would keep the earth warm for a much longer period: the heat produced by radioactivity. Moreover, there is growing agreement among investigators that the earth did not originate as a molten body. More likely, the earth, as well as the other planets, was accumulated through the gradual accretion of smaller bodies and subsequently warmed by heat from radioactive decay.

It may be difficult for today's reader to realize the impact of those various estimates—not only upon others but also upon such scientists as Darwin, Lyell, and Kelvin themselves. They were forced to readjust their thinking from the widespread Christian assumption of a 6,000-year-old world to one that could be several tens of millions of years old. When figures reach such an astronomical scale, it does not require a grand readjustment in thinking to conceive of the world's age as 200 million or even several billion years. Moreover, grasping the meaning of such vast amounts of time is beyond the capacity of most of us. Probably as a consequence of this challenge, twentieth-century geochronologists appear to have been more interested in the accuracy of their calculations than in their antiquity.

Methods depending on radioactive processes have been by far the most fruitful and accurate for determining the age of the earth. Only a few of the more informative of these methods can be briefly summarized here, nor does space allow much detailing of the many intricate processes and complex considerations by which geochemists and geophysicists now date rocks. Generally speaking, progress in the past five or ten years has made it possible to obtain accurate dates for the formation of many of the main types of rocks.

The origins of geochemical dating go back to 1905, when Bertram Boltwood pointed out the universal presence of lead in uranium-bearing rocks and observed that the ratio between lead and uranium was often more or less uniform for uranium minerals from the same region. He also expressed his belief that this must indicate lead of some kind is the end product of the spontaneous breakdown of the radioactive element uranium. Later he proposed that if lead was a product of this process of decay, the ratio of lead to uranium should be the same for minerals of equal age. Consequently, if in a given sample we learn both the amount of

Comparative Geologic Time Scale

Era	Period	Epoch	Distinctive features	Kulp (1961)	Symposium (1964)
Cenozoic	Quaternary	Pleistocene	Early man: continental ice sheets	1	7
	Tertiary	Pliocene	Large carnivores	13	1.5-2
		Miocene	First abundant grazing mammals	25	26
		Oligocene	Large running mammals	36	37 / 38
		Eocene	Many modern types of mammals	58	53 / 54
		Paleocene	Diversified hoofed mammals	63	65
Mesozoic	Cretaceous	Maestrichtian	First primates (?)		
		Upper	First placental mammals		
			First flowering plants: climax of dinosaurs and ammonites followed by extinction	110	
		Lower		135	136
	Jurassic	Upper	First birds, first mammals; dinosaurs and ammonites abundant	166	
		Middle lower		181	
	Triassic	Upper	First dinosaurs, abundant cycads and conifers	200	190 / 195
		Middle lower		(230)	225
Palezoic	Permian	Upper middle	Extinction of many kinds of marine animals, including trilobites. Southern glaciation.	260	
		Lower		280	280
	Carboniferous	Pennsylvanian	Great coal forests, conifers. First reptiles		
		Mississippian	Sharks and amphibians abundant. Large and numerous scale trees and seed ferns	320	
				345	345
	Devonian	Upper	First amphibians and ammonites; fishes abundant	(365)	
		Middle		390	
		Lower		405	395
	Silurian		First terrestrial plants and animals	(425)	430
	Ordovician	Upper	First fishes; invertebrates dominant	445	440
		Middle lower		500	ca. 500
	Cambrian	Upper	First abundant record of marine life; trilobites dominant	530	
		Middle lower			570
	?		?	600?	

Millions of years

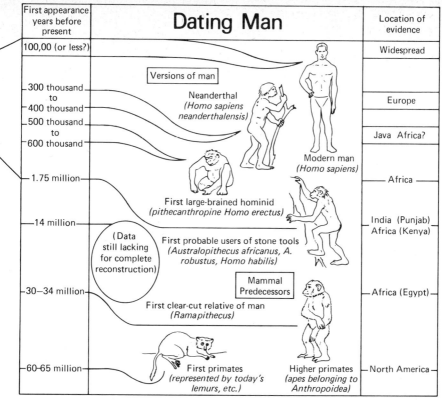

First appearance years before present	Dating Man	Location of evidence
100,00 (or less?)		Widespread
300 thousand to 400 thousand	Versions of man — Neanderthal (Homo sapiens neanderthalensis)	Europe
500 thousand to 600 thousand	Modern man (Homo sapiens)	Java Africa?
1.75 million		Africa
14 million	First large-brained hominid (pithecanthropine Homo erectus)	India (Punjab) Africa (Kenya)
	(Data still lacking for complete reconstruction) First probable users of stone tools (Australopithecus africanus, A. robustus, Homo habilis)	
30–34 million	Mammal Predecessors First clear-cut relative of man (Ramapithecus)	Africa (Egypt)
60–65 million	First primates (represented by today's lemurs, etc.) Higher primates (apes belonging to Anthropoidea)	North America

The detailed chart of primate evolution, above, shows that even when man is defined as any hominid who makes stone tools, evidence dates his arrival at less than two million years ago. This explains why the complete geologic calendar on the facing page allows little space for late stages leading to modern man.

radioactivity-produced lead and the amount of still-unchanged uranium, the age of the rock can be determined by the relative proportions—if we know the rate at which the uranium disintegrates.

It happens that there are several types of lead, each differing slightly in atomic weight. Such varieties of the same element are called isotopes. The isotopes of lead include: (1) primal lead, atomic weight 204—a lead believed originally formed as that element; (2) uranium lead, at. wt. 206—resulting from the breakdown of uranium's predominant isotope (U-238); (3) actinium lead, at. wt. 207—end product of a breakdown sequence beginning with another uranium isotope (U-235); (4) thorium lead, at. wt. 208—from the breakdown of thorium (Th-232). The rate for uranium is so slow that it takes over four and a half billion (4.51 x 10^9) years for half of the atoms in a starting amount of it to disintegrate. This figure is the "half-life" of uranium. It would take even longer, 13.9 x 10^9 years, for half of thorium's atoms to "decay."

But a timing method should be accurate. How can science be so sure that the disintegration of uranium and other radioactive elements has

163

continued at the same rate throughout geologic time? Here is how the rocks themselves indicate that the rates of decay have remained unaltered. Dark mica, or biotite, acts as a sort of photographic plate for the alpha particles released during the radioactive breakdown. If a microscopic particle of uranium is trapped in a piece of this mica, examination under polarized light will show small darkened rings around the central particle. Such microscopic rings form the so-called pleochroic halo and are known from rocks of all geologic ages. They are caused by the discharge of alpha particles, and the distance these particles will travel in the mica depends on the energy of atomic transformation. With uranium, the pleochroic halo will consist of eight separate rings. Very accurate measurements have shown that the energy of transformation, and therefore the rate of breakdown, is constant in all samples. In addition, numerous experiments have confirmed that the constant rate cannot be changed by an outside influence such as pressure or temperature. There is no danger of confusing primal lead (204), which might also be present in a radioactive sample, with the products of atomic breakdown, because the leads of different origin differ in atomic weight.

So far, the oldest rocks carefully dated (in North America) by the uranium method are Precambrian ones in the Minnesota River Valley. The Morton and Montevedes granite deposits in this valley contain zircons indicating they were crystallized at least 3,300 million years ago. In some parts of the world, rocks giving ages of the same order of magnitude are intruded into or overlay older rocks. It has not yet been possible to date these older rocks, because they have been seriously altered by more recent events. Possibly they were more immediately derived from the original crust of the earth, but geologists have not yet found a rock that could qualify as part of the earth's original crust.

However, some scientists believe that meteorites that contain lead may have the same age as the earth and the rest of our solar system. If so, then the age of the earth, as indicated by recently calculated uranium-lead ages of meteorites, increases to approximately 4,500 million years. This figure can be considered within the framework provided by other probing for dates. For instance, recent studies based on radioactive potassium indicate that about ten billion years ago such potassium began breaking down into an isotope of calcium. Here may be a clue to the age of our part of the universe.

Work with this potassium illustrates the usefulness for earthly data of radioactive materials other than uranium and thorium. Specifically, accurate determinations can be made on rock containing carbon, potassium, or rubidium. The two last-named, both members of the alkali metals family, are especially significant. In the past seven to nine years improvement in dating geologic events has come through analysis of the radioactive decay of potassium (at. wt. 40) to argon 40, and rubidium (at. wt. 87) to strontium 87.

These two geochronometric methods have widely expanded the dating possibilities because radioactive potassium and rubidium occur in many more rock types than datable uranium and thorium. (And uranium dates can be used as a countercheck on potassium or rubidium dates if two or more analytical methods can be run on the same rock.) Potassium-to-argon dates are particularly useful to paleontology because they may be determined from rocks often found associated with fossils, such as glauconites (greensands), ash falls, and lava flows. The principle underlying determination of potassium/argon and rubidium/strontium ages of rocks is basically the same as that of the uranium/lead and thorium/lead dating methods, but the laboratory procedures for analyzing the very small traces of these isotopes are considerably more complex.

Combining all these various dating methods, a reasonably accurate chronology or geologic time scale is now available for the time during which development of plant and animal life can be traced on earth from the Cambrian Period to the present. This is valuable because, although traces of simple life—perhaps residues of single-celled organisms—have been found and dated back to three billion years ago, the Cambrian is the oldest geologic period in which fossil organisms are abundant and can be traced in detail through successive evolutionary stages. From then on, the geologic calendar can be dated with considerable accuracy. The table on page 162 shows dating done in 1961 by J. Laurence Kulp, eminent American geochemist, and revisions made by a Geologic Society of London symposium in 1964.

Now let us apply the foregoing techniques to a problem of general interest—to dating the main stages of evolution in the order of Primates, climaxed by the arrival of modern man. About six years ago, Dr. Kenneth Oakley, noted anthropologist of the British Museum, outlined an approximate dating of some of these main steps, but so rapid has been the recent advance of knowledge—not only of dates, but of fossil primates as well—that another analysis seems warranted.

Until recently, the earliest known primates were found in deposits of Middle Paleocene age in western North America, but in 1965, what is probably a primate tooth was reported from latest Mesozoic deposits in Montana. This means that documentation of the emergence of this mammalian order should be sought in the faunas contemporary with the last of the dinosaurs. The date of the close of the Mesozoic Era, when those earliest primates existed, is indicated by several geochemical dates. One was obtained from ash falls in Alberta deposits of the latest Cretaceous substage—on the order of 63 to 66 million years ago. A potassium/argon date, calculated by J. F. Evernden and his associates at the University of

California, was derived from a basalt rock near Denver, Colorado. The basalt, which lies above shales or mudstones containing mammals typical of the earliest kinds of Paleocene fauna, shows an age in excess of 58 million years. Consequently, primates apparently came into existence as a group between 60 and 65 million years ago.

The primate fossils so far obtained from rocks laid down many millions of years following that time all resemble the present-day lower primates, or prosimians—including the lemurs, lorises, and tarsiers. These were widespread in the Early Tertiary, particularly in the Northern Hemisphere. They are abundant in the Eocene rocks of western North America and Europe, and are beginning to turn up increasingly in China, in rocks of similar age. Members of the order Primates did not reach South America until the Oligocene Epoch. Although Eocene land mammals are unknown in Africa, it is probable that there were also early primates on that continent, as the ancestors of the Malagasy lemurs of today presumably reached Madagascar early in the Cenozoic Era by way of the African continent.

A second main stage in the evolution of primates was the development from prosimians of the higher primates, or Anthropoidea (apes and monkeys as well as men). Among the many common features of the Anthropoidea that might signal the appearance of this level of primate organization in the fossil record are the characteristic closure of the orbit behind the eye into an eye socket and the reduction of the number of premolar teeth from four to two on each side of each jaw. Both of these features are first observed in fossil "apes" from the Early Oligocene in the Qatrani formation of the Fayum province of Egypt. No geochemically datable rocks have been found in these Egyptian fossil fields, but they are overlain by a volcanic rock that may soon be dated. The animal fauna with which these ancient apes (*Propliopithecus, Aeolopithecus,* and *Aegyptopithecus*) were recovered indicate an earliest Oligocene land mammal age. In North America, Early Oligocene rocks are associated with Chadronian fauna, which Evernden, *et al.,* have dated by the potassium/argon method with ages ranging from 33 to 36 million years. By that time the higher primates had definitely differentiated in Africa, and perhaps elsewhere as well.

At least since Darwin's time, a great problem confronting evolutionists has been: How close are apes and men taxonomically, and when did their ancestry split? Although these questions remain partly unanswered, we can now determine the age of the oldest clear-cut relatives of man, as opposed to apes, from data concerning species of the Mio-Pliocene genus *Ramapithecus.*

This primate, like both modern and primitive man, had small canines and front teeth, a short face, an arched palate, and massive jaws with

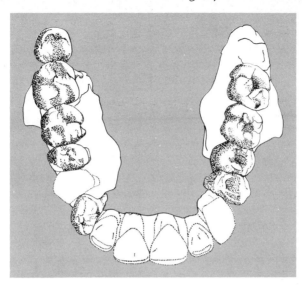

Prehuman *Ramapithecus,* dated at 14 million years ago, was probably the first clear-cut relative of man. It had an arched palate and crowded teeth in a semi-circular instead of U-shaped array.

crowded teeth arranged in a semi-circle, rather than in a U-shaped dental arcade with parallel cheek tooth rows. I have recently placed species of *Ramapithecus* in the taxonomic family of man: Hominidae. The oldest dated members of this genus come from near Fort Ternan, Kenya—from a deposit that the potassium/argon method has shown to be 14 million years old. How much prior to this date the families of man and apes separated is still unknown.

Much more is understood today about the time when modern man's forerunners first began to use stone tools. Paleoanthropologists have usually adopted the definition of man as any hominid that makes stone tools to a set and regular pattern. So defined, the earliest "men" are evidenced by living floors in Bed I of the now-celebrated Olduvai Gorge sequence of Early Pleistocene deposits in Tanzania, East Africa. The deposits of Bed I, in which stone tools first occur, have been dated by the University of California group at 1.75 million years. This figure sums up more than sixty separate potassium/argon analyses. Its accuracy therefore correlates with the accuracy of this dating method.

Admittedly, these oldest dated human cultural sites at Olduvai are not necessarily those of the first actual users of tools. But it can be assumed that tool using probably did not become general much earlier. If it had, archeologists would presumably be able to document this from sites of tool use and manufacture elsewhere in Africa (apart from East Africa) and in Eurasia as well. Controversy about the toolmakers at Olduvai has

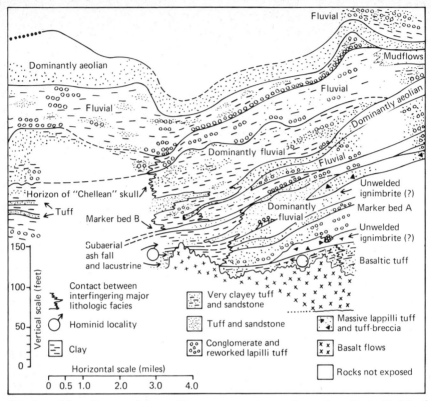

Fluvial

Dominantly aeolian

Mudflows

Fluvial

Fluvial

Dominantly aeolian

Dominantly fluvial

Fluvial

Horizon of "Chellean" skull

Unwelded ignimbrite (?)

Tuff

Marker bed A

Marker bed B

Dominantly fluvial

Unwelded ignimbrite (?)

Subaerial ash fall and lacustrine

Basaltic tuff

Contact between interfingering major lithologic facies

Very clayey tuff and sandstone

Massive lappilli tuff and tuff-breccia

Hominid locality

Tuff and sandstone

Clay

Conglomerate and reworked lapilli tuff

Basalt flows

Horizontal scale (miles)

Rocks not exposed

Vertical scale (feet)

150 — 100 — 50 — 0

0 0.5 1.0   2.0   3.0   4.0

Stratigraphic evidence of the earliest true man comes from deposits in Bed I of Olduvai Gorge, East Africa. Radioisotopes have dated its tool users as living 1.75 million years ago. Depicted here is bed sequence prior to faulting of rock.

been stirred by the unsound taxonomic procedures that have been used to name them. Although the names *Zinjanthropus boisei* and *Homo habilis* have been proposed for them, there is no clearcut evidence that hominids other than two species of *Australopithecus—A. africanus* and *A. robustus* —earlier named in South Africa, existed in Tanzania in Bed I times.

The next important stage in human evolution was signaled by the appearance of large-brained hominids of *Pithecanthropus* type—those with brain volumes of more than 700 cc., as opposed to the earlier *Australopithecus* brain volumes on the order of 400 to 600 cc. The first of these pithecanthropines, from Java, was described in 1892 as *Pithecanthropus erectus*. Investigators usually now consider it to be a species of *Homo* —*H. erectus*. The evolutionary emergence of *H. erectus* appears to be documented also at Olduvai, but is not as precise or certain as we wish because the rock samples from this horizon do not have as suitable a composition for potassium/argon (K/A) dating as do those from Bed I.

Drs. Curtis and Evernden have recently rechecked their K/A date from sediments at the top of Bed II Olduvai and have determined an approximate age of 500,000 years. Inasmuch as the Oldowan pithecanthropines were all collected below this horizon, they must be older than half a million years—which correlates reasonably with the recent K/A determination of 0.6 million years on tcktites from the Djetis beds of Java, from which the earliest *H. erectus* come.

For the time period within which the completely modern *H. sapiens* type of men evolved, another geochronometric method can be used—radiocarbon dating. But because the half-life of radioactive carbon is only about 5,000 years, dates in excess of 40,000 or 50,000 years are inaccurate. Also, the K/A dating method is increasingly inaccurate for dates of less than one million years. Consequently, there is a period during Early and Middle Pleistocene times when dating human remains is difficult and uncertain. During this time, *H. erectus* populations presumably graded into the Neanderthal type of primitive man, a variety to which most modern taxonomists would give only subspecific status, *H. sapiens neanderthalensis*. Neanderthalers have been dated as persisting in western Europe for two or three hundred thousand years. They seem to overlap with human fossils that appear to be more like those of modern man, of which the ones from Fontechevade and La Ferassie in France are examples. These most probably represent only the more progressive individuals and not a different race. The more abundant skeletal finds of Stone Age men are about 30,000 to 40,000 years old. Many have been dated accurately by the radiocarbon method and, consequently, the structural and cultural evolution of Late Paleolithic man is well documented and dated. Sufficient to say that within the period in which radiocarbon dating of fossil man is possible, only slight alterations in skeletal anatomy have occurred; as far as the main stages of locomotor adaptation and evolutionary change are concerned, man's ancestors of 10,000 years ago were essentially the same as man is today.

And what may the future reveal about our prehistoric calendar? As more and more fossil-bearing sediments are dated, multiple cross-checks on individual dates will be possible, and this should eventually lead to a nearly absolute chronology for life of the past. Nevertheless, some sediments will probably never be datable, because many lack sufficient concentrations of any of the known unstable isotopes of common elements. Even so, if such rocks contain traces of past organisms, they can be correlated in time with the same organisms in dated rocks.

Some scientists have suggested that when, and if, this geochronological "millennium" arrives there would be great value in rearranging or renaming the divisions of geologic time to correspond with fixed time durations. Thus the Tertiary, which lasted roughly sixty million years, could be divided into six subdivisions of ten million years each. Whatever the course adopted, it is clear that geochemical dating will eventually provide an accurate calendar for placement of fossil species in their proper temporal relationships. Until this is done, science cannot remain satisfied: uncertainties in interpreting the exact course of evolutionary history will remain among almost all groups of organisms.

10

THE HOMINIZATION PROCESS

F. Clark Howell

Although most anthropologists read the fossil record to signify the transformation of lower-Pleistocene australopithecine populations into mid-Pleistocene Homo, the details of the evolutionary sequence are sufficiently unclear so as to admit dissent. Some scientists even hold that Homo originated and evolved side by side with the australopithecines.

F. Clark Howell, now at the University of California, Berkeley, subscribes to the prevailing view of the phyletic evolution of Homo from the australopithecines. In this article, he discusses the evolutionary transformation of infrahominid populations into full hominid status in the early stages of human evolution. He draws on recently accumulated evidence from the fossil record and the archeological record, as well as data about ancient Pleistocene animal and plant populations and climatic fluctuations. Howell has made important contributions to our understanding of the status of the "Neanderthal" fossils and the emergence of modern man.

Modern paleoanthropological studies seek to understand, in both biological and cultural perspective, those factors which effected the evolution of man. The biologically oriented anthropologist is especially concerned

Reprinted from *Horizons of Anthropology*, Sol Tax, Editor. Chicago: Aldine Publishing Company, 1964. Copyright © 1964 by Aldine Publishing Company. By permission of the author and the publishers.

with the nature and adaptive significance of major anatomical and physiological transformations in the evolution of the body from an apelike higher primate to the single variable species, *Homo sapiens.* He must equally concern himself with the origin and evolution of distinctively human patterns of behavior, especially capabilities for culture and the manifestations of such capacities, and not only with their biological bases.

The fossil record of man and his higher primate relatives is still far from adequate. However, in the last several decades significant discoveries have been made which considerably expand our knowledge of ancient human and near-human populations. There is not now a single major range of Pleistocene time from which some one or more parts of the world has not at last yielded some hominid skeletal remains. Hence, there is now some pertinent evidence to suggest the general sequence and relative order of those bodily transformations during this process of hominization. In this process major changes were effected at quite unequal rates, in: the locomoter skeleton, the teeth and their supporting facial structures, the size and proportions of the brain, and the enveloping skull bones. And there were equally significant and accompanying changes in behavior. In the course of the last decade the earlier phases of this process have received considerable investigation. Some significant aspects of that work are to be considered here.

Man is a primate and within the order Primates is most closely related to the living African anthropoid apes. How immediate the relationship, or to put it another way, how far removed in time the point of common ancestry prior to divergence, is still unsettled. Except under special circumstances skeletal remains are not readily preserved in the acid soils of forested habitats; hence fossil remains of anthropoid apes from the requisite late Tertiary time range, some twenty to a few million years ago, are uncommon, and when found are often very inadequately preserved. Nevertheless, apelike higher primates are known to have had a widespread Eurasiatic distribution (up until five to ten million years ago, by which time such creatures had disappeared from increasingly temperate Europe); they were presumably also common in parts of Africa, although there, fossiliferous beds of that age are singularly rare. Fragmentary jaws and teeth of such creatures indicate their higher primate—indeed, specifically ape—affinities. They also suggest substantial diversity in anatomical structure as well as in over-all size. The rare and fortunate occurrence of other skeletal parts (such as limb bones) indicate that some distinguishing characteristics of modern apes were later evolutionary "specializations" rather than the "primitive" ancestral condition. Several specimens of jaws and teeth, from regions as widely separate as northern India and eastern Africa, and some ten to fourteen million years in age, show some hominid resemblances. Until more adequately preserved skeletal remains are recovered, these few provocative fragments will remain enigmatic. The antecedents of the hominids, the so-called proto-hominids, are still really unknown, and one can only speculate about the very early formative phases in the process of hominid emergence.

The anatomical-physiological basis of the radiation of the hominids is generally acknowledged to have been a major transformation in structure and function of the locomotor system. The lower limb skeleton and associated musculature were modified under selection pressures eventually to permit a fully erect posture and efficient, habitual bipedal gait. The changes effected in the lower limb were extensive and revolutionary. The characteristic curvature of the loins, the short, broad and backwardly shifted hip bones and their displaced and strengthened articulation with the sacrum, their sinuous distortion to form a basin-like structure about the lower abdomen, as well as the shortened ischial region, were all part of a complex of largely interrelated modifications adaptive for terrestrial bipedalism. There were related changes in the musculature of the hip and thigh, in relative proportions and in structure and function of specific muscle groups, all to afford power to run and to step off in walking, to maintain the equilibrium of the upright trunk during the stride, and to extend fully and to stabilize the elongated lower limb at hip and knee—an impossible stance for any ape. And the foot was fully inverted, with the lateral toes shortened and the hallux enlarged and immobilized, the rigidity of the tarsus enhanced through the angularity of joints and strengthened ligaments, with the development of prominent longitudinal and transverse arches, and the heel broadened to become fully weight-bearing.

The singularity of the erect posture was long ago recognized from comparative anatomical studies of man and the nonhuman primates. Its priority in the hominization process has been fully confirmed by the discovery in Africa of the still earliest known hominids, the australopithecines (genus *Australopithecus*), creatures with small brains, but with lower limbs adapted to the erect posture and bipedal gait, or at least for upright running. Some evidence suggests that the full-fledged adaptation to bipedalism, that which permitted leisurely and prolonged walking, was not yet wholly perfected. The hominid type of dental structure, with small incisor teeth, reduced and spatulate-shaped canines, and noninterlocking canines and anterior premolar teeth all set in a parabolic-shaped dental arch, was also fully differentiated. Brain size, as estimated from skull capacities, was only about a third to two-fifths that of the size range of anatomically modern man. There are several distinct forms of the genus *Australopithecus*, surely distinct species (and probably valid subgenera), with consistent differences in skeletal anatomy as well as in body size. One larger form attained a body weight of some one hundred twenty to one hundred fifty pounds, whereas another was much smaller with a body weight of only some sixty to seventy-five pounds.

Although thus far restricted to Africa, these earliest known hominids were nonetheless fairly widely distributed over substantial portions of that continent. Their ecological adjustments are now known in some measure and can even be paralleled among certain present-day African environments within the same regions. One small South African species is recorded under rainfall conditions some 50 per cent less than that of the present-day (now twenty-eight to thirty inches) in the same region—a rolling, high veld country of low relief and little surface water. Other occurrences testify to more favorable climatic and environmental situations. Generally speaking, the environments were relatively open savanna. In southern Africa remains of these creatures, along with other animals, occur in caves in a limestone plateau landscape where caverns, fissures, and sink holes probably afforded fairly permanent sources of water which was otherwise scarce. But there, and also in eastern Africa, sites were in proximity to more wooded habitats fringing shallow water courses, or mantling the slopes of adjacent volcanic highlands. It is just such transitional zones, the "ecotones" of the ecologist, which afford the greatest abundance and diversity of animal and plant life.

The absolute age of some of these creatures can now be ascertained as a consequence of refinements in the measurement of radioactivity (potassium/argon or K/A) in some constituent minerals of volcanic rocks. Their temporal range extends back nearly two million years with some representatives apparently having persisted until less than a million years ago. Their discovery has therefore tripled the time range previously known for the evolution of the hominids.

Culturally patterned behavior appears concurrently with these creatures. In several instances there is direct association with some of their skeletal remains. The field investigation of undisturbed occupation places, maximizing the possibility for the recovery of evidence in archeological context, has culminated in these significant discoveries. Traditional prehistoric archeological studies, on the other hand, were largely preoccupied with the sequential relationships of relics of past human endeavors, often in secondary contexts. The careful exposure of undisturbed occupation places has permitted wholly new inferences into the nature of past hominid adaptations and patterns of behavior. This work has broader implications for it forces complete rejection of the traditional viewpoint of some anthropologists which envisioned the sudden appearance of human behavior and culture at a "critical point" in man's phylogeny. . . .

This most primitive cultural behavior is manifest in several ways. There was a limited capability to fashion simple tools and weapons from stone (and presumably in other media, although perishable materials are not preserved). These objects, the raw material of which was not infrequently brought from sources some distance away, include deliberately collected, sometimes fractured or battered, natural stones or more substantially modified core (nodular) and flake pieces fashioned to produce chopping, cutting, or piercing edges. Several undisturbed occupation places with associated animal bones attest also to the acquisition of a meat-eating diet. It was limited, however, to the exploitation of only a narrow range of the broad spectrum of a rich savanna and woodland fauna. It comprised predominantly various freshwater fish, numerous sorts of small amphibians, reptiles (mostly tortoises and lizards), and birds, many small mammals (rodents and insectivores), and some infants (or the very young) of a few moderate-sized herbivorous ungulates. Vegetal products doubtlessly constituted a very substantial part of the diet of these predaceous-foragers, but the conditions of preservation prohibit other than inferences as to what these may have been. At any rate carnivorous behavior of these earliest hominids contrasts markedly with the essentially vegetarian proclivities of recent apes (and monkeys).

Such food remains and associated stone artifacts are concentrated over occupation surfaces of restricted extent—in part at least seasonally exposed mud flats around ephemeral lakes adjacent to periodically active volcanoes. These occupational concentrations have a nonuniform distribution over the occupation surfaces; there are dense central clusters of tools and much broken-up and crushed bones (presumably to extract the marrow), and peripherally more sparse occurrences of natural or only battered stones and different, largely unbroken skeletal parts of their prey. In one case a large ovoid-shaped pattern of concentrated and heaped-up stony rubble, with adjacent irregular piles of stone, suggests a structural feature on the occupation surface. These uniquely preserved sites in eastern Africa, sealed in quickly by primary falls of volcanic ash, afford some tantalizing glimpses into the activities of these primitive creatures. Such occupation places may well represent an ancient manifestation of the adjustment to a "home base" within the range, a unique development within the hominid adaptation.

We can now delineate some of the basic features of the early radiation of the hominids to include: (1) differentiation and reduction of the anterior dentition; (2) skeletal and muscular modifications to permit postural uprightness and erect cursorial bipedalism; (3) effective adjustment to, and exploitation of, a terrestrial habitat; (4) probably a relatively expanded brain; (5) extensive manipulation of natural objects and development of motor habits to facilitate toolmaking; and (6) carnivorous predation adding meat protein to a largely vegetal diet.

The adaptation was essentially that of erectly bipedal higher primates adjusting to a predaceous-foraging existence. These adaptations permitted or perhaps were conditioned by the dispersal into a terrestrial environment and the exploitation of grassland or parkland habitats. The African apes (and also the Asiatic gibbon), especially the juvenile individuals, show occasional though unsustained efforts at bipedalism; it is highly probable that this preadaptive tendency, which developed as a consequence of the overhand arboreal climbing adaptation of semierect apes, was pronounced in the still unknown proto-hominids of the Pliocene. Wild chimpanzees are now recognized sometimes to eat meat from kills they have made, and also to manipulate inanimate objects, and even to use and occasionally to shape them for aid in the food quest. This would surely suggest that such tendencies were at least equally well developed among the closely related proto-hominids.

Terrestrial environments were, of course, successfully colonized long previously by other primates. These are certain cercopithecoid monkeys, the secondarily ground-dwelling quadrupedal patas monkeys and baboons of Africa and the macaques of Asia (and formerly Europe). Hence their adaptations, social behavior, and troop organization provide a useful analogy for inferences into the radiation of the proto-hominids. Comparative investigations of the nonhuman primates, including the increasingly numerous and thorough behavioral and ecological studies of monkeys and apes in natural habitats, . . . have substantially broadened our understanding of the primate background to human evolution. These studies serve to emphasize those particular uniquenesses of the human adaptation.

A half million or probably nearly a million years ago, hominids were in the process of dispersal outside the primary ecological zone exploited by the australopithecines. In part, this dispersal can be understood only in respect to the opportunities for faunal exchange between the African and Eurasiatic continents, and the prevailing paleogeographic and paleoecological conditions of the earlier Pleistocene. The diverse Saharan zone failed to constitute a barrier to this dispersal, or to that of Pliocene and early Pleistocene mammal faunas for that matter. Moreover the extensive seas of the Pliocene and earliest Pleistocene were sufficiently lowered, either due to continental uplift, or, less likely, as a consequence of the incorporation of oceanic waters in extensive arctic-subarctic ice caps, so as to afford substantial intercontinental connections.

Probably within a hundred thousand years, or less, representatives of the genus *Homo* were dispersed throughout most of the Eurasian subtropics and had even penetrated northward well into temperate latitudes in both Europe and eastern Asia. This dispersal involved adjustment to a diverse new variety of habitats. Cultural and perhaps physiological adaptations permitted, for the first time, man's existence outside the tropics under new and rigorous climatic conditions, characterized by long and inclement winters. It was unquestionably facilitated by anatomical-physiological modifications to produce the genus *Homo* including prolongation of growth and delayed maturation, and behavioral changes favoring educability, communication, and over-all capabilities for culture.

The fully human pattern of locomotion was probably perfected by this time. These final transformations in the hip, thigh, and foot permitted a fully relaxed standing posture, with the body at rest, as well as sustained walking over long distances. The skeletal evidence is unfortunately still incomplete, but some four to six hundred thousand years ago the lower limb skeleton appears not to have differed in any important respect from that of anatomically modern man. Brain size, and especially the relative proportions of the temporal-parietal and frontal association areas, were notably increased to some one-half to two-thirds that of *Homo sapiens* (and to well within that range which permits normal behavior in the latter species). And some further reduction and simplification also occurred in the molar (and premolar) teeth and the supporting bony structures of the face and lower jaw.

Hunting was important as a basis for subsistence. Meat-eating doubtless formed a much increased and stable portion of the normal diet. Much of the mammalian faunal spectrum was exploited, and the prey included some or all of the largest of herbivorous species, including gregarious "herd" forms as well as more solitary species, and a variety of small mammals. Several occupation places of these early and primitive hunters, some of which are quite undisturbed, are preserved and have been excavated in eastern Africa and now also in Europe. These localities preserve prodigious quantities of skeletal remains of slaughtered and butchered mammals. The famous and enormous cave locality (Locality I) of Choukoutien (near Peking) in eastern Asia is a unique occurrence of occupation of a site of this type at such an early time. At Choukoutien, although other ungulate and carnivorous mammals are also present, about 70 per cent of the animal remains are represented by only two species of deer. In Africa the impressive quarry included a number of gigantic herbivorous species, as well as other extinct forms. In two such occupation sites in eastern Africa, over five hundred thousand years old, the very abundant fauna included species of three simians, two carnivores, two rhinos, eight pigs, two to three elephants, sheep and buffalo, two hippos, three giraffids, a chalicothere, six horses, as well as numerous antelopes and gazelles, and other remains of small mammals (rodents), birds and some reptiles (tortoises). Preferential hunting of certain herd species is recognized at several somewhat younger occupation sites in Europe. At one of two sites in central Spain recently worked by the writer only five large mammalian species are represented, and of these a woodland elephant and wild horse are most numerous, with infrequent wild oxen (aurochs) and stag (red deer), and very rare rhinoceros. The remains of some thirty individual elephants, many of which were immature, are represented in an area of approximately *three hundred* square meters! At another such open-air site, on the edge of the Tyrrhenian Sea north of Rome, remains of horse predominated over all other species. Some indication of the level of cultural capability and adaptation, as well as requisite plasticity for local ecological adjustment, is afforded by the diversity of game species which were exploited and the corresponding distinctions in occurrence, habitat of preference, size of aggregation, and their species-specific patterns of behavior.

Toolmaking capabilities are notably improved along with the establishment of persistent habits of manufacture. These reflect, in part at least, more dexterous and effective control of manual skills. Corresponding evolutionary changes in the structure and function of the hand, especially development of the fully and powerfully opposable thumb, with expansion and complication of the corresponding sector of the cerebral motor cortex and interrelated association areas, were all effected under the action of natural selection.

Not only was the over-all quantity and quality of the stone tools increased. New techniques were developed for the initial preparation as well as for the subsequent fashioning of diverse and selected sorts of stone into tools (and weapons). New types of stone tools make their appearance, including in particular sharply pointed and cutting-edged tools of several sorts, seemingly most appropriate for butchery of tough-skinned game. Certain stones already of favorable form were deliberately trimmed into a spheroidal shape, it is thought, as offensive missiles. These and other forms of tools subsequently become remarkably standardized. This fact, and the very broad pattern of geographic distribution throughout Africa, southern and western Europe, and through western into southern Asia and the Indian subcontinent, suggest also a sophisticated level of communication and conceivably even the capability of symbolization.

More perishable stuffs, such as wood and fiber, are unfortunately very rarely preserved. However, several such early sites in Europe attest the utilization and working of wood, fashioned into elongate, pointed, and spatulate shapes. The discovery had doubtless been made of the thrusting spear, a major offensive weapon in the pursuit of large, thick-skinned mammals. Again, although traces of the utilization of fire are nearly equally as rarely preserved, there is incontrovertible evidence of its discovery and utilization (whether for heat or cookery is uncertain), both in Europe and in eastern Asia.

The development of a hunting way of life, even at a very unsophisticated level of adaptation, it has been argued, set very different requirements on early human populations. It led to markedly altered selection pressures and was, in fact, responsible for profound changes in human biology and culture. Many workers regard this adaptation as a critical factor in the emergence of fundamentally human institutions. Some of those changes which represent the human (*Homo*) way of life would include: (1) greatly increased size of the home range with defense of territorial boundaries to prevent infringement upon the food sources; (2) band organization of interdependent and affiliated human groups of variable but relatively small size; (3) (extended) family groupings with prolonged male-female relationships, incest prohibition, rules of exogamy for mates, and subgroups based on kinship; (4) sexual division of labor; (5) altruistic behavior with food-sharing, mutual aid, and cooperation; and (6) linguistic communities based on speech.

It may appear impossible ever to obtain direct evidence of this sort from the fossil and archeological record. Yet an approach which combines the field and laboratory study of the behavior of living nonhuman primates with analysis of basic patterns of adaptation and behavior of human hunter-gatherer populations can enhance enormously the sorts of inferences usually drawn from the imperfect evidence of paleoanthropological investigations. The favorable consequences of active cooperation between students concerned with the origin and evolution of human behavior, however diverse in background and orientation, is already evident and has considerably advanced understanding of the process of hominization. In the coming years it may be comparable with those advances in paleoanthropological studies effected through the fullest cooperation with colleagues in the natural sciences.

BIBLIOGRAPHICAL NOTE

The point of view expressed here is also discussed by Washburn and Howell (in Tax 1960) and at greater length in Howell (1961) where pertinent references may be found. A colloquium volume (Colloques Internationaux 1958, Centre National de la Recherches Scientifiques, Paris, 1958), with chapters by various authorities, provided the title for my own essay, and is rewarding reading for anyone seriously interested in the subject. Much basic information, as well as ecological viewpoints which will surely provide the basis for much further research, will be found in the recently published symposium volume (Howell and Bourlière, eds., 1963). Much of the research work from which I have drawn conclusions and generalizations is still in progress, in both field and laboratory, and only preliminary reports are available at most. Many of my colleagues have shared with me their field experiences, basic data and preliminary interpretations, and to each of them I am deeply grateful.

Washburn, S. L. and F. Clark Howell. 1960 Human evolution and culture in Tax, Sol (Ed.) *Evolution after Darwin.* 3 Vols. Chicago: University of Chicago Press.

Colloques Internationaux Sciences Humaines 1958 *Les Processus de l'Hominisation.* Paris: Centre National de la Recherche Scientifique.

Howell, F. Clark 1961 Ismilia: a Paleothic site in Africa. *Sc. Am.,* October 1961.

Howell, F. Clark and François Bourlière (Ed.) 1963 *African Ecology and Human Evolution.* (Viking Fund Publications in Anthropology, No. 38). Chicago: Aldine Publishing Company.

11

THE TAXONOMIC
EVALUATION
OF FOSSIL HOMINIDS

Ernst Mayr

*Even the most enthusiastic student of human evolution can be
discouraged by the simply staggering array of "scientific" names by
which the various fossils are identified.*

*Should he venture timidly beyond the assigned classroom text,
the student may encounter a fossil by the name* Sinanthropus
pekinensis; *in another place the same fossil appears under the alias*
Homo erectus, *or even simply* Choukoutien. *What is to be made
of this confusing array of terminology?*

*The names assigned to specimens of animals (and plants as
well), whether living or extinct, should ostensibly follow the rules,
the basis of which is that the names assigned should reflect the
evolutionary relationships which exist between known forms and
newly discovered forms. The names should be based on "natural
taxonomy" for fossil hominids as for any other kinds of specimens.*

*But zoologists have often criticized the cavalier manner in which
paleoanthropologists have named and renamed human fossils without
regard to the principles of zoological nomenclature. They object
to the stress anthropologists have placed on minor differences in
classifying fossils and in building genera and species of fossil man.
Population geneticists have scoffed at the ways in which anthropologists disregard population models in making classifications. It
has been claimed that attempts have been made to construct
relationships without taking into consideration either variation
within populations or the polymorphic and polytypic character of
most human populations.*

Ernst Mayr is a zoologist who in recent years has specialized in the evolution of birds. Like many other zoologists he has begun applying his experience to problems of human evolution and fossil classification. In this article, he suggests a taxonomy for fossil hominids that is more in line with zoological thinking.

INTRODUCTION

The concepts and methods on which the classification of hominid taxa is based do not differ in principle from those used for other zoological taxa. Indeed, the classification of living human populations or of samples of fossil hominids is a branch of animal taxonomy. It can only lead to confusion if different standards and terminologies are adopted in the two fields. The reasons for the adoption of a single, uniform language for both fields, and the nature of this language, have been excellently stated by G. G. Simpson [in the symposium *Classification and Human Evolution*]. . . .

There is, perhaps, one practical difference between animal and hominid taxonomy. Hominid remains are of such significance that even rather incomplete specimens may be of vital importance. An attempt must sometimes be made to evaluate fragments that a student of dinosaurs or fossil bovids would simply ignore. But, of course, even a rather complete specimen is only a very inadequate representation of the population to which it belongs, and most specimens are separated by large intervals of space and time. Yet, it is the task of the taxonomist to derive from these specimens an internally consistent classification.

The non-taxonomist must be fully aware of two aspects of such a classification: first, that it is usually by no means the only possible classification to be based on the available evidence, so that a taxonomist with a different viewpoint might arrive at a different classification; and second, that every classification based on inadequate material is provisional. A single new discovery may change the picture rather drastically and lead to a considerable revision.

The material of taxonomy consists of zoological objects. These objects are individuals or parts of individuals who, in nature, were members of populations. Our ultimate objective, then, is the classification of populations as represented by the available samples.

OBJECTS VERSUS POPULATIONS

The statement that we must classify populations rather than objects sounds almost like a platitude. . . . Yet, it is not so many years ago that the study of fossil man was in the hands of strict morphologists who arranged specimens in morphological series and based their classification almost entirely on an interpretation of similarities and differences without regard to any other factor. He who classifies specimens as representatives of populations knows that populations have a concrete distribution in space and time and that this provides a source of information that is not available to the strict morphologist. Any classification that is inconsistent with the known distribution of populations is of lowered validity.

THE APPLICATION OF
TAXONOMIC PRINCIPLES
IN CONCRETE CASES

There have been several previous attempts to apply the principles of systematic zoology to some of the open problems of hominid classification. A great deal of new evidence has since accumulated and there has been some further clarification of our concepts. The time would seem proper for a new look at some of these problems.

SUBSPECIES OR SPECIES

The decision whether to rank a given taxon in the category "subspecies" or in the category "species" is often exceedingly difficult in the absence of conclusive evidence. This is as true in the classification of living populations (geographical isolates) as it is for fossils. A typical example as far as the hominids are concerned is the ranking of Neanderthal Man. I have found three interpretations of Neanderthal in the literature.

(1) "Neanderthal Man is a more primitive ancestral stage through which *sapiens* has passed." This we might call the classical hypothesis, defended particularly during the period when the interpretation of human evolution was based primarily on the evaluation of morphological series. This classical hypothesis had to be abandoned when it was found to be in conflict with the distribution of classical and primitive Neanderthals and of *sapiens* in space and time.

(2) "Neanderthal is an aberrant separate species, a contemporary of early *sapiens* but reproductively isolated from him."

(3) "Neanderthal is a subspecies, a geographic race, of early *sapiens*."

What evidence is there that would permit us to come to a decision as to the relative merits of alternatives (2) and (3)? We must begin by defining rigidly what a species is and what a subspecies [is]. As clearly stated by Simpson and Mayr, degree of morphological difference per se is not a decisive primary criterion. A species is reproductively isolated from other species coexisting in time and space, while a subspecies is a geographic subdivision of a species actually or potentially (in the case of geographical isolates) in gene exchange with other similar subdivisions of the species.

The difficulty in applying these concepts to fossil material is obvious. It can be established only by inference whether two fossil taxa formed a single reproductive community or two reproductively isolated ones. In order to draw the correct inference, we must ask certain questions:

Does the distribution of Neanderthal and *sapiens* indicate that they were reproductively isolated? Not so many years ago Neanderthal was considered by many as a Würm "eskimo," but he is now known to have had an enormous distribution, extending south as far as Gibraltar and North Africa and east as far as Iran and Turkestan. There is no evidence (but see below) that Neanderthal coexisted with *sapiens* anywhere in this wide area. Where did *sapiens* live during the Riss-Würm Interglacial and during the first Würm stadial? No one knows. Ethiopia, India and southeast Asia have been suggested, but these will remain wild guesses until some properly dated new finds are made. All we know is that at the time of the first Würm interstadial Cro-Magnon Man suddenly appeared in Europe and overran it in a relatively short time.

THE SAPIENS PROBLEM

There has been much talk in the past of the "Neanderthal problem." Now, since the average morphological differences between the classical Neanderthal of the first Würm stadial and the earlier Neanderthals of the Riss-Würm Interglacial have been worked out, and since the distribution of Neanderthal has been mapped, *sapiens sapiens* has become the real problem. Where did he originate and how long did it take for pre-*sapiens* to change into *sapiens*? Where did this change occur?

All we really know is that *s. sapiens*, as Cro-Magnon, suddenly appeared in Europe. Sufficient remains from the preceding period of unmixed Neanderthal in Europe and adjacent parts of Africa and Asia (and a complete absence of any blade culture) prove conclusively that *s. sapiens* did not originate in Europe. The rather wide distribution of types with a strong supraorbital torus (e.g., Rhodesia, Solo-Java) suggests that *s. sapiens* must have originated in a localized area. The sharpness of distinction between Neanderthal and *s. sapiens* (except at Mt. Carmel) further indicates that Neanderthal, as a whole, did not gradually change into *sapiens*, but was replaced by an invader.

There is a suspicion that evolutionary change can occur the faster (up to certain limits), the smaller and more isolated the evolving population is. If *s. sapiens* lost his supraorbital torus very quickly (and the various other characters it has before becoming *sapiens*) then it can be postulated with a good deal of assurance that *sapiens* evolved in a rather small, peripheral, and presumably well isolated population. Even if we assume that the rate of change was slow and the evolving population large, we must still assume that *sapiens* was rather isolated. Otherwise, one would expect to find more evidence for intergradation with late Neanderthal.

It is obvious that the available evidence is meager. Let us assume, however, for the sake of the argument, that Neanderthal and *sapiens* were strictly allopatric, that is, that they replaced each other geographically. Zoologists have interpreted allopatry in the past usually as evidence for conspecificity, because subspecies are always allopatric. We are now a little more cautious because we have discovered in recent years a number of cases where closely related species are allopatric because competitive intolerance seems to preclude their geographical coexistence. The rapidity with which Neanderthal disappeared at the time Cro-Magnon Man appeared on the scene would seem to strengthen the claim for competitive intolerance and consequently for species status of these two entities. Yet, here is clearly a case where it is perhaps not legitimate to apply zoological generalizations to man. The Australian Aborigines and most of the North American Indians disappeared equally or perhaps even more rapidly, and yet no one except for a few racists would consider them different species.

We are thus forced to fall back on the two time-honored criteria of species status, degree of morphological difference and presence or absence of interbreeding. Our inference on the taxonomic ranking of Neanderthal will be based largely on these two sets of criteria, supplemented by a third, available only for man.

1. *Degree of Morphological Difference*   The amount of difference between the skulls of Neanderthal and *sapiens* is most impressive. There are no two races of modern man that are nearly as different as classical Neanderthal and *sapiens*. And yet one has a feeling that the differences are mostly of a rather superficial nature, such as the size of the supraorbital and occipital torus and the general shape of the skull. The cranial capacity, on the other hand, is remarkably similar in the two forms. The gap between Neanderthal and *sapiens* is to some extent bridged by two populations, Rhodesian Man and Solo Man, which are widely separated geographically from Neanderthal. Although sharing the large supraorbital torus with Neanderthal, these two other populations differ in many details of skull shape and cranial capacity from Neanderthal as well as from *s. sapiens*. Whether or not these peripheral African and Asiatic types acquired their Neanderthaloid features independently, can be established only after a far more thorough study and the investigation of additional material. It seems quite improbable that they are directly related to Neanderthal. In view of their small cranial capacity they may have to be classified with H. *erectus*.

As it now stands, one must admit that the inference to be drawn from the degree of morphological difference between Neanderthal and *sapiens* is inconclusive.

2. *Interbreeding between Neanderthal and* Sapiens   Cro-Magnon Man, on his arrival in western Europe, seems to have been remarkably free from admixture with the immediately preceding Neanderthal. There is, however, some evidence of mixture in the material from the two caves of Mt. Carmel in Palestine. Both caves were inhabited early in the Würm glaciation. The older cave (Tabun) was inhabited by almost typical Neanderthals with a slight admixture of modern characters, the younger cave (Skhul) by an essentially *sapiens* population but with distinct Neanderthaloid characters. The date is too late to consider these populations to have belonged to the ancestral stock that gave rise both to Neanderthal and modern man. It seems to me that the differences between Tabun and Skhul are too great to permit us to consider them as samples from a single population coming from the area of geographical intergradation between Neanderthal and modern man, although this could be true for the Skhul population. Hybridization between invading Cro-Magnon Man and Neanderthal remnants is perhaps a more plausible interpretation for the Skhul population, while there is no good reason not to consider Tabun essentially an eastern Neanderthal, particularly in view of its similarity to the Shanidar specimens.

Repeated re-examinations of the Mt. Carmel material have thus substantiated the long-standing claims that this material is evidence for interbreeding between Neanderthal and *sapiens*.

3. *The Cultural Evidence*   As our knowledge of human and hominid artifacts increases, it becomes necessary to include this source of evidence in our considerations. My own personal knowledge of this field is exceedingly slight, but when I look at the implements assigned to Neanderthal and those assigned to Cro-Magnon, I feel the differences are so small that I can not make myself believe they were produced by two different biological species. I realize that the history of human or hominid artifacts goes back much further than we used to think, yet this is not in conflict with my hunch that there was no opportunity for the simultaneous existence of two separate hominid species of advanced tool-makers.

I would like to add some incidental comments on tools and human evolution. The history of peoples and tribes is full of incidences of a secondary cultural deterioration, *vide* the Mayas and their modern descendants! Most of the modern native populations with rudimentary material cultures (e.g., certain New Guinea mountain natives) are almost surely the descendants of culturally more advanced ancestors. This must be kept in mind when paleolithic cultures from Africa and western Eurasia are compared with those of southern and eastern Asia. Stone tools and the hunting of large mammals seem to be closely correlated. Could such peoples have lost their tool cultures after they had emigrated into areas poor in large game? Could this be the reason for the absence of stone tools in Javan *Homo erectus*?

Conclusion The facts that are so far available do not permit a clear-cut decision on the question whether Neanderthal was a subspecies or a separate species. It seems to me, however, that on the whole they are in better agreement with the subspecies hypothesis. It would seem best for the time being to postulate that Neanderthal (*sensu stricto*) was a northern and western subspecies of *Homo sapiens* (*sensu lato*), [one] which was an incipient species but probably never reached species level prior to its extinction.

POLYTYPIC SPECIES AND EVOLUTION

All attempts to trace hominid phylogeny still deal with typological models. "*Australopithecus* gave rise to *Homo erectus*," etc. In reality there were widespread polytypic species with more advanced and more conservative races. One or several of the advanced races gave rise to the next higher grade. It may happen in such a case that the descendant species lives simultaneously (but allopatrically) with the more conservative races of the ancestral species. This is often interpreted to indicate that the ancestral species could not have given rise to the descendant species. *True*, as far as the ancestral species in a typological sense is concerned, but *not* true for the ancestral species as a polytypic whole.

The concept of most polytypic species being descendants of ancestral polytypic species creates at once two formidable difficulties. One of these is caused by unequal rates of evolution of the different races. Let us say that there was an ancestral species 1 with races 1a, 1b, 1c and 1d. Race 1a evolved into species 2, absorbing in the process much of race 1b, and now forms races 2a and 2b. Race 1c became extinct and race 1d persisted in a relic area without changing very drastically. We now have 2 (a and b) and 1 (d) existing at the same time level, even though they represent different evolutionary stages (morphological grades). It is thinkable, for instance, that Heidelberg Man was the first population of *Homo erectus* to reach the *sapiens* level, and that as *Homo sapiens heidelbergensis* it was contemporary with *Homo erectus* of Java and China. (This is purely a thought model, as long as only a single mandible of Heidelberg Man is available.) It is possible that *Homo erectus* persisted in Africa as *rhodesiensis* and in Java as *soloensis* at a time when European populations clearly had reached *Homo sapiens* level. Such a possibility is by no means remote, in view of the many polytypic species of recent animals in which some races are highly advanced and others very primitive.

I am calling attention to this situation to prevent too far a swing of the pendulum. The late Weidenreich arranged fossil hominids into morphological series strictly on the basis of morphology without regard to distribution in space and time (*e.g.,* Neanderthal—Steinheim—*H. sapiens*). Some modern authors tend to swing to the other extreme by classifying fossil hominids entirely on the basis of geological dating without paying any attention to morphology. The unequal rates of evolutionary change in widely dispersed and partially isolated races of polytypic species make it, however, necessary to take morphology and distribution equally into consideration. Even though *Homo sapiens* unquestionably descended from *Homo erectus*, it is quite possible, indeed probable, that some races of *Homo erectus* still persisted when other parts of the earth were already populated by *Homo sapiens*.

The same argument is even more true for genera. The fact that *Homo* and *Australopithecus* have been found to be contemporaries does not in the least invalidate the generally accepted assumption that *Homo* passed through an *Australopithecus* stage. The Australopithecines consisted of several species (or genera, if we recognize the generic distinction of *Paranthropus*) and each of these species, in turn, was polytypic. Only a segment of this assemblage gave rise to *Homo*. Much of the remainder persisted contemporaneously with *Homo*, for a longer or shorter period, without rising above the Australopithecine grade. The modern concepts of taxonomy and speciation do not require an archetypal transformation (*in toto*) of *Australopithecus* into *Homo*.

The second great difficulty caused by the evolution of polytypic species is a consequence of the first one. It is the difficulty to determine what part (which races) of the ancestral species has contributed to the gene pool of the descendant species. This in turn depends on the amount of gene flow between the races of the ancestral species while it passed from the level of species 1 to the level of species 2. The amount of gene flow is determined by the nature of the interaction between populations in zones of contact. Unfortunately, the situation in the near-human hominids (*Homo erectus* level) was probably different from both the anthropoid condition and the condition in modern man. A number of possibilities are evident in an area of contact between races:

(1) Avoidance
(2) Extermination of one by the other
(3) Killing of the men and absorption of the women
(4) Free interbreeding

There is much evidence that all four processes have occurred and it becomes necessary to determine their relative importance in individual cases. The Congo pygmies, the bushmen, and various negritoid pygmies in the eastern tropics illustrate avoidance by retreating into inferior environments. The Tasmanians and some Indian tribes illustrate extermination.

The white invaders in North America and Australia absorbed extremely few genes of the native peoples. The frequently made assertion that invaders kill off the men and take the women is often contradicted by the facts. The sharpness of the difference between classical western European Neanderthal and invading Cro-Magnon indicates to me that Cro-Magnon did not absorb many Neanderthal genes (some contrary opinions notwithstanding). Language and cultural differences must have militated at the *Homo erectus* level against too active a gene exchange between different races. The distinctness of the negro, mongoloid, and caucasian races supports this assumption. Gene flow obviously occurred, but against considerable obstacles.

SYMPATRIC SPECIES OF HOMINIDS

When one reads the older anthropological literature with its rich proliferation of generic names, one has the impression of large numbers of species of fossil man and other hominids coexisting with each other. When these finds were properly placed into a multi-dimensional framework of space and time, the extreme rarity of the coexistence of two hominids became at once apparent. We have already discussed the case of Neanderthal and *sapiens*, but there are others in the Middle and early Pleistocene.

At the *Homo erectus* level, we have Java Man and Pekin Man, originally described as two different genera, but so strikingly similar that most current authors agree in treating them as subspecies. Ternifine Man in North Africa may be another representative of this same polytypic species. The existing material is, however, rather fragmentary. A further contemporary is Heidelberg Man, whose massive mandible contains teeth that appear smaller and more "modern" than those of a typical *erectus*. Was this a second species or merely a deviant peripheral isolate? This can only be settled by additional discoveries.

In Africa we find incontrovertible evidence of contemporaneity of several species of hominids. *Australopithecus* and *Paranthropus* apparently differed considerably in their adaptations. Perhaps this is the reason why they are not found together in most South African deposits. Yet the degree of difference between them and the time span of their occurrence leaves no doubt that they must have been contemporaries. Here, then, we have a clear case of the contemporaneity of two species of hominids. The fragments of the small hominid (*Telanthropus*) found at Swartkrans with A. *robustus*, which may belong to an Australopithecine or *Homo*, supply additional proof for the coexistence of two hominids in South Africa. In Java, in the Djetis layers, there is also the possibility of the coexistence of two hominids, "*Meganthropus*" and *Homo erectus*.

By far the most exciting instance of the coexistence of two hominids is that established by Leakey in East Africa. In layer 1 of Olduvai *"Zinjan-thropus,"* an unmistakable Australopithecine of the *Paranthropus* type, is associated with remains of an advanced hominid, "co-Zinjanthropus," that—when better known—may well turn out to be closer to *Homo* than to *Australopithecus*. Whether the tools of this layer were made by both hominids or only the more advanced, can be determined only when the two types are found unassociated at other sites. This will also influence the decision on the identity of the maker of the Sterkfontein tools in South Africa.

The picture that emerges from all these new discoveries is that only one species of hominids seems to have been in existence during the Upper Pleistocene, but that there is much evidence from the Middle and Lower Pleistocene of several independent lines. Some of these gave rise to descendant types, others became extinct. The coexisting types were, so far as known, rather distinct from each other. This is what one would expect on the basis of *a priori* ecological considerations The principle of "competitive exclusion" would prevent sympatry if there were not considerable ecological divergence. *Australopithecus* and *Paranthropus*, or *Zinjanthropus* and the associated hominid, "co-Zinjanthropus," were able to coexist only because they utilized the resources of the environment differently. Whether one of them was more of a hunter, the other more of a gatherer (or hunted), whether one was more carnivorous, the other more of a vegetarian, whether one was more of a forest creature, the other a savanna inhabitant, all this still remains to be investigated, when better evidence becomes available.

It is important to emphasize that nothing helped more to make us aware of these problems and to assist in the reconstruction of evolutionary pathways than an improvement of the classification of fossil hominids both on the generic and on the specific levels. It is here that the application of principles of zoological taxonomy has been particularly fruitful. Indeed, the earlier morphologists never appreciated the biological significance of the problem of coexistence or replacement of closely related species.

GENERIC PROBLEMS

The category "genus" presents even greater difficulties than that of the species. There is no non-arbitrary yardstick available for the genus as reproductive isolation is for the species. The genus is normally a collective category, consisting of a group of species believed to be more closely related to each other than they are to other species. Yet, every large genus includes several groups of species that are more closely related to each other than to species of other species groups within the same genus. For instance, in the genus *Drosophila* the species belonging to the *virilis* group are more closely related to each other than to those belonging to the *repleta* group, yet both are included in *Drosophila*. They are not separated in different genera because the species groups have not yet reached the degree of evolutionary divergence usually associated with generic rank. As Simpson has pointed out, the genus usually has also a definite biological significance, indicating or signifying occupation of a somewhat different adaptive niche. Again, this is not an ironclad criterion because even every species occupies a somewhat different niche, and sometimes different genera may occupy the same adaptive zone.

It is particularly important to emphasize again and again that the function of the generic and the specific names in the scientific binomen are different. The specific name stresses the singularity of the species and its unique distinctness. The generic name emphasizes not a greater degree of difference but rather the belonging-together of the species included in the genus. To place every species in a separate genus, as was done by so many of the physical anthropologists of former generations, completely stultifies the advantages of binomial nomenclature. As Simpson has stated correctly . . . the recognition of a monotypic genus is justified only when a single isolated known species is so distinctive that there is a high probability that it belongs to a generic group with no other known ancestral, collateral or descendant species. The isolated nature of *bamboli*, the type species of *Oreopithecus*, justifies the recognition of this monotypic genus.

Of the literally scores of generic names proposed for fossil hominids, very few deserve recognition. More and more students admit, for instance, that the degree of difference between *Homo erectus* and *H. sapiens* is not sufficient to justify the recognition of *Pithecanthropus*.

There are a number of reasons why it would seem unwise to recognize the genus *Pithecanthropus* in formal taxonomy. First of all, *Homo* would then become a monotypic genus and *Pithecanthropus* contain at most two or three species. This is contrary to the concept of the genus as a collective category. More importantly, the name *Pithecanthropus* was first applied to an actual fossil hominid when only a skull cap was known and the reconstruction envisioned a far more anthropoid creature than *erectus* really is. When the teeth and other body parts were discovered (or accepted, like the femur) it was realized that the total difference between *erectus* and *sapiens* was really rather small and certainly less than is normally required for the recognition of a zoological genus. The recognition of *Pithecanthropus* as a genus would lead to an undesirable heterogeneity of the genus category.

The genus *Australopithecus* has already many of the essential characters of *Homo*, such as a largely upright posture, bicuspid premolars, and reduced canines. For this reason I suggested previously that "not even *Australopithecus* has unequivocal claims for generic separation." I now agree with those authors who have since pointed out not only that the upright locomotion was still incomplete and inefficient, but also that the tremendous evolution of the brain since *Australopithecus* permitted man to enter a new niche so completely different that generic separation is fully justified. The extraordinary brain evolution between *Australopithecus* and *Homo* justifies the generic separation of these two taxa, no matter how similar they might be in many other morphological characters. Here, as in other cases, it is important not merely to count characters but to weight them.

Whether or not one wants to recognize only a single genus for all the known Australopithecines or admit a second genus, *Paranthropus*, is largely a matter of taste. The species (*robustus*) found at Swartkrans and Kromdraai is larger and seems to have more pronounced sexual dimorphism than *A. africanus*. Incisors and canines are relatively small, while the molars are very large and there are pronounced bony crests on the skull, particularly in adult males. These differences are no greater than among species in other groups of mammals. *Zinjanthropus* in East Africa seems to belong to the more massive *Paranthropus* group. The two Australopithecines (*africanus* and *robustus* seem to represent the same "grade" as far as brain evolution is concerned, but the differences in their dental equipment and facial muscles indicate that they may have occupied different food niches. It may well depend on future finds whether or not we want to recognize *Paranthropus*. The more genuinely different genera of hominids are discovered, the more important it may become to emphasize the close relationship of *Australopithecus* and *Paranthropus* by combining them in a single genus. It depends in each case to what extent one wants to stress relationships. We have a similar situation among the pongids. I have pointed out earlier that gorilla and chimpanzee seem to me so much nearer to each other than either is to man or to the orang or the gibbons, that degree of relationship would seem to be expressed better if the gorilla were included in the genus *Pan* rather than to be recognized as a separate genus. The decision on generic status is as always based on somewhat arbitrary and subjective criteria. One cannot prove that gorilla and chimpanzee belong to the same genus, but neither can one prove that they belong to different genera.

DIAGNOSTIC CHARACTERS

The collective nature of the categories above the species level show clearly why it is often so difficult to provide an unequivocal diagnosis for taxa belonging to the higher categories. Those who think that "the characters make the genus" have little difficulty in characterizing differences between species and calling them generic differences. It is much easier to characterize the species "chimpanzee" and the species "gorilla" than to find diagnostic characters that clearly distinguish the chimpanzee-gorilla group from man, the orang and the gibbons. Higher categories often can be diagnosed only by a combination of characters, not by a single diagnostic character. The definition of the genus *Homo* presented here is an example of such a combinational diagnosis.

The problem of the relation between taxonomic ranking and diagnostic characters will become increasingly acute as new Pliocene and Miocene fossils are found. Nothing would be more short-sighted than to base the classification of such finds on isolated "diagnostic" characters. We must ask ourselves each time whether relationship will be expressed better by including such new taxa in previously established ones, or by separating them as new taxa. If we combine them with previously established taxa, that is, if we include them in previously recognized genera, subfamilies, or families, we may have to modify the diagnosis of such taxa. We must always remember that the categories above the species are collective categories and subjectively delimited. The pronouncement made by Linnaeus, "It is the genus that gives the characters, and not the characters that make the genus," is true not only for the genus but for the categories at the family level. Diagnostic characters are a convenient tool of the working taxonomist, [and] should never become a strait jacket.

NOMENCLATURE AND COMMUNICATION

Superimposed on all the taxonomic difficulties are some purely nomenclatural ones. Simpson . . . has already pointed out that it is altogether inadmissible to change a scientific name because it is considered inappropriate. A scientific name is, so to speak, merely a formula and its etymological meaning is quite irrelevant. What is important is to avoid arbitrary changes, because the words of a language lose all usefulness if they are shifted around or replaced by new ones. If an anthropologist wants to play around with names, let him concentrate on the vernacular names. No one will care whether he talks of Heidelberg Man or Man of Mauer, or of Pekin Man rather than Man of Choukoutien.

As soon as an anthropologist employs zoological nomenclature that has very definite rules, he must obey these rules. In particular, I would like to call the attention of anthropologists to Article 35b of the Rules of Zoological Nomenclature, which states that the names of families and subfamilies must be based on the name of an included genus. Since there is no genus

Euhomo, there can be no family name Euhominidae. If a subfamily is recognized for the Australopithecines, it can only be Australopithecinae. Not only is this system the only valid one, but it also has the advantage of being simple and unambiguous. I should hope that such confusing terms as Praehominidae would soon disappear from the literature.

Those who give names to fossil Hominidae might also be more careful in the choice of specific names. To have several *africanus* and *robustus* in this family is confusing, particularly during this period of rearranging of genera. It would not seem an impossible demand that only such new specific names be given that had not been given previously to other species in the family Hominidae.

THE CLASSIFICATION OF
THE MISSING LINK

Nothing characterized the early study of human evolution as much as the search for the missing link. When one looks at early reconstructions of the missing link, one realizes how strongly the concept was dominated by the ancient idea of the *scala naturae*. If evolution were limited to a single lineage, as thought for instance by Lamarck, the missing link would simply be the halfway stage between the anthropoids and man. Now as we realize that there is no single line of descent but a richly branching phylogenetic tree, the search for the missing link has become somewhat illusory. It is now evident that there is not just one missing link but a whole series of missing links. There is first the species which was at the branching point between the Pongidae and the Hominidae. On the hominid branch there were those species that first acquired such essentially human traits as making tools, making fire, and possessing speech. There is the first species to be referred to the genus *Homo*, and there is the species that acquired a brain capacity about halfway between the anthropoids and modern man. We may already have representatives satisfying most of these qualifications, and rather than searching for *the* missing link we are now beginning to classify kinds of missing links.

It is now clear that we must distinguish between two essential phenomena. There is on one hand the phylogenetic branching of the hominid line from the pongid line. Yet even after this branching had taken place, which presumably was sometime during the Miocene, there was no sign of Man on the new hominid line. The hominids throughout the Miocene and Pliocene were still apes, and even the Australopithecines of the early Middle Pleistocene can hardly be classified as human. It does appear that *Homo erectus* qualifies better as representative of the stage between prehuman hominids and Man than any other form. It is almost certainly the stage at which the hominids became Man.

THE HIGHER CATEGORIES

The Pleistocene hominids present no problem at the family level. They all clearly belong to the Hominidae. Whether or not to separate the Australopithecines in a subfamily Australopithecinae is essentially a matter of taste. Matters are more difficult when it comes to Pliocene and Miocene fossils. Not only are most of them known from insufficient fragments, but the criteria on which to base the decision pongid or hominid become increasingly elusive as we go back in time. Furthermore, there is considerable probability of the existence of additional equivalent branches of anthropoids or near-anthropoids which have since become extinct. *Oreopithecus* seems to represent such a branch.

The evidence concerning the branching-off point of the hominid from the pongid line seems on first sight contradictory. Schultz finds that all the great apes agree with each other in very numerous characters of general morphology, in which they differ from Man. Yet, the African apes (*Pan* sensu lato) are closer to Man than to orang or gibbon in hemoglobin structure, in serum proteins and in chromosomal morphology. What can be the explanation of this apparent conflict in the evidence? Perhaps the simplest interpretation would be to assume that Man's shift into the niche of the bipedal, tool-making and speech-using hominid necessitated a drastic reconstruction of his morphology, but that this reconstruction did not, in turn, require a complete revamping of his biochemical system. Different characters and character complexes thus diverged at very different rates. If one assumes this to be correct, one will conclude that the *Homo*-line branched from the *Pan*-line well after the line of their common ancestor had separated from the orang (*Pongo*)-line.

Full awareness of mosaic evolution is particularly important for the correct placing of early hominid fossils. As Le Gros Clark and I have emphasized, the Hominidae are a classical example of mosaic evolution. Every character or set of characters evolved at a different rate. Even *Australopithecus* is still essentially anthropoid in some characters while having considerably advanced toward the hominid condition in other characters, for instance with respect to upright posture and the general shape of the tooth row. We must furthermore be aware of the fact that evolution is not necessarily irreversible and that temporary specializations may secondarily be lost again. The Pliocene forms *Ramapithecus* from India and *Kenyapithecus* from Africa may well belong to the hominid line. Even more difficult is the allocation of Miocene genera. To place them correctly one may have to use "prophetic characters" (a preevolutionary term of Agassiz), that is, characters which foreshadow future evolutionary trends. It is quite certain that the Miocene hominoids lacked some of the characteristic specializations of both the pongid and the hominid lines.

They were not the extreme brachiators that some of the modern pongids are, nor had they reached the completeness of upright posture and the special features of dentition of the later hominids. Such seemingly irrelevant characters as the shape of tooth cusps may be more revealing in such forms than the relative size of the canines or the development of a simian shelf. The recent arguments about *Oreopithecus* show how difficult it is to reach an objective evaluation of the evidence.

CONCLUSIONS

Fossil hominids are samples of formerly existing populations distributed in space and time. Their classification must be consistent with the generalizations derived from the study of polytypic species in animals. Whenever the anthropologist uses the terminology of subspecies, species, and genera, such terminology must be consistent with the meaning of these categories as developed in modern systematics. Application of the principles of systematics has helped to clarify the formerly bewildering diversity of morphological types. Pleistocene hominids display much geographic variation, but the number of full species coexisting at any one time is not known to have exceeded two. The relatively wide distribution of some of the fossil hominids indicates that there has been a considerable amount of gene flow as early as the lowest Middle Pleistocene.

REFERENCES

Clark, W. E. Le Gros 1950 "New palaeontological evidence bearing on the evolution of the Hominoidea." *Quart. J. Geol. Soc. London*, 105:225–264.

Coon, C. 1962 *The Origin of Races*. New York: A. A. Knopf.

Dobzhansky, Th. 1944 "On species and races of living and fossil man." *Amer. J. Phys. Anthr.*, 2 (n.s.):251–265.

Goodman, M. 1963 "Man's place in the phylogeny of the primates as reflected in serum proteins." In: *Classification and Human Evolution*. S. L. Washburn, ed. Chicago: Aldine.

"International code of zoological nomenclature" 1961 London: *Int. Trust Zool. Nomenclat.*

Klinger, H. P., J. L. Hamerton and D. Mutton 1963 "The chromosomes of the Hominoidea." In: *Classification and Human Evolution*. S. L. Washburn, ed. Chicago: Aldine.

Mayr, E. 1944 [Horizontal and vertical subspecies]. N.R.C. Committee on Common Problems of Genetics, Paleontology and Systematics. *Bull.* No. 2 (June):11–16. 1950 "Taxonomic categories in fossil hominids." *Cold Spring Harbor Symp. Quant. Biol.* 25:109–118. 1957 "Species concepts and definitions." In *The Species Problem*, E. Mayr, ed., *Amer. Assoc. Adv. Sci. Publ.* No. 50:1–22.

Simons, E. L. 1961 "The phyletic position of *Ramapithecus*." *Postilla.* Yale Peabody Mus. No. 57:1–9.

Simpson, G. G. 1961 *Principles of animal taxonomy.* New York: Columbia Univ. Press. 1963 "The meaning of taxonomic statements." In: *Classification and Human Evolution.* S. L. Washburn, ed. Chicago: Aldine.

Stewart, T. D. 1960 "Form of the pubic bone in Neanderthal man." *Science,* 131:1437–1438.

Zuckerkandl, E. 1963 "Perspectives in molecular anthropology." In: *Classification and Human Evolution.* S. L. Washburn, ed. Chicago: Aldine.

12

WHY MAN IS SUCH A SWEATY AND THIRSTY NAKED ANIMAL: A SPECULATIVE REVIEW

Russell W. Newman

Using data on the characteristics and the behavior of animals as the basis for assertions about human characteristics and behavior has become a familiar literary device in recent years. Such books as Robert Ardrey's African Genesis *and* The Territorial Imperative, *Desmond Morris's* The Naked Ape *and* The Human Zoo, *Lionel Tiger's* Men in Groups, *and popular books by Nikko Tinbergen and Konrad Lorenz have all acustomed us to the notion that the comparative method can be used to establish almost any proposition concerning mankind, just as long as the animals or behaviors are selected to fit the argument. It is almost as if scientists outside the field of anthropology have decided to take over the subject matter of anthropology—human behavior in evolutionary perspective—for themselves; perhaps because anthropologists themselves have been so cautious in making statements about how man has come to be the way he is.*

Reprinted from *Human Biology,* Vol. 42, February 1970, pages 12–27. By permission of the author and the Wayne State University Press. Copyright 1970 by Wayne University Press.

Nevertheless, what the author of the following selection terms "judicious speculation based on comparison" is often all we have to go on in reconstructing evolutionary events in man's biological and behavioral history. Fossil documentation is often impossible. Dr. Newman's article is speculative, but in his careful use of data from infrahuman forms, in his care in limiting the scope of the problems he is discussing, and in the tentativeness of his conclusion, he offers us a model for evolutionary speculation. Human physiological adaptation to heat load in evolutionary perspective is his topic, and we are treated to the extra dividend of a conclusion about the origin of human "nakedness" that runs counter to the usual one about the evolution of clothing.

Dr. Newman is a physical anthropologist with the U.S. Army Research Institute of Environmental Medicine at Natick, Massachusetts.

Our present view of the earliest differentiation of man's ancestors from the primate stock assumes a gradual shift in habitat from forest to open grasslands in the tropics and sub-tropics of the Old World during the Pliocene epoch. This period encompassed some critical innovations in: morphology (complete upright posture and gait, increased cranial capacity, etc.), economy (increasingly carnivorous), behavior (tool making, speech), and probably others not integrated into the central theory of evolution. This paper presents the argument that this shift from forest to grassland resulted in certain physiological adaptations to environmental heat loads that still characterize our species; Barnicot (1959) suggested just such a relationship a decade ago.

Specifically, the physical avenues of heat exchange between an animal and its environment will be reiterated to establish a common vocabulary; the interplay between these avenues and selected tropical environments will then be examined with reference to human responses to point up the specializations which characterize our species. Water requirements and the associated patterns of thirst and drinking habits are reviewed since they may have influenced the daily regimen. Finally, the experimental data now available on non-human primates under heat stress will be presented to see whether the human response is characteristic of the primate order.

AVENUES OF HEAT EXCHANGE

There are four channels of energy exchange between the animal and its environment: conduction, convection, radiation, and evaporation. Modern man uses all of them in varying combinations and proportions, as did his ancestors. The first three channels operate toward either a net loss or gain of heat for the organism. The fourth channel, evaporation, is generally considered to result only in heat loss.

Conduction is the flow of heat without displacement of the material and can occur in gases, liquids, or solids. The most familiar example of conduction occurs when touching an object which is either hot or cold; in the first instance we conduct heat towards us, in the second we lose heat. The importance of conduction is tied to the habits of a given species; in man it is not a major source of heat exchange. Conduction is often considered simultaneously with convection as a dual avenue of heat exchange, and this combination will be used here.

Convection is a flow of heat by the physical movement of the gas or liquid in contact. The best example is the increase in cooling (or heating) produced by wind. This convective air current cools by carrying away the heat conducted from the skin and by presenting a steady supply of air which is cooler than the skin surface. Convective exchange thus occurs at the outer boundary of a thin layer of still air which surrounds the body. It acts in the absence of wind because of the air movement produced by the expansion of the warmed air *per se*, but convective exchange increases dramatically with movement of either the air or the body.

Radiation is an exchange of heat from a warmer to a cooler object which is independent of the intervening medium. It includes the direct acquisition of heat from the sun as ultra-violet, visible, and infra-red energy and the more subtle exchanges by which we radiate heat to, or receive it from our surroundings. This is an exceedingly important channel for heat exchange in man, and some of its complexities will be examined later.

The fourth method, evaporation, results in the loss of "latent" heat required to vaporize water from the skin and from the membranes of the respiratory tract. The total evaporative heat loss, comprised of three parts, includes respiratory loss—which is relatively constant in man since he does not generally pant, diffusional skin loss—which is also rather constant, and sensible perspiration, i.e. active thermal sweating, which supplements the diffusional phase when either the environment imposes a sufficient heat load (at rest at about 30° C), [86° F] or when exercise increases body heat production.

ENVIRONMENTAL CONDITIONS OF THE FOREST AND SAVANNA

Man's ancestors probably did not step from tropical forest to sunbaked grasslands in one short bound. However, these two extremes will be contrasted, with the understanding that there are many intermediate conditions. Except for periods of storms and rain, tropical forests at ground level in the daytime are typified by warm temperatures averaging 28-32° C [82.4°-89.6°F] as the daily maximum, very little air movement, a very high humidity (the air almost saturated with moisture), and little radiant heat from the sun (Richards 1952, Read 1968). This is only marginally stressful to an inactive animal because this combination of an air temperature slightly below skin temperature with a high ambient humidity and low air movement makes it just physically possible for the animal to lose the heat generated by metabolism. An active animal has increased difficulty because it will produce two or three times as much heat and, although the higher skin temperature accompanying the physical activity of an exercising animal increases the heat transferred by radiation slightly, little real increase from convection or evaporation is possible because of the relatively small temperature and vapor pressure differences between the skin and air. We can characterize a forest-dwelling animal as being almost solely dependent on radiational heat loss for thermoregulation with a minor contribution from conductance when lying or sitting on the forest floor.

Open country (savanna or steppe) at comparable latitudes presents a different set of meteorological conditions for mammals. The most obvious difference is that solar radiation becomes an important element since it can impinge directly on the animal and its immediate surroundings. If an animal is not in shade, it will receive radiation directly from the sun; that energy which is not reflected (50-60%) will be absorbed as heat. Even in the shade, air temperature rises, usually about 5° C [9° F], primarily from the heat re-radiated by the ground and all nearby objects which are in sunlight. The radiant heat load absorbed cannot be simply described in terms of surface temperatures since animals have defensive mechanisms which modify this load. Under high radiant heat loads, exposed sand or rock may reach surface temperatures of 60° C [140° F], a level painful to touch; the tips of the wool in Merino sheep can reach 85° C [185° F] in desert sun (MacFarlane 1964), but their skins never reach this level for various reasons. The effect of solar insolation in mammals is best described either in terms of the energy received and absorbed or by some measure of the resulting strain (reaction to the stress). A nude man sitting in the sun in the desert may absorb 200 kcal/m²/hr (Adolph 1947), while Lee (1963) equates desert solar radiation to a rise of 7° C [12.6° F] in air temperature. In the tropical savanna, direct radiation is reduced below that in the desert by the increased amount of water vapor in the air which absorbs some of the ultra-violet and visible portions of the light spectrum; a figure of about 100 kcal/m²/hr has been estimated for such an area (Roller and Goldman 1967). This reduces the 7° C [12.6° F] equivalent air temperature increase, but nevertheless still represents a stressful condition for a quiet animal. The other two meteorological factors mentioned for the forest environment, air movement and humidity, also change. Air movement increases many-fold outside the forest although it obviously varies from time to time; in fact, the term "windspeed" is only appropriate outside the forest proper. Humidity in the tropics is always higher than that which we associate with higher latitudes, but relative humidities of less than 50% are common outside the forest. These humidities provide much greater potential for evaporative heat loss than the saturated air under the forest canopy.

This "open country" combination of meteorological conditions imposes a higher stress on the animal than did the tropical forest. Both conductive and radiant heat exchanges now flow from the environment to the animal; they have become avenues of heat gain. The only way to reestablish a heat-loss relationship (except by retreating to caves or burrows) is for peripheral vasodilation to raise skin temperature to levels above ambient temperature and, indeed, above the surface temperatures of everything nearby. There are definite limits to this approach since skin temperature must remain below deeper body temperature for metabolic heat to be dissipated from its sites of origin and since the upper tolerable level of skin (as well as deep body) temperature is relatively low. Convective heat loss is now more important, because of increased air movement, but can only assist heat transfer from the skin if the air temperature is lower than skin temperature; if the reverse is true, the warmer air simply increases the heat load as it moves over the skin. Fortunately, the fourth avenue of heat transfer, evaporative heat loss, is much greater in the relatively dry open country than it was under the higher humidities of the forest, if sufficient moisture can be made available at the surface of the animal by panting or sweating or the spreading of saliva or urine on the skin. Finally, it is noteworthy that under conditions where heat storage from radiant insolation is a problem, as in open country, larger body size becomes advantageous since it minimizes the surface area (cm^2) per unit of body volume (cm^3) and thus spreads the radiant heat load throughout a greater mass, sparing the critical centers.

ADAPTATIONS OF HEAT STRESS

Many specific adaptations to environmental stresses have been observed in mammals. Only a few appear relevant to human evolution, and these are observed primarily in ungulates and carnivores. Almost all of these findings are from domestic animals although many of the principles must also apply to related wild species.

Some small animals in hot environments (especially desert forms) have become nocturnal and spend the day underground; this was not man's adaptation. Many of the larger carnivores (especially the cats) kill and feed at intervals of as much as 2-3 days, resting between kills in the shadiest and coolest locations in their territory. On the other hand, man seems to have followed the general primate pattern of frequent feeding, and our ancestors must have spent many of their waking hours in pursuit of food even under conditions of high heat stresses. There may have been some behavioral adaptations in daily activity and feeding habits, but it does not appear that major changes in food acquisition patterns took place specifically to avoid stressful environmental conditions. There is one peculiarity of man that needs to be mentioned although it is not a true heat adaptation. Being erect *per se*, in open grassland, substantially reduces the solar heat load by minimizing the amount of surface area exposed to direct sunlight. For example, a standing man receives only two-thirds as much direct solar radiation as a standing sheep of equivalent size on a daily average and less than one-quarter as much at the noontime peak loads (Lee 1950). However, the assumption of an upright posture was so fundamental to human evolution that this thermal advantage can only be considered as a minor and fortuitous byproduct.

One of the most important mammalian defenses against radiant heat loads is a dense and highly reflective coat of body hair. This serves multiple purposes: reflecting up to a half of the solar energy, absorbing some of the unreflected portion at a distance from the skin for dissipation by convection and re-radiation, and providing an insulative space against the conductance of the absorbed heat toward the skin. A dense coat may be a dual purpose thermal protective system since it is most conspicuous in large desert mammals who face a high radiant heat input during the day and an equally impressive radiant heat loss at night. Man's evolution obviously did not include this specialization; in fact, man has gone in the opposite direction, toward virtual absence of body hair.

Keeping a body surface covered with a film of moisture, thus insuring a relatively high heat loss, is the only alternative to reducing a high radiant heat load by means of insulative fur or wool. Man's most important thermal adaptation is increased evaporative heat loss through thermal sweating. Of course, not all mammals sweat to regulate their body temperature; the dog dissipates body heat by panting. Even well adapted tropical animals such as goats may depend primarily on respiratory heat losses for cooling. However, respiratory heat loss offers much less effective surface area and is not as efficient as sweating. Animals which depend on panting usually show other defensive adaptations and, unlike man, have developed tolerance to hyperventilatory sequelae. Sweat glands have been observed in many animals which do not appear to use this type of evaporative heat loss as their principal defensive mechanism. Man is noted among the mammals not for either the number or size of his sweat glands (Weiner and Hellman 1960), but only for the very high secretory level at which they operate. Man has been observed to sweat at the rate of three liters per hour for short periods of heavy work in high heat (Eichna et al. 1945). Two liters per hour is quite common for the combination of exercise and heat and about the human limit for sustained sweating (Ladell 1964). One liter per hour is a reasonable figure for many hot conditions. No other mammal is known to sweat as much per unit surface area as man. The nearest competitors on a surface area basis are the donkey and the camel (Schmidt-Nielson 1964). Some breeds of cattle are heavy sweat producers but also simultaneously utilize panting (Mac-Farlane 1968). None of these can produce more than half of the sweat per unit of surface area (500 gm vs. 1000 gm/m^2/hr) that can be achieved by man. Sweating and panting are in general complementary, and for a sweating animal such as man panting is a secondary line of defense, used only when sweating is inadequate (Bianca 1968). Panting in man is apt to produce physiological problems of respiratory instead of thermal nature (Goldman et al. 1965).

Since all other primates have considerable hair covering, it has always been accepted that our ancestors must once have had a respectable amount of body hair. The question has been when and why did they lose it. Most of the explanations offered have implied that nakedness was an advantage in hot conditions. For example, Coon (1955) states: ". . . the absence of this covering . . . must be considered adaptive, on the one hand to hot dry conditions in which the surface of the skin must be free to permit the breezes to evaporate sweat, . . ." La Barre (1964) links nakedness with a need for "diffusion of metabolic heat from the rapid spurts of energy required in hunting"; Montagu (1964) postulates a loss of hair with increased sweating capacity as a mechanism for avoidance of overheating from a hunting way of life. This concept has been perpetuated in recent textbooks (Campbell 1967) and in a much more qualified format in a recent work for the lay audience (Morris 1967). One of the purposes of this review is to point out that this explanation does not fit the available data from man and other mammals. Unless we postulate an ancestral condition of dense, long fleece such as in wool-bearing sheep or the winter coat of camels, body hair is no bar to convective heat loss and has nothing to do with the radiation of long-wave infra-red heat to cooler objects. There is no evidence that a hair coat interferes with the evaporation of sweat; what can be said is that exposure to the sun after the removal of the body hair increases sweating in cattle (Berman 1957) and panting in sheep (MacFarlane 1968), because the total heat load has been increased. Many men have been studied in heat with a solar load while nude and wearing light clothing which is roughly equivalent to body hair. At rest or light work the clothed man gained about two-thirds as much heat as the nude man and sweated commensurately less, and this is roughly the same savings in heat gain that one gets from going out of the sunshine into shade (Adolph 1947, Lee 1964). It seems obvious that man's present glabrous state is a marked disadvantage under high radiant heat loads rather than the other way around, and that man's specialization for and great dependence on thermal sweating stems from his increased heat load in the sun.

It is much more difficult to evaluate the "metabolic-heat-generated-from-hunting" suggestion with comparative data since there is no other predatory species which seems to have had to lose its hair to be able to catch its prey. This might indicate that our ancestors either pushed their prey to exhaustion in what Krantz (1968) has termed persistence hunting, or that they were woefully inefficient hunters. Sweat rate is not a very good measure of physiological strain under conditions of short and uneven bursts of exercise although it is an excellent indicator for long-term, moderate work in the heat. This is because sweating is a somewhat delayed response, usually requiring at least 20 minutes to reach peak production. There certainly are no data on the energy requirements of our hypothetical early ancestors, but there are well established limits on modern man which place a work limit of approximately 300 kcal/hr of heat production for 8 hours per day if the level is to be maintained day after day without a problem of accumulated fatigue (Passmore and Durnin 1955). This is only about one half the limit of man's capacity for evaporative heat loss by sweating.

WATER REQUIREMENTS AND THIRST

Any discussion of evaporative heat losses must include a consideration of the replacement of the lost body fluid. Man is the most dependent on thermal sweating among the mammals thus far investigated and may well be the most dependent on a continuing source of water; his principal competitor would be the horse. The only two aspects of this problem to be considered will be man's drinking requirements and the effects of dehydration.

Every liter of water that an average 70 kg man loses represents nearly a 1½% loss in body weight. Sweating in the heat can reach the level of 2 liters per hour under conditions of sustained work; respiratory water loss in man is small, seldom over 15 grams per hour; and diffusional water loss through the skin probably never exceeds 1½% of weight per day (Ladell 1965). These losses must be replaced to avoid progressive dehydration. A man who has accumulated a 2% weight loss from sweating is thirsty, by 10% he is helpless, and death occurs at 18-20% (Folk 1966). There is much less information on other species, but the cat and dog die at about this same level of dehydration while camels, sheep, and donkeys can survive over a 30% weight loss of water (Whittow 1968). The mechanisms by which species tolerate dehydration vary and lie outside the scope of this survey, but it is interesting that man does not diminish sweating in the heat to fit his state of hydration until he is dangerously dehydrated (Adolph 1947).

Total daily water requirements for an animal in the heat is a complicated problem and many of its facets are not particularly relevant to our consideration of evolutionary forces. Assuming no marked seasonal differences in the water content of the vegetable foods consumed, and that the animal portion of the diet is very constant in water content, then body water losses from increased requirements for evaporative heat loss have to be replaced by additional drinking. In theory, this could be accomplished by either more at one time or by frequent watering. All domestic animals and probably most wild forms increase the frequency of drinking under heat stress provided water is available. Many have a remarkable ability to rehydrate themselves each time they drink. This may involve large quantities in remarkably short periods; camels ingest up to 100 liters within 10 minutes (Schmidt-Nielson et al. 1956), donkeys, 20 liters in 3 minutes (Schmidt-Nielson 1964), guanacos, 9 liters in 8 minutes (Rosemann and Morrison 1963), and sheep, 9 liters in 10 minutes (MacFarlane 1964). Carnivores generally cannot ingest as large a quantity of water as the ruminants for anatomical reasons, although the dog can hold almost 2 liters (Adolph 1947), and the cat has been reported to drink over 7% of its body weight in 10 minutes (Wolf 1958). Except for some small rodents and lagomorphs, man must be the least capable of rapidly ingesting water among all the mammals. Man reaches satiety after rapidly consuming 1 liter of water and cannot imbibe over 2 liters in a 10 minute period (Folk 1966). It is obvious, therefore, that man must resort to frequent rather than copious drinking to prevent even moderate dehydration. It is strange to find, in the same animal, both the least capacity to ingest water and the greatest dependence on thermal sweating. Its occurrence in a species which has in modern times been so successful in the tropics represents a triumph of technology (the ability to carry and store water outside the body) over biological limitations.

Not only is man unable to consume more than a small amount of water at one time, but he does not generally replace his water loss during the heat of the day. This "voluntary dehydration" in man can exceed 2 liters, but such a temporary negative water balance is not unique to man, probably occurring in all species. Presumably, in voluntary dehydration the free circulating water of the gut is being withdrawn and tissue dehydration does not start until the gut water has been largely utilized. This provides an initial buffer against dehydration which may be quite important in ruminants (12-15% of body weight) but of more limited potential in non-ruminants (5% of weight or less) (Chew 1965). The significant point in man is that any voluntary dehydration incurred during the day is routinely replaced at night, with and following his evening meal.

PROBLEMS OF CHANGING PATTERNS OF THIRST

Thirst in humans is a personal imperative which is only satisfied by individual drinking. The term "thirst" is difficult to use with the same connotations for other species because our subjective sensations cannot be verified in non-humans. Many gregarious species have group watering periods, widely spaced in time and probably related to both the water content of the diet and the ambient daytime temperature. Individual members of such groups rarely wander off alone in search of a drink. In domestic animals this may be an ancient trait, carried over and perhaps even intensified by breeding selection. In many wild species including the terrestrial primates it must stem from the increased vulnerability of individuals (cf. Washburn and DeVore 1961 on baboons). Even the time of day for group drinking is quite consistent. This implies either that the members do not normally develop marked individual differences in thirst or that these are subordinated to the daily regimen of the group. Both alternatives appear to be the antithesis of human thirst behavioral patterns. In our species the young have a propensity for becoming thirsty at inopportune times and are vociferous about satisfying this thirst. In fairness, it must be admitted that daily water requirements for the young are higher than adults per unit of body weight (Wolf 1958), while their capacity per drink is commensurately smaller. A gradual shift from group patterns to individual drinking in accordance with physiological needs would appear first in sub-adults and lactating females, the two most vulnerable portions of a mammalian group. Perhaps this shift could have occurred when, with sufficient economic specialization, part of the group did not wander with the adult male or males but remained relatively sedentary and near the water source. Even adult hominoid males might have been under similar constraints until they developed defensive weaponry to inhibit predators.

Cattle which are normally daytime drinkers increase the frequency and volume of night drinking under hot conditions (Yousef et al. 1968), but cattle are protected domestic animals. It was pointed out earlier that modern man rehydrates by drinking during the evening when evaporative heat loss is minimal. Modern day, terrestrial, non-human primates have not been reported as nocturnal drinkers in the wild. Before water containers were invented this might have posed quite a problem for our ancestors and been just as important as security in dictating the location of campsites.

HEAT RESPONSES IN NON-HUMAN PRIMATES

Human reaction to various levels of heat stress is by far the best known of all animals, but the rest of the primate order are among the least studied. From what little is known, it is obvious that we cannot safely extrapolate what we have learned about man to his primate relatives. Some data are available in the cebus, rhesus, baboon, and chimpanzee. A pertinent question raised by this review is do non-human primates utilize thermal sweating as their important avenue of heat loss under hot conditions? This cannot be answered from the available information; it is not even certain from laboratory experiments whether monkeys can tolerate heat stresses well within the capability of many animals.

A series of cebus monkeys (Hardy 1954) and chimps (Dale, et al. 1965) have been measured in calorimeters which distinguish between evaporation, conduction-convection, and radiation as sources of heat loss. Unfortunately for this discussion the chimps were only exposed to one temperature, 24° C [75.2° F]. A comparison of these two primates with man is given in Table 12-1.

Obviously if the percentage for one avenue of heat loss rises the others must decrease. At 24° C [75.2° F], both the cebus and chimp showed greater conductive-convective losses than man, but the cebus were in a metal chair and the chimps in a metal cage; these provided abnormal surface areas for the conduction-convection losses. This plus the hair covering on the cebus and chimp and the relatively high evaporative loss explain why the radiant exchange was so much less than that of a naked man. The evaporative heat losses at 24° C [75.2° F] were absolutely as well as relatively high in the non-human subjects. This temperature is too low for us to expect thermal sweating in any of the species so the high values must have been caused by higher diffusional skin water losses or increased respiratory exchange. Man and cebus were losing more heat than they were producing at 24° C [75.2° F], but the chimp's were very close to thermal balance.

At 34° C [93.2° F], radiation exchange was virtually eliminated since skin and wall temperatures in the calorimeters were practically identical. Evaporative loss increased three-fold in man from sweating. The two-fold increase in evaporative loss in the cebus has been ascribed to sweating because skin temperature remained relatively cool as calorimeter temperature rose, and this requires a moist skin surface. On the other hand the total evaporative cooling in the monkeys was not sufficient to prevent rectal temperature from rising rapidly to 40° C [104° F] when the experiments were terminated to avoid heat stroke.

Experiments with macaques and baboons in heat present a very con-

TABLE 12-1
Relative Mechanisms of Heat Loss Utilized in Three Primate Species

	TEMP.	RADIATION	CONDUCTION AND CONVECTION	EVAPO-RATION
Man[1]	[75.2° F] 24° C	66%	15%	19%
	[93.2° F] 34° C	0	45%	55%
Cebus[1]	[75.2° F] 24° C	40%	25%	35%
	[93.2° F] 34° C	0	35%	65%
Chimp[2]	[75.2° F] 24° C	30%	35%	35%

[1] From Hardy 1954.
[2] From Dale, et al. 1965.

fused picture. One young rhesus was exposed to 40° C [104° F], 55% RH presumably in a cage and with water (Robinson and Morrison 1957). This monkey's rectal temperature rose very slightly, and the animal showed no evidence of panting. The investigators did not check for sweating but inferred that it must have been present. A group of rhesus were exposed to two temperatures, 29° C [84.2° F] and 38° C [100.4° F], with comparable humidity, and two different positions of arm restraint while the animals were prone (Frankel et al. 1958). At 29° C [84.2° F], the type of restraint did not make any difference, and the animals maintained a steady, slightly depressed rectal temperature for over 4 hours. At 38° C [100.4° F], those with arms held back along the sides had slightly elevated but steady rectal temperatures while a position with the arms extended resulted in an almost explosive rise in rectal temperature which required premature termination of exposure. As if this were not confusing enough, no evidence of sweat production could be observed on these rhesus during the heat exposures (Folk 1965). Four young male savanna baboons exposed to 45° C [113° F], 50% RH (Funkhauser et al. 1967) in a primate restraint chair showed both sweating and panting but with insufficient total evaporative heat loss to prevent a dangerous elevation of body temperature. This combination of heat and humidity was very stressful and would have caused some panting in man (Goldman et al. 1965). The baboons increased their respiratory rate very rapidly, a response expected from a panting species. Unfortunately, these subjects were deprived of water for 24 hours prior to the heat exposure, and most animal species show effect of dehydration in their responses; the oryx, which is primarily a sweating species, shifts entirely to panting for heat loss when dehydrated (Taylor 1969).

Our knowledge of the non-human primates' reaction to heat is obviously fragmentary and contradictory. Whether the contradictions arise from problems in techniques, e.g. the resistance of primates to physical restraint, or for other reasons is as yet unknown.

CONCLUSIONS

Man's specialized dependence on thermal sweating appears to be an evolutionary adaptation to the tropics. The marked limitations on the amount of water which man can consume at one time, and the lack of any of the water conserving adaptations observed in desert mammals argues for a well-watered tropical habitat, or, at least certainly not a tropical desert. On the other hand profuse sweating does not in itself provide efficient evaporative heat loss under the high humidity and lack of air movement which characterize the tropical forest. An environmental niche of tropical parklands and grasslands, best corresponds to the observed thermoregulatory responses.

It has been argued here that man's propensity for sweating came about because his nakedness increased the total radiant heat load received and not because loss of hair somehow enhanced the efficiency of sweat evaporation. Therefore, either the two processes developed simultaneously or the decline in body hair preceded the increase in sweating. If nakedness was a disadvantage in a savanna environment which required a compensatory adaptation (sweating), loss of hair must have stemmed from other causes or preceded the occupation of the habitat in question, at least for its inception. Our traditional ideas of what our ancestors looked like has included far more body hair than our own species until forms very similar and closely related to *Homo sapiens* appear. The illustrations (with one exception) for as recent a work as Howell's "Early Man" (1968) show hirsutism as a rather conservative trait over millions of years. If one had to select times and habitats when progressive denudation was not a distinct environmental disadvantage, the choices would seem to be between a very early period when our ancestors were primarily forest dwellers or a very recent period when primitive clothing could provide the same protection against either solar heat or cold.

why recent?

The primary difficulty in arguing for the recent loss of body hair is that there seems to be no single and powerful environmental driving force other than recurrent cold that is obvious after the Pliocene epoch. Furthermore, the developing complexity and efficiency of even primitive man's technology would have decreased the probability of a straightforward biological adaptation. Finally, since all modern men lack an effective sun-screen of body hair and all sweat profusely in the heat, this specialization either must have occurred at a time and place when the evolving species was geographically compact, or at least contiguous, or represents a highly improbably parallel evolution in a variety of noncomparable environments.

The obvious time and place where progressive denudation would have been least disadvantageous is the ancient forest habitat. Radiant energy does penetrate the forest canopy in limited amounts, and that portion of the spectrum which is primarily transmitted through vegetation, the near infrared wavelengths of 0.75 to 0.93 microns, is exactly the energy best reflected by human skin (Gates 1968). This may be purely coincidental, but it lends a little support to this suggestion that loss of body hair in man was somehow stimulated much earlier in time than our present estimates.

In conclusion, man suffers from a unique trio of conditions: hypotrichosis corpus, hyperhydrosis, and polydipsia. The appearance and development of this combination must have been the result of some of the habits of our ancestors, or in more popular phraseology, of the ecological niche of the survivors who gave rise to our evolving ancestors. There is little hope that the osseous remains in the fossil record will provide much in the way of clues to the time of initiation and rate of change in these characteristics. There is hope that judicious speculation based on comparative studies which focus on the problems can help us to define the possible limits within which one may safely theorize.

ABSTRACT AND SUMMARY

A shift in habitat from forest to tropical grasslands by man's early ancestors in Tertiary times resulted in a rather distinctive reaction to external heat loads. If our ancestors had already lost most of their body hair or were in the process of doing so, the combination of increased solar exposure in open country and greater absorption of solar energy by the naked skin magnified the total heat load. This must have constituted a disadvantage which required some compensation; in man it has taken the form of dependence on thermal sweating for heat dissipation to the point where *Homo sapiens* has the greatest sweating capacity for a given surface area of any known animal. Heavy sweating without simultaneous drinking inevitably leads to tissue dehydration. Man is very intolerant of dehydration and has limited capacity to rehydrate rapidly. He is dependent on frequent and relatively small drinks of water under hot conditions. This must have influenced species behavior, at least until water containers were invented. Unfortunately, our present knowledge of how our non-human primate relatives react to heat stress is fragmentary and contradictory. We can't be sure that our modern specializations are uniquely human or are part of a general primate pattern.

LITERATURE CITED

Adolph, E. F. 1947 Physiology of man in the desert. Interscience, New York.

Barnicot, N. A. 1959 Climatic factors in the evolution of human populations. Cold Spring Harbor Symposia on Quantitative Biology, 24:115–129.

Berman, A. 1957 Influence of some factors on the relative evaporation rate from the skin of cattle. Nature, 179:1256.

Bianca, W. 1968 Thermoregulation. *In*: Adaptation of Domestic Animals, ed. by E. S. E. Hafez, pp. 97–118. Lea and Febiger, Philadelphia.

Campbell, B. G. 1967 Human Evolution. An Introduction to Man's Adaptations. Aldine, Chicago.

Chew, R. M. 1965 Water metabolism of mammals. *In*: Physiological Mammalogy, vol. II, Mammalian reactions to stressful environments, ed. by W. V. Mayer and R. G. Van Gelder, pp. 43–148. Academic Press, New York.

Coon, C. S. 1955 Some problems of human variability and natural selection in climate and culture. Amer. Naturalist, 89:257–280.

Dale, H. E., M. D. Shankline, H. D. Johnson and W. H. Brown 1965 Energy metabolism of the chimpanzee. A comparison of direct and indirect calorimetry. Tech. Report ARL–TR–65–17, 6571st Aeromed. Res. Lab., Holloman AFB, New Mexico.

Eichna, L. W., W. F. Bean, W. B. Bean and W. B. Shelley 1945 The upper limits of heat and humidity tolerated by acclimatized men working in hot environments. J. Indust. Hygiene and Toxicol., 27:59–84.

Folk, G. E., Jr. 1966 Introduction to Environmental Physiology. Lea and Febiger, Philadelphia.

Frankel, H. M., G. E. Folk, Jr. and F. N. Craig 1958 Effects of type of restraint upon heat tolerance in monkeys. Proc. Exp. Biol. Med., 97:339–341.

Funkhauser, G. E., E. A. Higgins, T. Adams and C. C. Snow 1967 The response of the savannah baboon (Papio cynocephalus) to thermal stress. Life Sciences, 6:1615–1620.

Gates, D. M. 1968 Physical environment. *In*: Adaptations of Domestic Animals, ed. by E. S. E. Hafez, pp. 46–60. Lea and Febiger, Philadelphia.

Goldman, R. F., E. B. Green and P. F. Iampietro 1965 Tolerance of hot, wet environments by resting men. J. Appl. Physiol., 20:271–277.

Hardy, J. D. 1955 Control of heat loss and heat production in physiologic temperature regulation. Harvey Lectures, 49:242–270. Academic Press, New York.

Howell, F. C. 1968 Early Man. Time-Life, New York.

Krantz, G. 1968 Brain size and hunting ability in earliest man. Current Anthropology, 9:450–451.

La Barre, W. 1964 Comments on The Human Revolution by C. F. Hockett and R. Ascher. Current Anthropology, 5:147–150.

Ladell, W. S. S. 1964 Terrestrial animals in humid heat: man. *In*: Hdbk.

of Physiology, 4, Adaptation to the Environment, ed. by D. B. Dill, pp. 625–659. Williams and Wilkins, Baltimore.

—— 1965 Water and salt intakes. In: The Physiology of Human Survival, ed. by O. G. Edholm and A. L. Bacharach, pp. 235–299. Academic Press, London.

Lee, D. H. K. 1950 Studies of heat regulation in the sheep with special reference to the Merino. Aust. J. Agric. Res., 1:200–216.

—— 1963 Physiology and the arid zone. In: Environmental Physiology and Psychology in Arid Conditions, Reviews of Research, 22, pp. 15–36. UNESCO, Paris.

—— 1964 Terrestrial animals in dry heat: man in the desert. In: Hdbk. of Physiology, 4, Adaptation to the Environment, ed. by D. B. Dill, pp. 551–583. Williams and Wilkins, Baltimore.

Macfarlane, W. V. 1964 Terrestrial animals in dry heat: ungulates. In: Hdbk. of Physiology, 4, Adaptation to the Environment, ed. by D. B. Dill, pp. 509–539. Williams and Wilkins, Baltimore.

—— 1968 Comparative functions of ruminants in hot evironments. In: Adaptations of Domestic Animals, ed. by E. S. E. Hafez, pp. 264–276. Lea and Febiger, Philadelphia.

Montagu, A. 1964 Comments on The Human Revolution by C. F. Hockett and R. Ascher. Current Anthropology, 5:160–161.

Morris, D. 1967 The Naked Ape. McGraw-Hill, New York.

Passmore, R. and J. V. G. A. Durnin 1955 Human energy expenditure. Physiol. Rev., 35:801–840.

Read, R. G. 1968 Evaporative power in the tropical forest of the Panama canal zone. J. Appl. Meteor., 7:417–424.

Richards, P. W. 1952 The Tropical Rain Forest, an Ecological Study. Cambridge Univ. Press, London.

Robinson, K. W. and P. R. Morrison 1957 The reactions to hot atmospheres of various species of Australian marsupial and placental animals. J. Cellular Comp. Physiol., 49:455–478.

Roller, W. L. and R. F. Goldman 1967 Estimation of solar radiation environment. Int. J. Biometeor., 11:329–336.

Rosenmann, M. and P. R. Morrison 1963 The physiological response to heat and dehydration in the guanaco. Physiol. Zool., 36:45–51.

Schmidt-Nielson, K. 1964 Desert Animals. Physiological Problems in Heat and Water. Clarendon Press, Oxford.

Schmidt-Nielson, K., B. Schmidt-Nielson, T. R. Houpt and S. A. Jarnum 1956 The question of water storage in the stomach of the camel. Mammalia, 20:1–15.

Taylor, C. R. 1969 The eland and the oryx. Sci. Amer., 220:88–95.

Washburn, S. L. and I. DeVore 1961 Social behavior of baboons and early man. In: Social Life of Early Man, ed. by S. L. Washburn, pp. 91–105. Aldine, Chicago.

Weiner, J. S. and K. Hellmann 1960 The sweat glands. Biol. Rev., 25: 141–186.

Whittow, G. C. 1968 Body fluid regulation. In: Adaptation of Domestic

Animals, ed. by E. S. E. Hafez, pp. 119–126. Lea and Febiger, Philadelphia.
Wolf, A. V. 1958 Thirst. Physiology of the Urge to Drink and Problems of Water Lack. C. C. Thomas, Springfield.
Yousef, M. K., L. Hahn and H. D. Johnson 1968 Adaptation of cattle. *In:* Adaptation of Domestic Animals, ed. by E. S. E. Hafez, pp. 233–245. Lea and Febiger, Philadelphia.

13

THE EVOLUTION OF HUNTING

S. L. Washburn and C. S. Lancaster

The fossil record of human evolution indicates that the Hominidae separated from the other hominoid lines 15 to 20 million years ago, and that by 3 to 4 million years ago, adaptations for bipedalism, tool-use, and cultural modes of behavior were already firmly established. The 3-million-year span of the Pleistocene epoch was marked by further transformations in human genotypes that were to equip populations of bipedal, toolmaking, and symbol-using humans to make a unique niche for themselves among the earth's living creatures. What were these prehistoric human beings really like? How did they live? What goals for living did their societies develop? What did they think about? What did they believe in? What were they afraid of? Such questions are not beyond conjecture or even investigation.

An illustration of how such reconstruction may be effected is provided in the following selection. Starting with the fact that man has lived as a hunter for almost his entire history, the authors explore the implications of the hunting way of life. They utilize data from the fossil record, ecology, primate studies, and ethnography of contemporary hunting cultures to present a picture of early human populations living as hunters.

S. L. Washburn is Professor of Anthropology at the University

Reprinted from *Man the Hunter*, Richard B. Lee and Irvin DeVore, editors (Chicago: Aldine Publishing Company, 1968). Copyright © 1968 by Wenner-Gren Foundation for Anthropological Research, Inc. By permission of the authors and the publishers.

of California, Berkeley. C. S. Lancaster is at Rutgers University, New Jersey.

. . . Human hunting is made possible by tools, but it is far more than a technique, or even a variety of techniques. It is a way of life, and the success of the hunting adaptation (in its total social, technical, and psychological dimensions) has dominated the course of human evolution for hundreds of thousands of years. In a very real sense our intellect, interests, emotions, and basic social life are evolutionary products of the success of the hunting adaptation. When anthropologists speak of the unity of mankind, they are stating that the selection pressures of the hunting and gathering way of life were so similar and the result so successful that populations of Homo sapiens are still fundamentally the same everywhere. In this essay we are concerned with the general characteristics of man that we believe can be attributed to the hunting way of life.

Perhaps the importance of the hunting way of life in producing man is best shown by the length of time hunting has dominated human history. The genus Homo[1] has been existent for some 600,000 years, but agriculture has been important only during the last few thousand years. Even 6000 years ago large parts of the world's population were nonagricultural, and the entire evolution of earliest Homo erectus into existing races took place during the period in which man was a hunter. The common factors that dominated human evolution and produced Homo sapiens were preagricultural. Agricultural ways of life have dominated less than 1 percent of human history, and there is no evidence of major biological changes during that period of time. The kinds of minor biological changes that occurred and which are used to characterize some modern races are not common to Homo sapiens. The origin of all common characteristics must be sought in preagricultural times. Probably all experts would agree that hunting was a part of the social adaptation of populations of the genus Homo, and many would regard Australopithecus[2] as an earlier hominid who was already a hunter, although possibly much less efficient than the later forms. If this is true and if the Pleistocene period had a duration of 3 million years, then pre-Homo erectus human tool-using and hunting lasted for at least four times as long as the duration of the genus Homo. No matter how the earlier times may ultimately be interpreted, the observation of more hunting among apes than was previously suspected and increasing evidence of hunting by Australopithecus strengthens the posi-

---

[1] The term Homo includes Java, Peking, Maur, and later forms.

[2] This term is used to include both the small A. africanus and A. robustus large forms. Simpson (1966) briefly and clearly discusses the taxonomy of these forms and of the fragments called Homo habilis.

tion that less than 1 percent of human history has been dominated by agriculture. It is for this reason that the consideration of hunting is so important for the understanding of human evolution.

When hunting and the way of life of successive populations of the genus *Homo* are considered, it is important to remember that there must have been both technical and biological progress during this vast period of time. Although the locomotor system appears to have changed very little in the last 500,000 years, the brain did increase in size, and the form of the face evolved. But for present purposes it is particularly necessary to direct attention to the cultural changes that occurred in the last ten or fifteen thousand years before agriculture. There is no convenient term for this period of time, traditionally spoken of as the end of the Upper Paleolithic and the Mesolithic, but Binford has rightly emphasized its importance.

During most of human history water must have been a major physical and psychological barrier and the inability to cope with water is shown in the archeological record by the absence of remains of fish, shellfish, or any object that would have required either going deeply into water or the use of boats. There is no evidence that the resources of river and sea were utilized until this late preagricultural period, and since the consumption of shellfish in particular leaves huge middens, the negative evidence is impressive. It is likely that the basic problem in utilization of resources from sea or river was that man cannot swim naturally and to do so must learn a difficult skill. The normal quadrupedal running motions of monkeys serve to keep them afloat and moving quite rapidly. A macaque, for example, does not have to learn any new motor habit in order to swim. But the locomotor patterns of gibbons and apes will not keep them above the water surface, and even a narrow, shallow stream is a barrier for the gorilla. For early man, water was a barrier and a danger, not a resource.

In addition to the conquest of water, there seems to have been great technical progress in the late preagricultural period. Besides a wide variety of stone tools of earlier kinds, the archeological record shows bows and arrows, grinding stones, boats, houses of more advanced types and even villages, sledges drawn by animals and used for transport, and the domestic dog. These facts have two special kinds of significance . . . First, the technology of *all* the living hunters belongs at the earliest to this late Mesolithic era, and many have elements borrowed from agricultural and metal-using peoples. Second, the occasional high densities of hunters mentioned as problems and exceptions . . . are based on a very late and modified extension of the hunting and gathering way of life. For example, the way of life of the tribes of the Northwest Coast, with polished stone axes for woodworking, boats, and extensive reliance on products of river and sea, should be seen as a very late adaptation.

The presence of the dog is a good index of the late preagricultural period, and domestic dogs were used by hunters in Africa, Australia, and

the Americas. Among the Eskimos, dogs were used in hunting, for trans-
portation, as food in time of famine, and as watchdogs. With dogs, sleds,
boats, metal, and complex technology, Eskimos may be better examples
of the extremes to which human adaptation can go than examples of
primitive hunting ways. . . . Dogs were of great importance in hunting for
locating, tracking, bringing to bay, and even killing. Lee mentions that one
Bushman with dogs brought in much more game than the other hunters.
Dogs may be important in hunting even very large animals; in the Amboseli Game Reserve in Kenya one of the present authors saw two small
dogs bring a rhinoceros to bay and dodge repeated charges.

With the acquisition of dogs, bows, and boats, it is certain that hunting
became much more complex in the last few thousand years before agriculture. The antiquity of traps, snares, and poisons is unknown, but it appears that until a few thousand years ago man was able to kill large game
close in with spear or axe. As Brues has shown, this kind of hunting limits
the size of the hunters, and there are no very large or very small fossil
men. Pygmoid hunters of large game are probably possible only if hunting
is with bows, traps, and poison. It is remarkable that nearly all the estimated statures of fossil men fall between 5 feet 2 inches and 5 feet 10
inches. This suggests that strong selection pressures kept human stature
within narrow limits for hundreds of thousands of years and that these
pressures relaxed a few thousand years ago, allowing the evolution of a
much wider range of statures.

Gathering and the preparation of food also seem to have become more
complex during the last few thousand years before agriculture. Obviously
gathering by nonhuman primates is limited to things that can be eaten
immediately. In contrast, man gathers a wide range of items that he
cannot digest without soaking, boiling, grinding, or other special preparation. Seeds may have been a particularly important addition to the human
diet because they were abundant and could be stored easily.* Since grinding
stones appeared before agriculture, grinding and boiling may have been
the necessary preconditions to the discovery of agriculture. One can easily
imagine that people who were grinding seeds would see repeated examples
of seeds sprouting or being planted by accident. Grinding and boiling were
certainly known to the preagricultural peoples, and this knowledge could
have spread along an arctic route, setting the stage for a nearly simultaneous discovery of agriculture in both the New and Old Worlds. It was
not necessary for agriculture itself to have spread through the arctic but
only the seed-using technology, which could have then led to the discovery
of seed planting. If this analysis is at all correct, then the hunting-gath-
ing adaptation of the Indians of California, for example, should be seen as

* For a view of an earlier utilization of seed resources by the earliest hominids, see
Jolly's paper in this volume. [Ed.]

representing the possibilities of this late preagricultural gathering, making possible much higher population densities than would have been the case in a pregrinding and preboiling economy.

Whatever the fate of these speculations, we think that the main conclusion, based on the archeological record, ecological considerations, and the ethnology of the surviving hunter-gatherers, will be sustained. In the last few thousand years before agriculture, both hunting and gathering became much more complex. The final adaptation, including the use of products of river and sea and the grinding and cooking of otherwise inedible seeds and nuts, was worldwide, laid the basis for the discovery of agriculture, and was much more effective and diversified than the previous hunting and gathering adaptations.

Hunting by members of the genus *Homo* throughout the 600,000 years that the genus has persisted has included the killing of large numbers of big animals. This implies the efficient use of tools, as Birdsell has stressed. . . . The adaptive value of hunting large animals has been shown by Bourlière, who has demonstrated that 75 percent of the meat available to human hunters in the eastern Congo was elephant, buffalo, and hippopotamus. It is some measure of the success of human hunting that when these large species are protected in game reserves (as in the Murchison Falls or Queen Elizabeth Parks in Uganda) they multiply rapidly and destroy the vegetation. As evidenced in the Masai Amboseli Reserve in Kenya, elephants alone can destroy trees more rapidly than can be replaced naturally. Since the predators are also protected in reserves, it appears that human hunters have been killing enough large game to maintain the balance of nature for many thousands of years. It is tempting to think that man replaced the saber-toothed cat as the major predator of large game, both controlling the numbers of the game and causing the extinction of Old World saber-tooths. We think that hunting and butchering large animals put a maximum premium on cooperation among males, a behavior that is at an absolute minimum among the nonhuman primates. It is difficult to imagine the killing of creatures such as cave bears, mastodons, mammoths—or dinotherium at a much earlier time—without highly coordinated, cooperative action among males. It may be that the origin of male-male associates lies in the necessity of cooperation in hunting, butchering, and war. Certainly butchering sites, such as those described by Clark Howell in Spain, imply that the organization of the community for hunting large animals goes back for many, many thousands of years. From the biological point of view, the development of such organizations would have been paralleled by selection for an ability to plan and cooperate (or reduction of rage). Because females and juveniles may be involved in hunting small creatures, the social organization of big-game hunting would also lead to an intensification of a sexual division of labor.

As noted before, it is important to stress that human hunting is a set of ways of life. It involves division of labor between male and female, co-operation among males, planning, knowledge of many species and large areas, and technical skill. Recently, the old idea has been revived that the way of life of our ancestors was similar to that of wolves, rather than that of apes or monkeys. But this completely misses the special nature of the human adaptation. Human females do not go out and hunt and then regurgitate to their young when they return. Human young do not stay in dens but are carried by mothers. Male wolves do not kill with tools, butcher, and share with females who have been gathering. In an evolutionary sense the whole human pattern is new, and it is the success of a particularly human way that dominated human evolution and determined the relation of biology and culture for thousands of years. Judging from the archeological record, it is probable that the major features of this human way, possibly even including the beginnings of language, had evolved by the time of *Homo erectus*.[3]

THE WORLD VIEW OF THE HUNTER

Lévi-Strauss has urged that we study the world view of hunters, and, perhaps surprisingly, some of the major aspects of this world view can be traced from the archeological record. We have already mentioned that boats and the entire complex of fishing, hunting sea mammals, and using shellfish developed late. With this new orientation, wide rivers and seas changed from barriers to pathways and sources of food, and the human attitude toward water must have changed completely. But many hundreds of thousands of years earlier, perhaps with *Australopithecus*, the relation

[3] In speculations of this kind, it is well to keep in mind the purpose of the speculation and the limitation of the evidence. Our aim is to understand human evolution. What shaped the course of human evolution was a succession of successful adaptations, both biological and cultural. These may be inferred in part from the direct evidence of the archeological record. But the record is very incomplete. For example, Lee has described how large game may be butchered where it falls and only meat brought back to camp. This kind of behavior means that analysis of bones around living sites is likely to underestimate both the amount and variety of game killed. If there is any evidence that large animals were killed, it is probable that far more were killed than the record shows. Just as the number of human bones gives no indication of the number of human beings, the number of animal bones, although providing clues to the existence of hunting, gives no direct evidence of how many animals were killed. The Pleistocene way of life can only be known by inference and speculation. Obviously, speculations are based on much surer ground when the last few thousand years are under consideration. Ethnographic information is then directly relevant and the culture bearers are of our own species. As we go farther back in time, there is less evidence and the biological and cultural differences become progressively greater. Yet it was in a remote time that the human way took shape, and it is only through speculation that we may gain some insights into what the life of our ancestors may have been.

of the hunters to the land must also have changed from an earlier relationship that may be inferred from studies of contemporary monkeys and apes. Social groups of nonhuman primates occupy exceedingly small areas, and the vast majority of animals probably spend their entire lives within less than four or five square miles. Even though they have excellent vision and can see for many miles, especially from the tops of trees, they make no effort to explore more than a tiny fraction of the area they see. Even for gorillas the range is only about fifteen square miles, and it is of the same order of magnitude for savanna baboons; they refuse to be driven beyond the end of their range and double back. The known area is a psychological reality, clear in the minds of the animals. Only a small part of even this limited range is used, and exploration is confined to the canopy, lower branches, and bushes, or ground, depending on the biology of the particular species. Napier has discussed this highly differential use of a single area by several species. In marked contrast, human hunters are familiar with very large areas. In the area studied by Lee eleven waterholes and several hundred square miles supported a smaller number of Bushmen than the number of baboons supported by a single waterhole and a few square miles in the Amboseli Reserve in Kenya. The most minor hunting expedition covers an area larger than most nonhuman primates would cover in a lifetime. Interest in a large area is human. The small ranges of monkeys and apes restrict the opportunities for gathering, hunting, and meeting conspecies, and limit the kind of predation and the number of diseases. In the wide area, hunters and gatherers can take advantage of seasonal foods, and only man among the primates can migrate long distances seasonally. In the small area, the population must be supported throughout the year on local resources, and natural selection has been for biology and behavior that efficiently utilize these limited opportunities. But in the wide area selection is for the knowledge that enables the group to utilize seasonal and occasional food sources. Gathering over a wide and diversified area implies a greater knowledge of flora and fauna, knowledge of the annual cycle, and a different attitude toward group movements. Clearly one of the great advantages of slow maturation is that learning covers a series of years, and the meaning of events in these years becomes a part of the individual's knowledge. With rapid maturation and no language the chances that any member of the group will know the appropriate behavior for rare events is greatly reduced.

Moving over long distances creates problems of carrying food and water. Lee has called to our attention that the sharing of food even in one locality implies that food is carried, and there is no use in gathering quantities of fruit or nuts unless they can be moved. If women are to gather while men hunt, the results of the labors of both sexes must be carried back to some agreed-upon location. Meat can be carried easily, but the development of some sort of receptacle for carrying vegetable products may have been one of the most fundamental advances in human evolution. Without a way of carrying, the advantages of a large area are greatly reduced. However that may be, the whole human pattern of gathering and hunting to share is unique to man. In its small range a monkey gathers only what it itself needs to eat at the moment; the whole complex of economic reciprocity that dominates so much of human life is unique to man. Wherever archeological evidence can suggest the beginnings of movement over large ranges, cooperation, and sharing, it is dating the origin of some of the most fundamental aspects of human behavior, of the human world view. We believe that hunting large animals may demand all these aspects of human behavior that separate man so sharply from the other primates. If this is so, then the human way appears to be as old as *Homo erectus*.

The price that man pays for his high mobility is well illustrated by the problems of living in the African savanna. Man is not adapted to this environment in the same sense that baboons or vervet monkeys are. Man needs much more water, and without preparation and cooking he can only eat a limited number of the foods on which the local primates thrive. Unless there have been major physiological changes, the diet of our ancestors must have been far more like that of chimpanzees than like that of a savanna-adapted species. Further, man cannot survive the diseases of the African savanna without lying down and being cared for. Even when sick, the locally adapted animals are usually able to keep moving with their troop; and the importance to their survival of a home base has been stressed elsewhere (DeVore and Washburn, 1963). Also, man becomes liable to new diseases and parasites by eating meat, and it is of interest that the products of the sea, which we believe were the last class of foods added to the human diet, are widely regarded as indigestible and carry diseases to which man is particularly susceptible. Although many humans die of disease and injury, those who do not, almost without exception, owe their lives to others who cared for them when they were unable to hunt or gather, and this uniquely human caring is one of the patterns that builds social bonds in the group and permits the species to occupy almost every environment in the world.

A large territory not only provides a much wider range of possible foods but also a greater variety of potentially useful materials. With tool use this variety takes on meaning, and even the earliest pebble tools show selection in size, form, and material. When wood ceases to be just something to climb on, hardness, texture, and form become important. Availability of materials is critical to the tool-user, and the interest of early men in their environment must have been very different from that of monkeys or apes. Thus, the presence of tools in the archeological record is not only an indication of technical progress but also an index of interest in inanimate objects and in a much larger part of the environment than is the case with nonhuman primates.

The tools of the hunters include the earliest beautiful man-made objects, the symmetrical bifaces, especially those of the Acheulian tradition. Just how they were used is still a matter of debate, but, as contemporary attempts to copy them show, their manufacture is technically difficult, taking much time and practice and a high degree of skill. The symmetry of these tools may indicate that they were swung with great speed and force, presumably attached to some sort of handle. A tool that is moved slowly does not have to be symmetrical, but balance becomes important when an object is swung rapidly or thrown with speed. Irregularities will lead to deviations in the course of the blow or the trajectory of flight. An axe or spear to be used with speed and power is subject to very different technical limitations from those of scrapers or digging sticks, and it may well be that it was the attempt to produce efficient high-speed weapons that first produced beautiful, symmetrical objects.

When the selective advantage of a finely worked point over an irregular one is considered, it must be remembered that a small difference might give a very large advantage. A population in which hunters hit the game 5 percent more frequently, more accurately, or at greater distance would bring back much more meat. There must have been strong selection for greater skill in manufacture and use, and it is no accident that the bones of small-brained men (*Australopithecus*) are never found with beautiful, symmetrical tools. If the brains of contemporary apes and men are compared, the areas (both in cerebellum and cortex) associated with hand skills are at least three times as large in man. Clearly, the success of tools has exerted a great influence on the evolution of the brain and has created the skills that make art possible. The evolution of the capacity to appreciate the product must evolve along with the skills of manufacture and use, and the biological capacities that the individual inherits must be developed in play and practiced in games. In this way the beautiful, symmetrical tool becomes a symbol of a level of human intellectual achievement, representing far more than just the tool itself.

In a small group the necessity of practice in developing skills to a very high level restricts the number of useful arts. Where there is little division of labor, all men must learn to use the weapons of the hunt and of war. In sports we take it for granted that one person will not achieve a very high level of performance in more than a limited set of skills. This limitation is in part biological, but it is important socially as well because great proficiency in a skill necessitates practice. In warfare a wide variety of weapons is useful only if there are enough men so that there can be division of labor and different groups can practice different skills. Handedness, a feature that separates man from ape, is a part of this biology of skill. To be ambidextrous might seem ideal, but in fact the highest level of skill is attained by concentrating both biological ability and practice primarily on one hand.

Hunting changed man's relationship to other animals and his view of what is natural. The human notion that it is normal for animals to flee and the whole concept of animals being wild, is the result of man's habit of hunting. In game reserves many different kinds of animals soon learn not to fear man, and they no longer flee. Woodburn took a Hadza into the Nairobi Park, and the Hadza was amazed and excited, because although he had hunted all his life, he had never seen such a quantity and variety of animals close at hand. His whole previous view of animals was the result of his having been their enemy, and they had reacted to him as the most destructive carnivore. In the park, the Hadza hunter saw for the first time the peace of the herbivorous world. Prior to hunting, the relationship of our ancestors to other animals must have been very much like that of the other noncarnivores. They could have moved close among the other species, fed beside them, and shared the same waterholes. But with the origin of human hunting the peaceful relationship was destroyed, and for at least half a million years man has been the enemy of even the largest mammals. In this way the whole human view of what is normal and natural in the relation of man to animals is a product of hunting, and the world of flight and fear is the result of the efficiency of the hunters.

Behind this human view that the flight of animals from man is natural lie some aspects of human psychology. Men enjoy hunting and killing, and these activities are continued as sports even when they are no longer economically necessary. If a behavior is important to the survival of a species (as hunting was for man throughout most of human history), then it must be both easily learned and pleasurable (Hamburg, 1963). Part of the motivation for hunting is the immediate pleasure it gives the hunter, and the human killer can no more afford to be sorry for the game than a cat can be for its intended victim. Evolution builds a relation between biology, psychology, and behavior, and, therefore, the evolutionary success of hunting exerted a profound effect on human psychology. Perhaps this is most easily shown by the extent of the efforts devoted to maintain killing as a sport. In former times royalty and nobility maintained parks where they could enjoy the sport of killing, and today the United States government spends many millions of dollars to supply game for hunters. Many people dislike the notion that man is naturally aggressive, that he naturally enjoys the destruction of other creatures. We all know people who, for example, although denying human aggressive tendencies, use the lightest fishing tackle to prolong the fish's futile struggle, to maximize the personal sense of mastery and skill. And until recently war was viewed in much the same way as hunting. Other human beings were simply the most dangerous game, and war has been far too important in human history for it to be other than pleasurable for the males involved. It is only recently, with the entire change in the nature and conditions of war, that this institution has been challenged and the wisdom of war as a normal part of national policy or as an approved road to personal social glory has been questioned.

Human killing differs from killing by carnivorous mammals in that the victims are frequently of the same species as the killer. In carnivores there are submission gestures or sounds that normally stop a fatal attack, but in man there are no effective submission gestures. It was the Roman emperor who might raise his thumb; the victim could make no sound or gesture that might restrain the victor or move the crowd to pity. The lack of biological controls over killing conspecies is a character of human killing that separates this behavior sharply from that of other carnivorous mammals. This difference may be interpreted in a variety of ways. From an evolutionary point of view, it may be that human hunting is so recent that there has not been enough time for controls to evolve. Or it may be that killing other humans was a part of the adaptation from the beginning, and the sharp separation of war and hunting is due to the recent development of these institutions. Or it may be simply that in most human behavior stimulus and response are not tightly bound. Whatever the origin of this behavior it has had profound effects on human evolution, and almost every human society has regarded killing members of certain other human societies as desirable. Certainly this has been a major factor in man's view of the world, and every folklore contains the tales of the cultural heroes whose fame is based on the human enemies they destroyed.

The extent to which the biological bases for killing have been incorporated into human psychology may be measured by the ease with which boys can be interested in hunting, fishing, fighting, and games of war. It is not that these behaviors are inevitable, but that they are easily learned, satisfying, and have been socially rewarded in most cultures. The skills for killing and the pleasures of killing are normally developed in play, and the patterns of play prepare the children for their adult roles. . . . Woodburn's excellent motion pictures show Hadza boys killing small mammals, and Laughlin describes how Aleuts train boys from early childhood so that they will be able to throw harpoons with accuracy and power while seated in kayaks. The whole youth of the hunter is dominated by practice and appreciation of the skills of the adult males, and the pleasure of the game motivates the practice that is necessary to develop the skills of weaponry. Even in monkeys rougher play and play fighting are largely the activities of the males, and the young females explore less and show a greater interest in infants at an early age. These basic biological differences are reinforced in man by the division of labor that makes adult sex roles differ far more in humans than they do in nonhuman primates. Again, hunting must be seen as a whole pattern of activities, a wide variety of ways of life, the psychobiological roots of which are reinforced by play and by a clear identification with adult roles. Hunting is more than a part of the economic system, and the animal bones in Choukoutien are evidence of the patterns of play and pleasures of our ancestors.

THE SOCIAL ORGANIZATION
OF HUMAN HUNTING

The success of the human hunting and gathering way of life lay in its adaptability. It permitted a single species to occupy most of the earth without biological adaptation to local conditions. The occupation of Australia and the New World probably occurred entirely after the end-Paleolithic-Mesolithic development discussed earlier, but even so there is no evidence that any other primate species occupied more than a fraction of the area of *Homo erectus*. Obviously this adaptability makes any detailed reconstruction impossible, and we are not looking for stages in the traditional evolutionary sense. However, using both knowledge of the contemporary primates and the archeological record, certain important general conditions of our evolution may be reconstructed. For example, the extent of the distribution of the species noted above is remarkable and gives the strongest sort of indirect evidence for the adaptability of the way of life, even half a million years ago. Likewise all evidence suggests that the local group was small—twenty-five to fifty individuals has been suggested as the average size. Such a group size is common in nonhuman primates and so we can say with some assurance that it did not increase until after agriculture. This means that the number of adult males who might cooperate in hunting or war was very limited, and this restricted the kinds of social organizations that were possible. Probably one of the great adaptive advantages of language was that it permitted the planning of cooperation between local groups, temporary division of groups, and the transmission of information over a much wider area than that occupied by any one group.

Within the group of the nonhuman primates the mother and her young may form a subgroup that continues even after the young are fully grown. This grouping affects dominance and grooming and resting patterns, and, along with dominance, is one of the factors giving order to the social relations in the group. The group is not a horde in the traditional sense, but it is ordered by positive affectionate habits and by strength, that is, dominance. Both these principles continued into human society, and dominance based on personal achievement must have been particularly powerful in small groups living dangerous physical lives. The mother-young group certainly continued and the bonds must have been intensified by the prolongation of infancy. In human society economic reciprocity was added, and this created a wholly new set of interpersonal bonds.

When males hunt and females gather, the results are shared and given to the young, and the habitual sharing among a male, a female, and their offspring is the human family. According to this view the human family is the result of the reciprocity of hunting—the addition of a male to the mother-plus-young social group of the monkeys and apes.

This view of the family offers a reason for incest tabus. According to Sade, young sexually mature males do not mate with their mothers. How long this avoidance continues is not known at the present time, but at least some elements of the mother-son incest tabu are present among the non-human primates. This emphasizes the point mentioned above that the primate social group is not a horde, but is an orderly organization with principles that are seen to be both more lasting and more complex as the data become richer. If the function of the addition of a male to the group is economic, then the male who is added must be able to fulfill the role of a socially responsible provider. In the case of the hunter this necessitates a degree of skill in hunting and a social maturity that is attained some years after puberty. The necessary delay in the assumption of the role of provider for female and young can be achieved only by an incest tabu because brother-sister mating would result in an infant while the brother was still years away from effective social maturity. Father-daughter incest would produce a baby without adding a male; this is quite different from taking a second wife, which, if permitted, is allowed only for those males who have shown they are able to provide for more than one female.

To see how radically hunting changed the economic situation, it is necessary to remember that in the case of monkeys and apes an individual simply eats what it needs. After an infant is weaned, it is economically on its own and is not dependent on adults. This means that adult males never have economic responsibility for any other animal, and adult females have it only when they are nursing. In such a system there is no economic gain in delaying any kind of social relationship. But when hunting makes females and young dependent on the success of male skills, there is a great gain in not allowing the male to take the role of provider until he has proved his full effectiveness. These considerations in no way alter the importance of the incest tabu as a deterrent to role conflict in the family and as the necessary precondition to all other rules of exogamy. The more functions a set of behaviors fulfills the more likely such behaviors are to be widespread, and the rule of parsimony is completely wrong when applied to the explanation of social situations. However, these considerations do alter the emphasis and the conditions of the discussion of incest. First, the existence of mother-son incest in monkeys requires a different explanation from that for brother-sister or father-daughter incest. Incest is not a single problem, nor is the tabu to be accounted for in any one way. Second, the central consideration is that incest produces pregnancies, and the most fundamental adaptive value of the tabu is the provision of situations in which infants are more likely to survive. . . .

That family organization may be attributed to the hunting way of life is supported by ethnography. Since the same economic and social problems under hunting continued under agriculture, the institution continued. The data on the behavior of contemporary monkeys and apes also show why this institution was not necessary in a society in which each individual gets its own food.[4] Obviously the origin of the custom cannot be dated in the sense that we can prove *Homo erectus* had a family organized in the human way. But it can be shown that the conditions making the family adaptive existed at the time of *Homo erectus*. The evidence of hunting is clear from the archeological record. A further suggestion that the human family is old comes from physiology: the loss of estrus is essential to the human family organization. It is unlikely that this physiology, which is universal in contemporary mankind, evolved recently.

If the local group is looked upon as a source of male-female pairs (an experienced hunter-provider and a female who gathers and who cares for the young), then it is apparent that a small group cannot produce pairs regularly, since chance determines whether a particular child is a male or female. If the number maturing in a given year or two is small, then there may be too many males or females (either males with no mates or females with no providers). (The problem of excess females may not seem serious in our society or in agricultural societies, but among hunters it was recognized and was regarded as so severe that female infanticide was often practiced.) How grave the problem of imbalance can become is shown by the following hypothetical example: In a society of approximately forty individuals there might be nine couples. With infants born at the rate of about one in three years, this would give three infants per year, but only approximately one of these three would survive to become fully adult. The net production in the example would be one child per year. Because the sex of the child is randomly determined, all children might be male for a three-year period once every eight years, and similarly all might be female once every eight years. Smaller departures from a fifty-fifty sex ratio would be very common.

In monkeys, because the economic unit is the individual, not a pair, a surplus of females causes no problem. Surplus males may increase fighting in the group, or males may migrate to other groups.

---

[4] The advantage of considering both the social group and the facilitating biology is shown by considering the "family" of the gibbon. The social group consists of an adult male, an adult female, and their young. But this group is maintained by extreme territorial behavior in which no adult male tolerates another, by aggressive females with large canine teeth, and by very low sex drive in the males. The male-female group is the whole society. The gibbon group is based on a different biology from that of the human family and has none of its reciprocal economic functions. Although the kind of social life seen in chimpanzees lacks a family organization, to change it into that of man would require far less evolution than would be necessary in the case of the gibbon.

For humans, the problem of imbalance in sex ratios may be met by exogamy, which permits mates to be obtained from another group. The orderly pairing of hunter males with females requires a much larger group than can be supported locally by hunting and gathering, and this problem is solved by reciprocal relations between several local groups. It takes something on the order of 100 pairs to produce enough children so that the sex ratio is near enough to fifty-fifty for society to function smoothly, and this requires a population of approximately 500 people. With smaller numbers there will be constant random fluctuations in the sex ratio significant enough to cause social problems. This argument shows the importance of a sizable linguistic community, one large enough to cover an area in which people may find mates. It does not mean that either the large community or exogamy does not have many other functions, as outlined by Mair. As indicated earlier, the more factors that favor a custom, the more likely it is to be geographically widespread and long lasting. What the argument does stress is that the finding of mates and the production of babies under the particular conditions of human hunting and gathering favor both incest tabus and exogamy for basic demographic reasons.

Assumptions behind the argument are that social customs are adaptive, and that nothing is more crucial for evolutionary success than the orderly production of the number of infants that can be supported. This argument also presumes that, at least under extreme conditions, these reasons are obvious to the people involved, as infanticide attests. Or, as Whiting has mentioned . . . when a boy could find no suitable mate locally, his mother might well have suggested he try her brother's group.

If customs are adaptive and if humans are necessarily opportunistic, it might be expected that social rules would be labile under the conditions of small hunting and gathering societies. . . . Murdock has pointed out the high frequency of bilateral kinship systems among hunters, and experts on Australia all seem to believe that the Australian systems have been described in much too static terms. Under hunting conditions, systems that allow for exceptions and local adaptation make sense, and surely political dominance and status must have been largely achieved.

CONCLUSION

While stressing the success of the hunting and gathering way of life with its great diversity of local forms and while emphasizing the way it influenced human evolution, we must also take into account its limitations. There is no indication that this way of life could support large communities or more than a few million people in the whole world. To call the hunters "affluent," as Sahlins has, is to give a very special definition to the word. During much of the year many monkeys can obtain enough food in only three or four hours of gathering each day, and under normal conditions baboons have plenty of time to build the Taj Majal. The restriction on population, however, is the lean season or the atypical year, and, as Sahlins recognized, building by the hunters was limited by motivation and technical knowledge, not by time. Where monkeys are fed, population rises, and Koford estimates the rate of increase on an island at 15 percent per year.

After agriculture human populations increased dramatically in spite of disease, war, and slowly changing customs. Even with fully human (*Homo sapiens*) biology, language, technical sophistication, cooperation, art, the support of kinship, the control of custom and political power, and the solace of religion—in spite of this whole web of culture and biology—the local group in the Mesolithic was no larger than that of baboons. Regardless of statements . . . on the ease with which hunters obtain food, it is still true that food was the primary factor in limiting early human populations, as is shown by the events subsequent to agriculture.

The agricultural revolution, continuing into the industrial and scientific revolutions, is now freeing man from the conditions and restraints of 99 percent of his history, but the biology of our species was created in the long gathering and hunting period. To assert the biological unity of mankind is to affirm the importance of the hunting way of life. It is to claim that, however much conditions and customs may have varied locally, the main selection pressures that forged the species were the same. The biology, psychology, and customs that separate us from the apes—all these we owe to the hunters of time past. Although the record is incomplete and speculation looms larger than fact, for those who would understand the origins and nature of human behavior, there is no choice but to try to understand man the hunter.

BIBLIOGRAPHY

Aberle, David F., Urie Bronfenbrenner, Eckard H. Hess, Daniel R. Miller, David M. Schneider, and James N. Spuhler 1963 The incest taboo and the mating patterns of animals. American Anthropologist (n.s.), 65:253–265.

Binford, Lewis R., and Sally R. Binford 1966a The predatory revolution: a consideration of the evidence for a new subsistence level. American Anthropologist (n.s.), 68(2), pt. 1:508–512.

Bourlière, François 1963 Observations on the ecology of some large African mammals. In F. C. Howell and F. Bourlière (Eds.), African ecology and human evolution. Chicago: Aldine Publishing Company.

Brues, Alice 1959 The spearman and the archer, an essay on selection in body build. American Anthropologist (n.s.), 61:457–469.

DeVore, Irven, and Sherwood L. Washburn 1963 Baboon ecology and human evolution. In F. C. Howell and F. Bourlière (Eds.), African ecology and human evolution. Chicago: Aldine Publishing Company.

Freeman, Derek 1964 Human aggression in anthropological perspective. *In*: J. D. Carthy and F. J. Ebling (Eds.), The natural history of aggression. New York: Academic Press.

Goldschmidt, Walter R. 1959 Man's way: a preface to the understanding of human society. New York: Henry Holt.

Goldschmidt, Walter R. 1966 Comparative functionalism: an essay in anthropological theory. Berkeley and Los Angeles: University of California Press.

Goodall, Jane, and Hugo Van Lawick 1965 My life with wild chimpanzees. (16mm film). Washington, D. C.: National Geographic Society.

Hamburg, David A. 1963 Emotions in the perspective of human evolution. *In*: P. H. Knapp (Ed.), Expression of the emotions in man. New York: International University Press.

Koford, Carl B. 1966 Population changes in rhesus monkeys: Cayo Santiago, 1960–1964. Tulane Studies in Zoology, 13:1–7.

Lancaster, Jane B. The evolution of tool-using behavior; primate field studies, fossil apes and the archaeological record. American Anthropologist, Vol. 70, (1968) pages 56–66.

Lee, Richard B. 1965 Page 131. Subsistence ecology of Kung Bushmen. Unpublished doctoral dissertation, University of California, Berkeley.

Lorenz, Konrad Z. 1966 On aggression. Trans. by Marjorie K. Wilson. New York: Harcourt, Brace and World.

Mair, Lucy 1965 An introduction to social anthropology. Oxford: Clarendon Press.

Napier, John R. 1962 Monkeys and their habitats. New Scientist, 15:88–92.

Sade, Donald S. 1965 Some aspects of parent-offspring and sibling relations in a group of rhesus monkeys, with a discussion of grooming. American Anthropologist (n.s.), 23(1):1–17.

1966 Ontogeny of social relations in a group of free ranging Rhesus monkeys (Macaca mulatta Zimmerman). Unpublished doctoral dissertation, University of California, Berkeley.

Schaller, Geroge B. 1963 The mountain gorilla: ecology and behavior. Chicago: University of Chicago Press.

Service, Elman R. 1962 Primitive social organization: an evolutionary perspective. New York: Random House.

Slater, Miriam K. 1959 Ecological factors in the origin of incest. American Anthropologist (n.s.), 61:1042–1059.

Tax, Sol 1937 Some problems of social organization. *In*: Fred Eggan (Ed.), Social anthropology of North American tribes. Chicago: University of Chicago Press.

Vallois, Henri V. 1961 P. 222. The social life of early man: the evidence of skeletons. *In*: S. L. Washburn, (Ed.), Social life of early man. Chicago: Aldine Publishing Company.

Yamada, Munemi 1963 A study of blood-relationship in the natural society of the Japanese macaque. Primates (Journal of Primatology), 4:43–66.

Zeuner, F. E. 1963 A history of domesticated animals. New York: Harper and Row.

PART FOUR

HUMAN POPULATIONS
AND THEIR GENETIC
DIVERSITY

It is regrettable that, with the vital need for reliable information about the biological bases of differences among human populations, there is no substantive agreement among anthropologists as to even an elementary definition of the word *race*. No concept in physical anthropology has been the focus of more heated controversy than that of race. The record of the use of the term itself would almost constitute an intellectual history of anthropology.

Although the traditional approach that retains the concept of racial "types" is maintained by few American or Western European anthropologists, there are several other respectable positions on race.

The most widely accepted concept of race is illustrated by the population approach of Dobzhansky, who, as a population geneticist, is interested in problems of human evolution. He sees race as an array of populations with similar distributions of hereditary characteristics, differing from other similar arrays of populations within the species. This definition is based on the acceptance of the Mendelian population as the basic unit of evolutionary change. Hence a race is viewed as a collection of populations characterized not by the absolute presence or absence of some hereditary traits but by the relative incidence of certain hereditary traits.

This definition is flexible as to where the lines that separate the races are to be drawn. One could easily distinguish the extremes—populations with profoundly different distributions of traits like blond hair or very dark skin or heavy brow ridges—but the boundaries between races are much more tenuous. From this genetic point of view, they are not "real" boundaries, since populations are distributed continuously in all geographic areas, and a certain amount of gene flow between neighboring populations is assumed to occur. Stanley Garn, who has written extensively on race and has proposed several racial taxonomies, shares this point of view.

Another, more limited, opinion as to what constitutes race is that held by the serologist, William Boyd. Using a similar definition, he has restricted his classification of races to blood-group frequency. Using the distribution of the frequencies for the genes that determine the A-B-O-AB blood groups, Rh-factor compatibility, and M-N-MN blood groups, he has proposed a classification of large clusters of human populations that is in accord with the traditional Caucasoid-Negroid-Mongoloid Australoid-American Indian classification most nonanthropologists assume to be adequate. However, the layman's roughhewn taxonomy is based on "ideal" phenotype differences whereas Boyd's taxonomy rests exclusively, if narrowly, on those few characteristics for which the genetic basis is known and for which population distributions are well documented.

Julian Huxley, the British biologist, has proposed no classification nor constructed any list of races as such. Yet he does defend the reality of human races; he believes that at the close of Pleistocene there were three sharply demarcated stocks (corresponding to Caucasoid, Negroid, and Mongoloid) and that, although population expansion and migration have meanwhile attenuated the differences through hybridization, these three stocks are still discernible in modern populations.

The race concept is not taken for granted by all authorities, however, Ashley Montagu has suggested that, in view of the kind of thinking the term "race" has engendered—"racist" attitudes about how people should be treated based on their "racial" characteristics—the term should be dropped entirely, and the term "ethnic group" be substituted. For Montagu, ethnic group means a group distinguishable by the possession of biological and cultural characteristics.

Frank Livingstone has gone even further in questioning the validity of the concept of race. Beginning with a view quite different than that of Montagu, he argues that each of the traits that make up the traditional racial "type" are distributed differently, and that the race concept is worthless as a tool in organizing or classifying what we have learned about the differences among human populations, and how these differences have arisen.

There is no disputing that there are biological differences among human populations, and there is general agreement that genetic differences among populations are, for the most part, differences in frequency of genes, or, phenotypically, differences in the incidence of observable characteristics. The epicanthic fold of the eye, which in former days was said to be "typical" of the Mongoloid race, can be observed in low incidence among European populations, the Bushmen, and the Hottentot peoples. Its highest incidence is, of course, among East Asian populations. But what distinguishes Asian populations from other arrays of population is not the *absolute* presence or absence of the epicanthic fold, but rather its observed *high frequency* among them.

For the most part, the hereditary differences that anthropologists study probably fall into the category of relative frequency of traits that are generally present in most populations, for example, hemoglobins, haptoglobins, blood groups, and peppercorn hair form. Nevertheless, occasionally characteristics are encountered by which populations are capable of being described on an all-or-nothing basis—extremely dark pigmentation, blue eyes and blond hair, presence of lip seam, and others.

Investigation of the differences among human populations follows the lines laid down by commitment to the theory of evolution: the degree to which human populations differ from one another is the degree to which the evolutionary agencies of mutation and recombination, natural selection, gene flow, and the random fluctuation of gene frequencies in small isolated populations operate and interact to differentiate one population from another and to maintain such differences. Giles' article illustrates this genetic approach.

Since hereditary differences are the result of feedback to the gene pool of a population from forces and factors outside itself—that is, the environment within which the population must carry on life—considerable inquiry has been directed to such environmental factors as climate, intensity of solar radiation, altitude, humidity, mineral concentrations in soil, disease vectors, and the like. Accordingly, several of the articles that follow discuss attempts to account for population differences in terms of natural selection for traits that are adaptive to climatic and other environmental factors.

Admittedly, we are far from definitive answers to questions about the sorts of differences that we have been taught to regard as significant in our culture. The adaptive value of human skin pigment, for instance, is still an open arena for disagreement. An example of active research to illuminate the question is provided by the paper by W. F. Loomis.

For all human populations, however, the major environmental factor in the ecological niche is culture itself—the sum total of patterns of social life, symbolic transaction, and technology, all of which, transmitted from generation to generation, distinguish the human species from all others. The greater part of human evolution, as we have seen, has been evolution toward greater adaptive flexibility for cultural life.

In the past there has been considerable question (mostly by writers who had the answer in mind even as they asked) as to whether human populations are equally endowed with the hereditary basis for culture-building and cultural life. Some authorities such as Ashley Montagu and Dobzhansky believe that the phyletic evolution of man extends to the evolution of the capacity for culture and that no population has ever been separated from other populations long enough for significant differences in this potential to arise and to be maintained. Other authorities, such as Carleton Coon, have tried to build theories about human evolution that provide for the appearances at the sapiens threshold of different human populations at different times, leading to differences in cultural abilities.

The cultural component of human behavior needs no emphasis here, nor do physical anthropologists of any persuasion deny it. There is agreement that many of the differences in performance and ability among different population samples are obviously maintained in different cultural contexts. The major point of disagreement is the degree to which cultural differences and hereditary differences are responsible for population differences in behavior.

This question has long been dormant in anthropological circles. Cultural anthropologists have occupied themselves with exploring the variety of human cultures and salvaging the more exotic ones in monograph form. On the other hand, physical anthropologists have pursued their love affair with population genetics and gene counts among populations everywhere.

The current political atmosphere, coupled with the resurgence of American Black political and social activity, has had effects on the nature of statements concerning race. The political controversy is perhaps more responsible than anthropologists are willing to concede for spirited defenses of the concept that human populations do not differ significantly in their distribution of genes for ability, or at least, that such differences are not scientifically demonstrable.

Several approaches to the study of population differentiation are represented here: their common thread is the evolutionary base on which each is built as well as the acceptance in each case of the fact of human differences and the tentativeness of the theories that are advanced to explain these differences.

Loomis discusses the possible selective interplay of human skin pigment and solar radiation. The paper by Ginsburg and Laughlin discusses the problems of identifying interpopulational differences in behavioral capability, from the twin vantage points of psychology and anthropology. Baker discusses his studies of human adaptation to altitude. Bayless discusses the differences in human physiology distributed differentially among human populations, suggesting a new avenue of "raciation." Giles provides concrete examples of microevolutionary studies on contemporary populations.

If the race concept is to have any substantive meaning for anthropologists, it must be couched in terms that are evolutionary, populational, and of course, genetic. A set of static categories fitting a neat taxonomic chart is useless at best and misleadingly dangerous at worst. It is the *process* of population differentiation that should ultimately be illuminated.

14

CULTURE AND GENETICS

Eugene Giles

Physical anthropologists represent many diverse interests in the study of human biology. One of the most vigorous efforts is in the study of evolutionary mechanisms and their impact on contemporary human populations. Such studies have much in common with zoological and population genetic research on plant and animal populations. Mutation, natural selection, drift and gene flow are the processes which act to change the frequencies of genetically-based traits in populations. The anthropologists have conducted such studies on human populations. How has natural selection adapted a current population to the ecological niche which it occupies? How does this population differ from other populations? What influences are acting to transform this population? What may be predicted about the future genetic make-up of the population? These are all questions which are appropriate for study by anthropologists: micro-evolutionary studies which catch evolution "in the act," rather than attempts to reconstruct past biological events by analysis of the fossil evidence.

In the following article, the author discusses some typical research topics currently under way, which represent how anthropologists work and think.

Dr. Eugene Giles is Associate Professor of Anthropology at the University of Illinois.

Anyone who is or will be a physical anthropologist ought to, it seems to me, satisfy himself that his profession is or will be relevant. Whatever definition of relevance is applied, I think few would disagree that personal satisfaction is no longer sufficient to justify an active career in academic anthropology. But, I am afraid, a number of anthropologists seem to operate from self-satisfaction: archeologists who go on to another dig rather than prepare and defend analyses of material already gathered;

Reprinted from *Bulletin of the American Anthropological Association*, 3:3, 1970. By permission of the American Anthropological Association and the author.

biological anthropologists who collect one set of blood samples after another; or the social anthropologist who never quite let go of his data so that he alone commands academic knowledge of a particular population . . .

I . . . reflect—I hope in a fair, accurate and deliberate way—on the relation of cultural anthropology to biological anthropology. I am not talking here of a few regions of long-established overlap, such as aspects of archeology and osteology, but of central tendencies in both fields. In some departments of anthropology, the formal unity of the field seems to be breaking down. But at the same time other anthropology departments are adding biological anthropologists to their staffs, often for the first time. For students as well as faculty this raises questions. Some are practical—how long will the current academic sellers' market last? But others, more intellectual, intrude. Students are faced not only with the choice of courses in the anthropology spectrum over a limited graduate period, but also with shifting requirements, imposed or optional, evolved by their faculty. The importance of the faculty's view of anthropology grows, or should grow, in part out of their concern for students. In the following remarks, strictly from one biological anthropologist, I will try to persuade by example that anthropology is still an intellectually productive entity. In the recent past the field's integrity has benefited the biological anthropologist more than the social; it is just possible the exchange in the future may even out.

The anthropological geneticist, a term used by at least one other person (Derek Roberts), is in the broadest sense concerned with the genetic structure of peoples ordinarily falling under the anthropological tent. Genetic structure means the totality of factors ultimately affecting the population's genotype frequencies. When he is situated more directly under his genetic hat, the anthropological geneticist worries about those factors that make for stability and for change in genetic structure. Let's look at some of these factors, keeping in mind the question: what is the importance of culture here? Genetic change, as just about everyone must know these days, can have its effectors conveniently categorized under four headings: natural selection, mutation, migration or gene flow and genetic drift.

From the evolutionist's point of view, natural selection is by far the most important factor (although mutations provide the ultimate source of heritable variation). Selection is most often thought of as a mechanism for changing gene frequencies—and changing population gene frequencies through time *is* evolution—but it is also a great stabilizer. A recent article in *Science* (Ehrlich and Raven 1969) went to some pains to suggest that stability in animal populations, in a genetic-morphological sense, may well be due more to selection and less to gene flow than usually believed; that is, within-species genetic homogeneity may be maintained despite a restricted amount of intraspecies randomness in breeding. The dispersal of mankind is so complete that the single species status of man maintained by intraspecies gene flow—continually spreading the genes around—

boggles the mind. Just what is the genetic continuity between Eskimos and Bushmen, between Basques and Tasmanians? On the other hand, in extending his domain to encompass the terrestrial extremes of the planet, man obviously is exposing himself, as it were, to the most severe selective pressures available. What then maintains our human oneness? We can, of course, say that culture is our adaptive niche, everywhere. To me at least, this is easier to say than to conceptualize. And there is, after all, as biological anthropologists are intent on demonstrating, an enormous amount of genetic variation present in human populations. The parameters of culture's stabilizing selective activity—the underpinning of the catch phrase—is one of the most important, if difficult to analyze, interactions of genetics and culture.

Selection is more often thought of as promoting change. While this is the obvious interpretation of the fossil record, and is equally obvious in certain experimental treatment of animals, natural selection in man remains difficult to actually catch in the act. Suspicions abound but demonstrations are rare. One of the better established genetic traits upon which natural selection appears to be acting is a condition producing a deficiency in the activity of the enzyme glucose-6-phosphate dehydrogenase in human red blood cells. This condition, more conveniently known as G6PD deficiency, is produced by alleles at a locus on the X chromosome. The modern period of investigation of this trait goes back to the time after World War II when it was discovered that this particular enzyme deficiency was the cause of "primaquine sensitivity," a hemolytic anemia that occurred in some individuals each time they ingested certain antimalarial drugs. At about the same time it was discovered that a disease known as favism had the same etiology. People subject to favism undergo an attack of hemolytic anemia when they eat the fava or Italian broad bean (*Vicia faba*) or breathe its pollen. The bouts are temporary but can be severe: 250 deaths annually in Sardinia have been attributed to favism (Crosby 1956).

It seems a fair assumption (Giles 1962) that this unique endowment of the fava bean is the reason it has become so embedded in early Indo-European mythology, more so than any other plant or animal. The strange effect fava beans have on certain individuals, and only certain individuals, coupled perhaps with—to a naive observer—the apparently random appearance in family lines of an X-linked genetic trait, might easily be regarded as supernatural. If this is true, than the antiquity of the genetic anomaly in European populations can to some extent be gauged by datable references to the powers of the bean.

The establishment and persistence in human populations of what appears to be a disadvantageous gene in the case of G6PD deficiency is plausibly attributed to an increased fitness it must provide under malarial conditions—in other words, natural selection. Unambiguous evidence of the sort favored by medical investigators, such as statistically significant malarial death rates between "normals" and those G6PD deficient, is still lacking. But there is an impressive correlation between areas of known recent malarial endemicity and the prevalence of the G6PD deficiency alleles.

The fact that G6PD deficiency is one of the approximately 60 traits known to be on the human X chromosome, and one of the few traits on that chromosome important in terms of its frequency in the population, has added impetus to its investigation. Traits whose genes are found on the X chromosome do not seem to fall into any special morphological or physiological type—color blindness and hemophilia are the two best known—but they willy-nilly reflect the chromosomal basis of our sexuality. All females have two X chromosomes, males have one X and one Y (barring rare developmental abnormalities). A mother gives her offspring one of her X chromosomes, just as she gives one from each of the other 22 pairs of non-sex chromosomes, or autosomes. The father likewise gives the child one each from his 22 pairs of autosomes. But the father in addition determines the sex of the child. If the father contributes a Y chromosome, the child will be male, if it gets the X it will be female. Thus there is no way for a son to inherit his father's X chromosome and no way for his daughter to avoid it. This accounts for the peculiar inheritance pattern of X-linked traits. In the more common example, a color-blind father cannot have a color-blind son, since the trait is on his X chromosome. All his daughters will have the color-blind gene, but because it is recessive it will be masked in its effect by the normal gene for color vision they received from their mother. The sons of these daughters, however, have a 50-50 chance of being color-blind. If they received the X chromosome with the color-blind gene they have no other X and thus no normal gene to mask it (they are *hemizygous* for the trait) and will express it. For an analogous recessive autosomal trait, two mutant genes would have to be present (a *homozygous* condition), one in each chromosome of the pair, for the character to be manifested. With only one such mutant gene in an autosomal pair of chromosomes (a *heterozygous* condition), the single normal allele would suffice to produce the normal condition. (If the mutant and its normal allele are not in a dominant-recessive relationship, then the heterozygous condition will differ from both the homozygous normal and the homozygous mutant state.)

A genetic hypothesis proposed early in this decade by Mary Lyon has become quite well established (incidentally, in large part by experimentation involving the G6PD deficiency). The Lyon hypothesis is that at some early embryonic stage in females one of the two X chromosomes—which one is a random matter—becomes deactivated, and although it "goes through the motions" in cell divisions, the deactivated X no longer contributes its message to cellular activities. Recent investigations (Luzzatto,

Usanga and Reddy 1969) in Nigeria have taken advantage of one consequence of the Lyon hypothesis. Females who are heterozygous for G6PD
deficiency, that is to say, have one X chromosome carrying a normal gene
and one X carrying a G6PD deficiency allele, produce two kinds of red
blood cells, normal and enzyme deficient, in about equal numbers. (A
heterozygote for such a trait which was located on an autosome would be
expected to produce red blood cells all of one kind, whatever that kind
might be. If, for example, the normal allele was dominant to the mutant
allele, then the red cells would all be the same as those produced by a
heterozygote normal; if it was not a dominant, then they would all be
different. The important thing is that the expectation is the cells will all be
alike, not some with the trait and some without in the same individual.)
The normal red blood cells in such a heterozygous individual are presumably produced by the descendants of cells in which the X chromosome
carrying the deficiency allele was the one deactivated. All cells in these
lines will be normal because the active X chromosome carries a normal
gene. The deficient red blood cells are produced by descendants of cells in
which by chance the other X chromosome was deactivated, leaving the
X chromosome with the G6PD deficiency allele to direct production of deficient red blood cells.

In the heterozygous females investigated in Nigeria, all of whom were
suffering from acute falciparum malaria, the individual normal red blood
cells had from 2 to 80 times the parasite infection of the G6PD deficient
red blood cells. If these results are sustained, we may presume that under
malarial conditions natural selection favors the presence of a G6PD deficiency allele, at least up to that population frequency at which the benefit
conferred upon heterozygous females is balanced by the disadvantage to
the poor males, who with their single X chromosome either have all
normal red blood cells or all deficient—and if the latter, are subject to
attacks of hemolytic anemia.

G6PD deficiency suggests for one fairly well-studied trait quite intricate
interactions between genetics and culture, with benefits to students of both
fields from cognizance of the other. What we do know about G6PD deficiency, the sickle-cell trait and others implies that not only certain single-
gene characters, but many inherited differences among human populations, including those with a polygenic base, may have a corresponding
interrelationship with culture.

Mutation is, of course, the basis for G6PD variants and for all other
heritable variation. For anthropology the significance of genetic variants
that have become established in a population is often taken not in their
putative relationship to selective pressures, but rather in the presumed
absence of these that allows the variants to be treated as markers in
population movements. Anthropology has a long history of seeking such
features; the current fashionable crop consists mainly of polymorphic

traits known to be controlled by one gene. Blood groups are exhibit A. A typical blood group system (like the ABO) is produced by two or more alleles at one locus on the chromosome. Although in Western European populations there are four common alleles, A_1, A_2, B and O, no individual can carry more than two of these alleles since this is a system controlled by a single gene. Try as they may, an A_1B father (genotype A_1B) and an O mother (genotype OO) can have neither A_1B nor O children, but only A_1 (genotype A_1O) or B (genotype BO). Not all populations are polymorphic for the same traits: it is generally conceded that South American Indians originally had only the O gene in the ABO system. Summing this up, we can say that for anthropological purposes polymorphic means the character exists in two or more forms, produced by two or more alleles at the same locus, in at least some populations at allele frequencies sufficiently high (usually 1 percent or more) to rule out merely repetitive mutation.

Biology as well as cultural anthropology can be used to delineate the migrations of preliterate populations. Take the possible connection between Melanesia and Black Africa. The connection meant here is that an already Negroid African population migrated from Africa to Melanesia either around or through some other population occupying the area in between; it does not mean that Melanesians and other *Homo sapiens* cannot ultimately trace their origins to Africa. The superficial similarities are patent in skin color, hair form, nose shape and the like. "These are our brothers" was the response in Nairobi when New Guineans first visited East Africa (Mboya 1965). The connections have been denied, however, on the basis of genetic traits, among others the high frequency of the Rh blood group allele R_0 (cDe) in Africa and the reverse in Melanesia.

Single-gene traits do not provide clear-cut evidence for or against the Africa-Melanesia connection. In the serum fraction of human blood there are a number of proteins with specialized functions which, like the antigens on red blood cells, have inherited variants. One such is called transferrin, an iron-binding glyco-protein acting to transfer iron where it is needed in the body. Some 17 single-gene variants of normal transferrin, called transferrin C, have been found. The variants are detected by the speed at which they move during electrophoresis. In electrophoresis the protein is subjected, in a very standardized way, to an electric charge; the overall charge of the protein molecules, which reflects internal structure, determines how far towards the positive or negative poles of the electric field the protein will migrate during a specified period of time.

Transferrin variants that move faster than the normal transferrin C are labeled B, those that move more slowly D. One of the D variants, D_1, is especially interesting because it occurs in population frequencies apparently only in Africans (and consequently in American Blacks) and in

Melanesians and indigenous Australians. This would seem to be a positive and significant link between African and Oceania, but there are reasons for exercising caution in such interpretations. One is that, although electrophoresis is a sensitive means for detecting molecular variation, the technique does not preclude two types of protein of different composition ending up with the same overall electric charge, or charges so similar that the differences cannot be observed. Hence the transferrin D_1 in the two groups may not be the same in every detail. The structural difference between transferrin C and D_1 is precisely known, however, so this argument does not apply here. D_1 differs by only one amino acid from transferrin C, specifically a substitution of glycine for aspartic acid, and furthermore the same amino acid substitution is found when D_1 of American Blacks is compared with the D_1 of indigenous Australians (Wang, Sutton and Scott 1967).

Even though the similarity of transferrin D_1 in the two regions is exact down to the level of amino acids, the building-blocks of protein, it is not possible yet to be sure that the genes directing the synthesis of the protein are identical. We cannot begin to review the mechanisms of inheritance uncovered by molecular biologists just for this one point, but here is the problem very briefly. Deoxyribonucleic acid (DNA) is the genetic information system in the chromosome. One aspect of the structure of DNA permits an almost infinite information storage potential: in the DNA molecule four bases, cytosine (C), guanine (G), adenine (A) and thymine (T), are linked together in some linear order, like ATGGACTAGGCTTTACGA. The information DNA has to "remember" is the amino acids and their sequence necessary to form a specific protein. First the coding problem. There are only 20 amino acids. Obviously, if each base coded for one amino acid, say G for valine, T for lysine, etc, DNA could only "remember" four. If two bases coded for each amino acid, then 4^2 or 16 amino acids could be taken care of, eg, AA = valine, AT = lysine, AG = alanine. It turns out that 3 bases, or triplets, are used to code for the amino acids: $4^3 = 64$, more than enough to accommodate the 20 amino acids. All codings for the 20 amino acids are now known. Putting the amino acids in sequence is easy: just read the code from one end of the DNA segment to the other. Thus, the imaginary DNA segment above might better be written ATG GAC TAG GCT TTA CGA. These triplets, left to right, code for the following sequence of amino acids: tyrosine-leucine-isoleucine-arginine-asparagine-alanine.

That there are too many codes for the number of amino acids is the problem. The "degeneracy" of the genetic code refers to the fact that more than one sequence of three bases—one triplet—codes for each amino acid. The single amino acid, glycine, that provides all the difference between transferrin D_1 and normal transferrin C can be coded for in no fewer than four ways: CCA, CCG, CCT and CCC. One could assume that glycine's appearance is the consequence of a mutation from aspartic acid in that precise position in normal transferrin C molecular structure. Aspartic acid is coded for by one of two triplets: CTA or CTG. At present it is unknown which codon represents glycine in transferrin D_1, or whether it is the same one in both African and Australian populations. A single base substitution of cytosine for thymine in either CTA or CTG would result in CCA or CCG, both of which would then code for the amino acid glycine.

But even if the DNA structure were known to be precisely the same for both Africa and Australia, the very real possibility exists that quite independent mutations and subsequent selective advantage of transferrin D_1 led to the establishment of the variant in both regions. Unfortunately, nothing is known about the relative fitness of individuals carrying transferrin variants. All of this should lead, I think, to the realization by cultural anthropologists that anthropological geneticists have their own problems in using biological data to infer population relationships. But often overall probabilities, based on examination of many genetic traits, can be strongly supportive of, even if not conclusive about, one hypothesis over another.

Professor Pollitzer has so clearly discussed anthropological aspects of gene flow that I will merely commend his comments to you and proceed to a final means of genetic change—drift. Genetic drift subsumes all the non-directional (stochastic, random) changes in population gene frequencies. You should not let the term "drift" mislead you into thinking there is anything directional about genetic drift—there isn't. The proportion of gene frequency variation between populations that should be attributed to a non-directional mechanism is not something that all geneticists agree on. Currently, for human populations, there is a resurgence of interest in drift as an explanation of a significant portion of interpopulation variation. This is a change from the time when the whole topic was in ill repute. Perhaps it is more cautious to say that now there is much less assurance that observable variation is evidence for on-going natural selection. Much of the argument for significant stochastic processes at work is theoretical; there is also from anthropological investigations much evidence which seems most reasonably interpreted as genetic drift.

One example I am particularly familiar with occurs in New Guinea. As you know, there is a fantastic linguistic diversification in New Guinea, with well over 800 separate languages on the island. The people speaking one of these, called Waffa, were the object of a genetic survey I conducted in 1962-63 and in 1968 under NIH sponsorship. The people lived in three large villages and two very small ones—the total population was just under 1,000. Blood samples were obtained in the three big villages, Kusing, Tumbuna and Siaga, which together constitute 88 percent of Waffa speakers. The proportion tested was 64 percent of the total census population of the three villages, or 56 percent of all the Waffa speakers in the world. There can be no reasonable doubt then that the data gathered give an adequate representation of Waffa speakers—in fact the problem lies more in the proper analytical approach to such material when most statistical methods for comparison assume random sampling rather than incomplete enumeration. Table 14-1 presents gene frequencies and sample size for each of the three major Waffa villages for three common blood group systems (Giles, Wyber and Walsh 1970). If we assume the frequencies are based on random samples, a procedure admittedly open to some question though unlikely to be too far off, then there are statistically significant differences among these 3 villages in each blood group system. Consider, for instance, the fluctuation in frequency for the allele B in the ABO system: in one village it is 27 percent, in another 14 percent, and in the third 2 percent. The problem here is obvious without statistics.

I should think that a preliterate group possessing a language distinct enough from neighboring languages to be called a separate language by linguists is a fair indication of a cultural entity persisting through time. The Waffa speakers regard themselves as such, live in one area, and interchange genes occasionally between village groups. There is no apparent cultural variation within the group defined by language, and it is equally difficult to see any environmental differences. Yet the fluctuation in the frequencies of the ABO blood group system from village to village within this small, bounded population is relatively enormous. It should give anyone pause when they feel impelled to characterize a whole region, or even the whole island of New Guinea (Cavalli-Sforza, Barrai and Edwards 1964), by results derived from what appears to be a quite generous sample of an unacculturated group—in this case any one of the 3 villages. Whether the allele B frequency would be reported as 27 percent,

TABLE 14-1

Waffa Language Blood Group Frequencies

		KUSING	TUMBUNA	SIAGA
Allele	A	.104	.131	.076
	B	.273	.143	.025
	O	.623	.725	.899
	R_1	.928	.938	.962
	R_2	.005	.004	.032
	R_0	.067	.058	.007
	Ms	.005	.046	.038
	Ns	.925	.824	.921
	NS	.070	.130	.041
Sample Size		215	195	143

14 percent or 2 percent would depend entirely on which village happened to be tested. In 1962-63 I thought I had a good idea of the Waffa from testing the two villages with frequencies of 27 percent and 14 percent. In 1968 I found out how wrong I was.

Studies such as these are strong indicators of genetic drift, but they do not "prove" it, nor at this level of analysis do they contribute much to a partitioning of all genetic variation causality into its constituent effectors. Beginning with Sewell Wright in the 1940s genetic models have been developed to aid in elucidating situations in which genetic drift effects are partially mitigated by gene flow consequent upon migration. All of these models have been recently reviewed by Bodmer and Cavalli-Sforza (1968). The simplest of these dealt with a situation in which each of a number of distinct populations received a set fraction m of its genes from a common gene pool with a constant frequency. Distance between the "islands" was not taken into account, and in effect each island exchanged equally with every other island. This "island model" has very serious limitations when application is attempted for man.

A modification of the "island model" was made to take into account distance between populations; this was called the "isolation by distance model." Mobility is defined in terms of a continuous distribution, like the normal distribution, to measure genetic exchange.

A further development towards reality in model building has been termed the "stepping stone model." Here the population is distributed in "colonies" of equal size with given rates of genetic exchange based on the number of steps apart. A simple example would be one in which each colony has two neighbors and exchanges a fraction $m/2$ genes with each.

The development of these models has proved theoretically interesting but it should be clear from this very brief description that population genetics models do not approach human reality with the fidelity one could hope for. After all, real human populations are irregualar in size, irregular in exchange of genes, and irregular in all of these things over time.

In the same paper Bodmer and Cavalli-Sforza have presented a "migration model" which goes further than any of the others towards an approximation of human practices. Their model aims at predicting the amount of variation expected between gene frequencies of a finite number of colonies of different sizes subjected to a specifiable pattern of intercolony migration. The interesting aspect of this model is the way in which the buffering affects of migration on genetic drift are handled. Migration is analyzed as a matrix with k rows and k columns where k represents the number of colonies. It is a help if the colonies form a circle. Luckily I happen to have data on just such a situation in New Guinea where 14 villages (all of the same language) can be substituted for the colonies of the model. The migration matrix represents the displacement over one generation among the k colonies. Table 14-2 should make clear what I mean when I say that

TABLE 14-2

Onga Migration Data
Number of Living Children Produced by Mothers from each Village in each Village

BIRTHPLACE OF OFFSPRING	BIRTHPLACE OF MOTHER: Gnarowein	Yanuf	Guruf	Itsingants	Yatsing	Puguap	Intoap	Singas	Awan	Onga	Naruboin	Siats	Antir	Bampa
Gnarowein	56	6	0	0	0	0	0	0	0	0	0	0	3	2
Yanuf	16	15	4	0	0	0	0	0	0	0	0	0	0	0
Guruf	0	1	76	2	0	0	0	0	0	0	0	0	0	2
Itsingants	0	0	3	31	4	1	0	0	0	0	0	0	6	0
Yatsing	0	0	0	4	15	0	0	0	0	0	0	0	0	0
Puguap	0	0	1	0	3	53	5	0	1	3	0	0	1	0
Intoap	0	0	0	0	0	1	34	6	5	0	0	0	0	0
Singas	0	0	0	0	0	0	3	22	5	0	2	5	0	0
Awan	0	0	0	0	2	1	1	6	56	17	6	0	0	0
Onga	0	0	0	0	0	0	2	6	8	107	24	0	0	0
Naruboin	0	0	0	0	0	0	2	2	0	33	37	3	5	0
Siats	0	0	0	0	0	1	0	0	0	5	5	33	7	0
Antir	3	0	0	8	0	11	0	0	1	0	0	8	78	4
Bampa	0	0	2	0	0	0	0	0	0	0	0	0	9	9

the matrix elements represent the number of children born in the *i*th village to parents born in the *j*th village. Table 14-2 represents data collected from the Onga district in northeastern New Guinea concerning the birthplace of mothers and children. I introduce it here only for illustrative purposes, to show one way in which cultural data are crucial for genetic analysis.

In some cases it may be wise to form one matrix from the birthplaces of both fathers and mothers on one hand and their offspring on the other. In any event from such matrices new stochastic matrices can be produced (in these rows or columns sum to one) which represent the transition probabilities from one generation to the next. These have been labeled "backward" stochastic matrices where the transition is against time, ie, the birthplaces of the parents (rows) sum to one, and as "forward" where the transition is with time, ie, the birthplaces of the offspring (columns) sum to one. One can say, for example, that in a "forward" matrix we have the probability that children of mothers born in village *j* go to village *i*. This distinction may not be clear from such a condensed exposition; I mention these manipulations here only to point out that the so-called "backward" matrix is involved in predicting the variation in gene frequencies within and between villages, and the "forward" matrix, projected through time, will eventually reach equilibrium. If the relative sizes of villages in the equilibrium state differ significantly from those observed, then in the past migration must have differed from the present.

I think this is quite enough detail to give you an idea of the approach. The migration model comes nearer to what we would like to think of as reality far more than its predecessors, but it still leaves things to be desired, e.g., mixed generations rather than requiring everybody to reproduce at the same time (overlapping vs. non-overlapping generations). However, it does take into account in a reasonable way migration, and it has provision for immigration from outside the particular set of villages under consideration by adding an extra column and row for this purpose.

An alternative approach to mathematical models, and one which may ultimately be more satisfactory, involves computer simulation to give numeric answers to specific problems rather than explicit mathematical formulations (Schull and Levin 1964:179). Simulation permits the simultaneous consideration of many parameters, directional (such as natural selection) as well as stochastic. It may well turn out that while the difficulties of the models lie in the necessary simplifying assumptions, the real problem in computer simulation will be in the quality of the demographic data used for the explicit numeric results.

We have thus worked our way around in our consideration of genetic drift to another culture/genetic interfact: the adequacy of the available cultural data for small populations in terms of the anthropological geneticist's needs. In the same way in which current studies of hunter-gatherer ecology or primate behavior can, with caution, provide insight regarding our evolutionary progress, so investigation of people living in fashions best approximating the mating patterns of our long precosmopolitan history are likely to be the best base for extrapolation to the relative importance of the different genetic mechanisms in our evolution. The input for a computer simulation of gene behavior can utilize quite specific information on such population features as age, reproductive span, age-specific mortality, directional aspects of marriage patterns, age differential at marriage, fertility, fecundity and much more. These are the sorts of data that would be obtained most reliably by a cultural anthropologist. Many are affected by cultural practices; the accuracy of all is likely to benefit from the rapport a social anthropologist builds up in the field.

Some of the population parameters simply have to be guessed if they are not known. Age, for example, is so important for any analysis of the sort we are talking about here that in practice age will be recorded even if it can only be estimated by the investigator from the subject's appearance. A physical means for independently assessing the age of an individual would be a wonderful thing, worth at least ten new genetic polymorphisms. So far there is no reliable method. Tooth eruption and epiphyseal union (the latter requiring X-rays) or the like provide estimates, but for real reliability the standards developed among cosmopolitan Europeans must be validated on the local populations. People who feel certain that they can judge age by appearance have usually developed this talent over a long period notably devoid of actual checking. I am rather skeptical of the use of remembered events which have a date attached to them: the big drought, the plane crash, World War II events, and this sort of thing. It sounds good in theory, and using this method may have merit, but there is absolutely no check on the reliability of the informant's memory, nor on the year equivalent of being "so tall" (hand above the ground) when such and such an event occurred. It seems quite possible that prestige of sorts may accrue to persons claiming to have been alive during exciting events: I have heard of two persons in one small group given ages over 100 because of some connection with happenings during Queen Victoria's reign. With other aspects of human biology the situation can be much worse. A sobering study (Potter, Wyon, New and Gordon 1965) in India's Punjab demonstrated that the actual abortion rate, as determined by monthly home visits, was more than 5 times that calculated from asking the women about it. I do not mean by all this to suggest that indirect ascertainment should be abandoned—far from it. But the need for better techniques, the defects of those in use, and the probable error incorporated into results based on them, should all be kept well in mind.

A sustaining thought in difficult situations is that there is something about the very nature of anthropologically interesting populations that makes accuracy in demographic parameters hard to attain, thus blunting criticism of estimates. Civil record-keeping in developing countries is usually bad when it is attempted despite the best intentions of the governments involved. For some situations in anthropology even the *possibility* of accurate records means a level of acculturation too high to be acceptable for the research design. The fact that mission birth records, for example, are available for certain segments of a population may in fact mean that that subgroup is no longer representative of the population as a whole.

Genetic heterogeneity in human populations has been amply documented. The current difficulties lie in interpreting the inter- and intrapopulation variation. I hope I have been able to show you just how valuable the interaction between culture and genetics is, and particularly, as students of the phenomena, indicate how dependent on largely cultural matters anthropological geneticists are in dealing with their biological material. Obviously a biological anthropologist is more likely to stress the ways in which cultural factors impinge upon genetic mechanisms of change. How does genetic structure impinge on social structure? This is not the same as asking if there are any genetic variations that have cultural concomitants. Clearly there are: skin color to name one, and G6PD deficiency. And there *may* be population variants in behavior (or at least perception) that *may* be attributable to genetic differences between populations. Cross-cultural evidence with regard to certain optical illusions can be viewed this way (Spuhler and Lindzey 1967). But as for social structure, I see no evidence yet of a feed-back from the genetic structure of a population. This may come, but today it is those of us interested in population genetic structure that must be concerned with social structure rather than vice-versa.

REFERENCES

Bodmer, W. F., and L. L. Cavalli-Sforza   1968   A migration matrix model for the study of random genetic drift. *Genetics* 59:565–92.

Cavalli- Sforza, L. L., I. Barrai, and A. W. F. Edwards 1964 Analysis of human evolution under random genetic drift. *Cold Spring Harbor Symposia on Quantitative Biology* 29:9–20.

Crosby, W. H.   1956   Favism in Sardinia. *Blood* 11:91–92.

Ehrlich, P. R., and P. H. Raven   1969   Differentiation of populations. *Science* 165:1228–32.

Giles, E.   1962   Favism, sex-linkage, and the Indo-European kinship system. *Southwestern Journal of Anthropology* 18:286–90.

Giles, E., S. Wyber, and R. J. Walsh   1970   Micro-evolution in New Guinea: additional evidence for genetic drift. *Archaeology and Physical Anthropology in Oceania* 5:60–72.

Luzzatto, L., E. A. Usanga, and S. Reddy   1969   Glucose-6-phosphate dehydrogenase deficient red cells: resistance to infection by malarial parasites. *Science* 164:839–42.

Mboya, T.   1965   These are our brothers. *New Guinea* 1(1):11–13.

Potter, R. G., J. B. Wyon, M. New, and J. E. Gordon   1965   Fetal wastage in eleven Punjab villages. *Human Biology* 37:262–73.

Schull, W. J., and B. R. Levin   1964   Monte Carlo simulation: some uses in genetic study of primitive man. In *Stochastic models in Medicine and Biology*. J. Gurland, ed. Madison: University of Wisconsin Press.

Spuhler, J. N., and G. Lindzey   1967   Racial differences in behavior. In: *Behavior-genetic analysis*. J. Hirsch, ed. New York: McGraw-Hill.

Wang, A. C., H. E. Sutton and I. D. Scott   1967   Transferrin D_1: identity in Australian Aborigines and American Negroes. *Science* 156:936–37.

15

SKIN-PIGMENT REGULATION
OF VITAMIN-D BIOSYNTHESIS
IN MAN

W. Farnsworth Loomis

Have differences in ultraviolet solar radiation served as selective agents differentiating human populations with respect to skin color? For almost 40 years evidence has been accumulating which strongly suggests that the body's requirements for vitamin D and the intensity of solar radiation are both involved in the selection of dark-skinnedness in tropical areas, and of depigmented human skin in northern areas. In this paper, a biochemist reviews the physiological and anthropological evidence for the role of natural selection in determining human skin color. Russell Newman's paper on heat stress resistance in human evolution comes to a similar conclusion about the timing of man's loss of his hairy coat, but on the basis of different sorts of evidence.

The author of the following paper, W. Farnsworth Loomis, is Professor of Biochemistry at Brandeis University in Waltham, Massachusetts.

Vitamin D mediates the absorption of calcium from the intestine and the deposition of inorganic minerals in growing bone; this "sunshine vitamin" is produced in the skin, where solar rays from the far-ultraviolet region of the spectrum (wavelength, 290 to 320 millimicrons) convert the provitamin 7-dehydrocholesterol into natural vitamin D (1) (Figure 15-1).

Unlike other vitamins, this essential calcification factor is not present in significant amounts in the normal diet; it occurs in the liver oils of bony fishes and, in very small amounts, in a few foodstuffs in the summer (see Table 15-1). Almost none is present in foodstuffs in winter.

Reprinted from *Science*, Vol. 157, August 4, 1967, pages 501–506. Copyright 1967 by the American Association for the Advancement of Science. By permission of the author and the publishers.

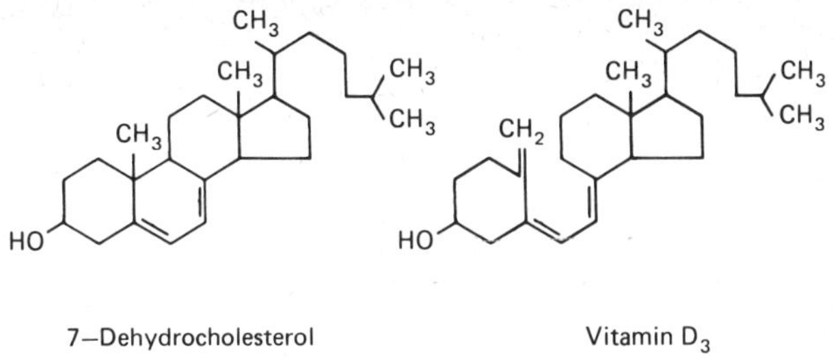

7–Dehydrocholesterol   Vitamin D_3

Fig. 15-1   Chemical structures of 7-dehydrocholesterol and vitamin D_3.

TABLE 15-1

Vitamin-D content of two fish-liver oils and of the only foodstuffs known to contain vitamin D.

FISH-LIVER OIL OR FOODSTUFF	VITAMIN-D CONTENT (I.U./GRAM)
Halibut-liver oil	2000–4000
Cod-liver oil	60–300
Milk	0.1
Butter	0.0–4.0
Cream	0.5
Egg yolk	1.5–5.0
Calf liver	0.0
Olive oil	0.0

From K. H. Coward, *The Biological Standardization of the Vitamins* (Wood, Baltimore, 1938), p. 223.

Chemical elucidation of the nature of vitamin D has made it possible to eradicate rickets from the modern world through artificial fortification of milk and other foods with this essential factor. Before this century, however, mankind resembled the living plant in being dependent on sunshine for his health and well-being, a regulated amount of vitamin D synthesis being essential if he were to avoid the twin dangers of rickets on the one hand and an excess of vitamin D on the other.

Unlike the water-soluble vitamins, too much vitamin D causes disease just as too little does, for the calcification process must be regulated and controlled much as metabolism is regulated by the thyroid hormone. The term *vitamin D* is, in fact, almost a misnomer, for this factor resembles the hormones more closely than it resembles the dietary vitamins in that it is not normally ingested but is synthesized in the body by one organ—the skin—and then distributed by the blood stream for action elsewhere in the body. As in the case of hormones, moreover, the rate of synthesis of vitamin D must be regulated within definite limits if both failure of calcification and pathological calcifications are to be avoided.

Synthesis of too little vitamin D results in the bowlegs, knock-knees, and twisted spines (scoliosis) associated with rickets in infants whose bones are growing rapidly. Similar defects in ossification appear in older children and women deprived of this vitamin; puberty, pregnancy, and lactation predispose the individual toward osteomalacia, which is essentially adult rickets. In osteomalacia the bones become soft and pliable, a condition which often leads to pelvic deformities that create serious hazards during childbirth. Such deformities were common, for example, among the women of India who followed the custom of purdah, which demands that they live secluded within doors and away from the calcifying power of the sun's rays (2). Cod-liver oil or other source of vitamin D is a specific for rickets and osteomalacia, the usual recommended daily dosage being 10 micrograms of 400 international units (1 I.U.=0.025 microgram of vitamin D).

Ingestion of vitamin D in amounts above about 100,000 I.U. (2.5 milligrams) per day produces the condition known as hypervitaminosis D, in which the blood levels of both calcium and phosphorus are markedly elevated and multiple calcifications of the soft tissues of the body appear. Ultimate death usually follows renal disease secondary to the appearance of kidney stones (3). Although this condition has been described only in patients given overdoses of vitamin D by mouth, similarly toxic results would probably follow the natural synthesis of equal doses of vitamin D by unpigmented skin exposed to excessive solar radiation. The body appears to have no power to regulate the amount of vitamin D absorbed from food and no power to selectively destroy toxic doses once they have been absorbed. These facts suggest that the physiological means of regulating the concentration of vitamin D in the body is through control of the rate of photochemical synthesis of vitamin D in the skin.

It is the thesis of this article that the rate of vitamin-D synthesis in the stratum granulosum of the skin is regulated by the twin processes of pigmentation and keratinization of the overlying stratum corneum, which allow only regulated amounts of solar ultraviolet radiation to penetrate the outer layer of skin and reach the region where vitamin D is synthesized. According to this view, different types of skin—white (depigmented and dekeratinized), yellow (mainly keratinized), and black (mainly pigmented)—are adaptations of the stratum corneum which maximize ultraviolet penetration in northern latitudes and minimize it in southern latitudes, so that the rate of vitamin-D synthesis is maintained within physiological limits (0.01 to 2.5 milligrams of vitamin D per day) throughout man's worldwide habitat.

Figure 15-2 provides evidence in support of this view, for it is apparent

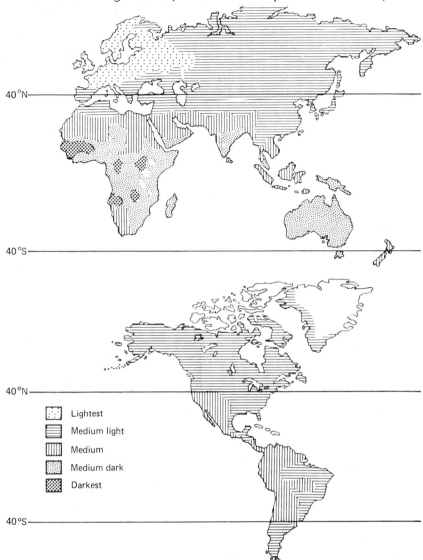

Figure 15-2   Distribution of human skin color before 1492. (Adapted from Brace Montague, *Man's Evolution* (Macmillan, New York, 1965), p. 272.

that there is a marked correlation between skin pigmentation and equatorial latitudes. In addition, the reversible summer pigmentation and keratinization activated by ultraviolet radiation and known as suntan represents a means of maintaining physiologically constant rates of vitamin-D synthesis despite the great seasonal variation in solar ultraviolet radiation in the northern latitudes.

ULTRAVIOLET TRANSMISSION AND VITAMIN-D SYNTHESIS

In 1958 Beckemeier (4) reported that 1 square centimeter of white human skin synthesized up to 18 I.U. of vitamin D in 3 hours. Using this figure, we calculate that an antirachitic preventive dose of 400 I.U. per day can be synthesized by daily exposure of an area of skin approximately equal to that of the nearly transparent pink cheeks of European infants (about 20 square centimeters). Perhaps this explains why mothers in northern climates customarily put their infants out of doors for "some fresh air and sunshine" even in the middle of winter.

From this high rate of synthesis by only a small area of thin unpigmented skin, one can calculate the daily amount of vitamin D that would be synthesized at the equator by the skin of adults who exposed almost all their 1½ square meters (22,500 square centimeters) of body surface during the whole of a tropical day. Such a calculation shows that the skin of such individuals would synthesize up to 800,000 I.U. of vitamin D in a 6-hour period if the stratum corneum contained no pigment capable of filtering out the intense solar ultraviolet radiation.

Direct evidence that pigmented skin is an effective ultraviolet filter was provided by Macht, Anderson, and Bell (5), who used a spectrographic method to show that excised specimens of whole skin from Negroes prevented the transmission of ultraviolet radiation of wavelengths below 436 millimicrons, while excised specimens of white skin allowed radiation from both the 405- and the 365-millimicron bands of the mercury spectrum to pass through.

These early studies with whole skin were refined by Thomson (6), who used isolated stratum corneum obtained by blistering the skin with cantharides. He found that the average percentage of solar radiation of 300- to 400-millimicron wavelength transmitted by the stratum corneum of 22 Europeans was 64 percent, while the average for 29 Africans was only 18 percent. There was no overlapping of values for the two groups (Figure 15-3), but there was considerable variation within each group, the values for the Europeans varying from 53 to as high as 72 percent and those for the Africans (who were mainly Ibos but also included men from most of the Nigerian tribes) varying from 36 to as low as 3 percent.

In his careful studies, Thomson measured skin thickness as well as pigmentation and found that the former was a minor variable. Studies on the degree of blackness of the various African specimens were made by skin-reflectance measurements. These showed that the darker the skin is, the lower is the percentage of ultraviolet radiation transmitted. One specimen from an albino African showed transmission of 53 percent—a value within the range for the European group. Thomson concluded from these

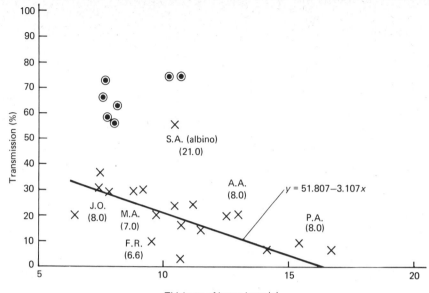

Fig. 15-3   Variation of transmission of solar ultraviolet light (3000 to 4000 angstroms) through the stratum corneum, plotted against thickness of this layer. ⊙, Europeans; ×, Africans. The numbers in parentheses after initials are percentages for reflectance of blue light on the forearm. [From M.L.T. Thomson (6)]

studies that skin pigmentation was mainly responsible for protecting the African from excessive solar ultraviolet radiation, the thickness of the horny layer in Africans playing only a minor role. Thomson did not mention the fact that skin pigmentation and thickening of the horny layer in Africans would protect against excessive vitamin-D synthesis as well as sunburn.

Thomson's results indicate that African stratum corneum filters out solar ultraviolet radiation equivalent to between 50 and 95 percent of that which reaches the vitamin-D-synthesizing region of the skin of Europeans. This explains "the fact, agreed to by all, that of all races the Negro is most susceptible to rickets" (7). It is clear from Thomson's figures that exposure of the face of Negro infants to winter sunlight in Scandinavia would result in synthesis of too little vitamin D to meet the infant's body requirements.

It was Hess who first proved that sunlight could cure rickets (8). Seeking experimental proof of a relationship between skin pigmentation and rickets, he took six white and six black rats and placed them on a rachitogenic diet containing low amounts of phosphorous. Exposing both groups to a critical amount of ultraviolet light, he found that all the white rats remained healthy while all the black rats developed rickets. He concluded (9), "It is manifest that the protective rays were rendered inert by the integumentary pigment."

To return now to Thomson's results and consider their bearing on hypervitaminosis, they explain why deeply pigmented Africans living near the equator and exposing almost all their body surface to the ultraviolet of the tropical sun do not suffer from kidney stones and other evidences of hyper-vitaminosis. Under conditions where untanned Europeans would synthesize up to 800,000 I.U. per day, deeply pigmented Africans would synthesize 5 to 10 percent as much; thus their daily production would fall within the acceptable range.

In this connection it is significant that Reinertson and Wheatley (10) found that the 7-dehydrocholesterol content of human skin does not vary significantly between Negroes and Whites. Skin from the back, abdomen, and thigh of adults of both races averaged 3.8 percent (standard deviation, 0.8 percent), the lowest result in their series being obtained in a specimen of the epidermis of the sole, an area that receives no radiation at all, while the highest result among adults was from a Negro. The highest content of all was found in a specimen from a 2-week-old infant that showed 8.8 percent of the provitamin, a fact that correlates well with the especially high need for vitamin D during the first 2 years of life.

In their paper and ensuing discussion, the above workers emphasize that 7-dehydrocholesterol is found almost entirely beneath the stratum corneum, thus establishing the fact that in man it is not present in the secretions of the sebaceous glands as it is in birds and some northern fur-covered animals which, respectively, obtain their vitamin D by preening or by licking their fur after the provitamin has been converted into the vitamin on the surface of the body. It would appear that vitamin D is made in man solely by the irradiation of the provitamin in the layers underneath the stratum corneum, a mechanism that would allow efficient regulation of the biosynthesis of this essential factor by varying the degree of ultraviolet penetration through differing amounts of pigmentation in the overlying stratum corneum.

ORIGIN OF WHITE SKIN

Having originated in the tropics where too much sunlight rather than too little was the danger, the first hominids had no difficulty in obtaining sufficient amounts of vitamin D until they extended their range north of the Mediterranean Sea and latitude 40°N (Fig. 15-2), where the winter sun is less than 20 degrees above the horizon (11) and most of the needed ultraviolet is removed from the sun's rays by the powerful filtering action of the atmosphere through which the slanting rays have to pass. Before the present century, for example, there was a very high incidence of rickets among infants in London and Glasgow, because in these latitudes the midday sun is less than 35 degrees from the horizon for 5 and 6 months, respectively, of the year; in Jamaica and other southern localities, on the other hand, the sun's midday altitude is never less than 50 degrees and rickets is almost unknown (2). The farther north one goes, the more severe becomes this effect of latitude on the availability of winter ultraviolet radiation, an effect compounded by cloudy winter skies.

Having evolved in the tropics, early hominids were probably deeply pigmented and covered with fur, as are most other tropical primates. The first adaptation one might expect therefore to lowered availability of ultraviolet light as they moved north of the Mediterranean would be a reduction of fur, for Cruikshank and Kodicek have shown (12) that shaved rats synthesize four times more vitamin D than normal rats do.

As early hominids moved farther and farther north, their more deeply pigmented infants must have been especially likely to develop the grossly bent legs and twisted spines characteristic of rickets, deformities which would cripple their ability to hunt game when they were adults. In this connection, Carleton Coon has written (13), "Up to the present century, if black skinned people were incorporated into any population living either north or south of the fortieth degree of latitude, their descendants would eventually have been selected for skin color on the basis of this vitamin factor alone." Howells agrees (14): "This variety of outer color has all the earmarks of an adaptation, of a trait responding to the force of sunlight by natural selection." The skin, he continues "admits limited amounts of ultraviolet, which is needed to form vitamin D, but presumably diminishes or diffuses dangerous doses by a screen of pigment granules."

Even in 1934 Murray clearly recognized the implications of these facts (15): "As primordial man proceeded northwards into less sunlit regions, a disease, rickets, accomplished the extinction of the darker, more pigmented elements of the population as parents and preserved the whiter, less pigmented to reproduce their kind and by progressive selection through prehistoric times, developed and established the white race in far northern Europe as it appears in historic times; its most extreme blond types inhabiting the interior of the northern-most Scandinavian peninsula."

It is a curious fact that Murray's thesis is almost unknown to the general public, including physiologists, biochemists, and physicians, and that it is not generally accepted by anthropologists, with the exception of Coon and Howells, quoted above, even though it fits the facts of Figure 15-2. Both in Europe and China, skin pigmentation becomes lighter as one goes north, and it is lighter in young children; in almost all races the skin is lighter in the newborn infant (16) and gradually darkens as the individual matures, a change that parallels the declining need for vitamin D.

WHEN DID EUROPEAN HOMINIDS BEGIN TO TURN WHITE?

On the basis of the conclusion that white skin is an adaptation to northern latitudes because of the lowered availability of winter ultraviolet radiation, it appears probable that the early hominids inhabiting western Europe had lost much of their body hair and skin pigmentation even half a million years ago. Anthropological evidence indicates that early hominids such as the Heidelberg, Swanscomb, Steinheim, Fontechevade, and Neanderthal men lived north of the Mediterranean Sea—particularly during warm interglacial periods (17). It is important to recognize that the effect of latitude on the availability of ultraviolet light in winter is not related to climate but operates steadily and at all times, through glacial and interglacial periods alike.

Hand axes and other early stone tools have been found throughout the tropics of the Old World and also in Europe as far north as the 50th degree of latitude (Figure 15-4). The presence of such stone tools as far

Fig. 15-4 Distribution of early stone tools throughout the tropics of the Old World Land in Europe as far north as 50°N. [From Brace and Montague, *Man's Evolution* (Macmillan, New York, 1965), p. 231]

north as England and France shows that some early hominids must already have adapted to the lowered level of ultraviolet radiation and consequent danger of rickets by partial loss of body hair and skin pigmentation, for without such adaptation they would have probably been unable to survive this far north.

England is at the same latitude as the Aleutian Islands, and no stone tools such as those found in southern England and France have ever been found in other areas at this latitude—for example, Mongolia and Manchuria. The unique combination of temperate climate and low levels of winter ultraviolet radiation in England and France is due to the powerful warming effect of the Gulf Stream on this particular northern area, which is unique in the world in this respect, for the Japan current in the Pacific is not as powerful as the Gulf Stream and warms only the Aleutian Islands, where no hominids existed until very recently.

Occupation of northern Europe and even Scandinavia up to the Arctic Circle seems to have taken place during the Upper Paleolithic, when presumably partially depigmented men already adapted to latitude 50°N lost nearly all their ability to synthesize melanin and so produced the blond-haired, blue-eyed, fair-skinned peoples who inhabit the interior of the northernmost part of the Scandinavian peninsula.

It has been held that the abundant appearance of stone scrapers in the Upper Paleolithic indicates that this far-northern extension of man's habitat followed his use of animal skins for clothing, a change that would select powerfully for infants with nearly transparent skin on their cheeks, who were thus still able to synthesize a minimum antirachitic dose of vitamin D even when fully clothed during the Scandinavian winter. Certainly the pink-to-red cheeks of northern European children are uniquely transparent; their color is due to the high visibility of the blood that circulates in the subepidermal region.

The one exception to the correlation between latitude and skin color in the Old World is the Eskimo; his skin is medium dark and yet he remains completely free of rickets (*18*) during the long dark arctic winters. Murray noted long ago that the Eskimo's diet of fish oil and meat contains several times the minimum preventive dose of vitamin D, concluding (*15*), "Because of his diet of antirachitic fats, it has been unnecessary for the Eskimo to evolve a white skin in the sunless frigid zone. He has not needed to have his skin bleached by centuries of evolution to admit more antirachitic sunlight. He probably has the same pigmented skin with which he arrived in the far north ages ago." Similar considerations would apply in the case of any coastal peoples of Europe and Asia, who would have been able to expand northward without depigmentation as long as they obtained sufficient vitamin D from a diet of fish; only when they ventured into the interior would antirachitic selection for blond types, as in Scandinavia, presumably have taken place.

YELLOW, BROWN, AND BLACK ADAPTATION

Human skin has two adaptive mechanisms for resisting the penetration of solar ultraviolet: melanin-granule production in the Malpighian layer and keratohyaline-granule production in the stratum granulosum. Melanin granules are black, whereas the keratohyaline granules produce keratin (from which nails, claws, horns, and hoofs are formed), which has a yellowish tinge. Particles of both types migrate toward the horny external layer, where they impart a black (melanin), yellow (keratin), or brown (melanin and keratin) tinge to the skin.

Thomson has shown (6) that, in Negroes, melanization of the stratum corneum plays the major role in filtering out excessive ultraviolet radiation, keratinization of the horny layer playing only a minor part. Mongoloids on the other hand have yellowish skin, since their stratum corneum is packed with disks of keratin (13) that allow them to live within 20 degrees of the equator even though their skin contains only small amounts of melanin (Figure 15-2). On the equator itself, however, even Mongoloid-derived peoples acquire pigmentation—for example, the previously medium-light-skinned Mongoloids who entered the Americas over the Bering Straits at latitude 66°N as recently as 20,000 to 10,000 years ago (Figure 15-2).

Even white-skinned peoples have to protect themselves against excessive doses of solar ultraviolet radiation in summer, for, as Blum has pointed out (19), on 21 June the solar ultraviolet is as intense in Newfoundland as it is at the equator, since at that time the two regions are at the same distance from the Tropic of Cancer (at 23°27′N). (At the equator, the solar ultraviolet is never less than on this date, while in Newfoundland it is never more.) In other words, adaptation to the variable intensities of solar ultraviolet in the north requires not only winter depigmentation but also the evolution of a reversible mechanism of summer repigmentation to keep the rate of vitamin-D synthesis constant throughout the year. It is significant that both the keratinization and melanization components of suntan are initiated by the same wavelengths which synthesize vitamin D, for it would be difficult to design a more perfect defense against excessive doses of vitamin D than this reversible response to ultraviolet light of these particular wavelengths—a pigmentation response that is further protected by the painful alarm bell of sunburn, which guarantees extreme caution against overexposure to solar ultraviolet in untanned individuals suddenly encountering a tropical sun.

Defenses against production of too much vitamin D therefore range from (i) reversible suntanning, as in Europeans, through (ii) constitutive keratinization, as in the Mongoloids of Asia and the Americas, to (iii) constitutive melanization, as in African and other truly equatorial peoples.

The physiological superiority of melanization as a means of protection against ultraviolet was demonstrated by the ability, historically documented, of imported Nigerian slaves to outwork the recently adapted American Indians in the sundrenched cane fields and plantations of the Caribbean and related tropical areas.

Additional evidence for the view that melanization of the stratum corneum is primarily a defense against the oversynthesis of vitamin D from solar ultraviolet is provided by the fact that the palms and soles of Negroes are as white as those of Europeans; *only* the palms and soles possess a thickly keratinized stratum lucidum under the external stratum corneum, which renders melanization of the latter unnecessary. The same reasoning explains the failure of the palms and soles of whites to sunburn during the summer.

Coon has written (13), "We cannot yet demonstrate why natural selection favors the prevalence of very dark skins among otherwise unrelated populations living in the wet tropics, but the answer may not be far away." Since overdoses of vitamin D administered orally are known to result in prompt and serious consequences, such as calcifications in the aorta and other soft tissues of the body, kidney stones, secondary renal disease, and death, it would appear that oversynthesis of vitamin D is sufficiently detrimental in young and old to favor the gradual selection for deeply pigmented skin near the equator, as seen, for example, in the repigmentation that has taken place among the equatorial American Indians during the last 10,000 years (Figure 15-2).

SECONDARY RESULTS OF PIGMENTATION AND DEPIGMENTATION

It is known that black skin absorbs more heat than white skin; the studies of Weiner and his associates (20) show that black Yoruba skin reflects only 24 percent of incident light whereas untanned European skin reflects as much as 64 percent. Of themselves, these facts would lead one to expect that reflective white skin would be found near the equator while heat-absorbing black skin would be found in cold northern climates.

Since the exact opposite is true around the world, it seems clear that man has adapted his epidermis in response to varying levels of ultraviolet radiation despite the price he has had to pay in being badly adapted from the standpoint of heat absorbance and reflectance of visible and near-infrared wavelengths. Similar considerations naturally apply to summer pigmentation due to suntan; ultraviolet regulation rather than heat regulation explains why Caucasians are white in the winter but pigmented in the summer.

In addition to being badly adapted for maximum heat absorbance, whiteskinned northern peoples are known to be particularly susceptible to skin cancer (21) and such skin diseases as psoriasis and acne. Therefore, only some powerful other advantage, such as relative freedom from rickets, would explain the worldwide correlation between high latitudes and white skin, for without some such factor it would seem that black or yellow skin would be the superior integument.

From this and other evidence, such as the fact that lion cubs and the young of other tropical animals develop rickets in northern zoos unless given cod-liver oil (2), it appears probable that depigmentation occurred north of latitude 40°N (a line marked by the Mediterranean Sea, the Great Wall of China, and the Mason-Dixon line) as an adaptation that allowed an increased penetration of winter ultraviolet radiation and consequent freedom from rickets. Certainly no other essential function of solar ultraviolet is known for man besides the synthesis of vitamin D.

SUMMARY

The known correlation between the color of human skin and latitude (Figure 15-2) is explainable in terms of two opposing positive adaptations to solar ultraviolet radiation, weak in northern latitudes in winter yet powerful the year around near the equator. In northern latitudes there is selection for white skins that allow maximum photoactivation of 7-dehydrocholesterol into vitamin D at low intensities of ultraviolet radiation. In southern latitudes, on the other hand, there is selection for black skins able to prevent up to 95 percent of the incident ultraviolet from reaching the deeper layers of the skin where vitamin D is synthesized. Selection against the twin dangers of rickets on the one hand and toxic doses of vitamin D on the other would thus explain the world-wide correlation observed between skin pigmentation and nearness to the equator.

Since intermediate degrees of pigmentation occur at intermediate latitudes, as well as seasonal fluctuation in pigmentation (through reversible suntanning), it appears that different skin colors in man are adaptations of the stratum corneum which regulate the transmission of solar ultraviolet to the underlying stratum granulosum, so that vitamin-D photosynthesis is maintained within physiological limits throughout the year at all latitudes.

REFERENCES AND NOTES

1. A. White, P. Handler, S. L. Smith, *Principles of Biochemistry* (McGraw-Hill, New York, ed. 3, 1964), p. 981.
2. C. H. Best and N. B. Taylor, *The Physiological Basis of Medical Practice* (Williams and Wilkins, Baltimore, ed. 3, 1943), pp. 1102, 1105.

3. F. Bicknell and F. Prescott, *The Vitamins in Medicine* (Grune and Stratton, New York, ed. 3, 1953), p. 578.

4. H. Beckemeier, *Acta Biol. Med. Ger.* 1, 756 (1958); ———— and G. Pfennigsdorf, *J. Physiol. Chem.* 214, 120 (1959).

5. D. I. Macht, W. T. Anderson, F. K. Bell, *J. Amer. Med. Assoc.* 90, 161 (1928); W. T. Anderson and D. I. Macht, *Amer. J. Physiol.* 86, 320 (1928).

6. M. L. Thomson, *J. Physiol. London* 127, 236 (1955).

7. A. F. Hess and L. J. Unger, *J. Amer. Med. Assoc.* 69, 1583 (1917).

8. ————, *ibid.* 78, 1177 (1922).

9. A. F. Hess, *ibid.*, p. 1177.

10. R. P. Reinertson and V. R. Wheatley, *J. Invest. Dermatol.* 32, 49 (1959).

11. F. Daniels, Jr., in *Handbook of Phyisology*, D. B. Dill, E. F. Adolph, C. G. Wilber, Eds. (American Physiological Society, Washington, D.C., 1964), pp. 969–88.

12. E. M. Cruikshank and E. Kodicek, *Proc. Nutr. Soc. Engl. Scot.* 14, viii (1955).

13. C. Coon, *The Living Races of Man* (Knopf, New York, 1965), pp. 232, 234.

14. W. W. Howells, *Mankind in the Making* (Doubleday, New York, 1959), p. 270.

15. F. G. Murray, *Amer. Anthropol.* 36, 438 (1934).

16. E. A. Hooton, *Up from the Ape* (Macmillan, New York, 1946), p. 466.

17. It is possible that the most northern or "classic" Neanderthal died out some 35,000 years ago in western Europe because of rickets which became severe when the arctic weather of the last glaciation made it necessary for him to dress his infants warmly in animal skins during the winter months, a change that would drastically reduce the area of their skin exposed to solar ultraviolet.

18. W. A. Thomas, *J. Amer. Med. Assoc.* 88, 1559 (1927).

19. H. F. Blum, *Quart. Rev. Biol.* 36, 50 (1961).

20. J. S. Weiner, G. A. Harrison, R. Singer, R. Harris, W. Jopp, *Human Biol.* 36, 294 (1964).

21. H. F. Blum, in *Radiation Biology*, A. Hollaender, Ed. (McGraw-Hill, New York, 1955), vol. 2, pp. 487, 509, 529.

16

INADEQUATE INTESTINAL DIGESTION OF LACTOSE

Theodore M. Bayless and Shi-Shung Huang

The article which follows will require more page fumbling and place-on-the-page losing than most, as the reader consults the glossary of technical terms at the back of this book. But (aside from the information provided) this paper on an enzyme deficiency which causes milk intolerance raises a number of interesting questions about the nature of populational, or "race" differences, and their relationship to culture facts.

In the folklore of "race" which Americans of all ethnic derivations absorb as they become enculturated, biological differences are thought to be on the level of very obvious characteristics, such as skin pigment, hair structure, and eye shape. To suggest that population differences are any more than "skin deep" is still in some quarters considered to be a heresy bordering on fascism. But evidence is accumulating that human populations do differ from one another in the incidence of many other genetically based physiological traits. The sickle-cell trait, for example, is practically nonexistent in populations of Asian or European derivation, but among African-derived populations it is found on the chromosomes of one person in 500.

There are many more such traits. They are not absolute markers, of course; not every man who is West Asian has the G6PD-enzyme deficiency gene, but the defect is found with higher incidence in Israel, Syria, and North Africa than anywhere else. So populations can be roughly differentiated on the basis of the frequency with which a gene is to be found in their gene pools, the proportion of individuals within the population showing the phenotypic effect of the gene.

The ability to break down lactose in digestion, like any other

Reprinted from the *American Journal of Clinical Nutrition*, Volume 22, March 1969, pages 250–256. By permission of the authors.

physiological digestive or metabolic function is genetically based. It has been assumed that this ability is distributed evenly through all human populations, even though there are individuals here or there who lack that ability. Does this paper by Bayless and Shi-Shung suggest otherwise?

Another significant idea illustrated by this paper is in the rapidly shrinking no man's land between biological anthropology and cultural anthropology. Milk is such an important food in American culture that it almost automatically comes to mind when Americans think of well-being, balanced diet, and good nutrition. Milk symbolizes the youthful bursting-with-health picture that Americans like to call to mind in thinking about American childhood and growing up. Is it possible that the milk = good syndrome may be another example of American ethnocentrism? What significance does this have for U.S. developed nutrition programs in Asia and Africa What is the relation between our pro-milk bias and programs of nutritional supplement for disadvantaged people in our own society? What about the TV commercials which exhort the viewers to drink "lots of milk"? Obviously, milk consumption is a likely candidate for loud arguments about social policy.

Interesting work is under way in lactose tolerances on the basis of different culture areas, with or without cattle. But with regard to the biological differences themselves, this paper is a useful, if difficult beginning for the reader in seeing the thread of argument from biological differences to the anthropology of culture change.

The authors are both members of the Medical School Staff at Johns Hopkins University. Dr. Bayless is in the Division of Gastroenterology, and Dr. Shi-Shung is an instructor in pediatrics.

Our understanding of the physiology of intestinal digestion and absorption of disaccharides has increased rapidly in the past 6 years. A group of syndromes based on inadequate levels of intestinal disaccharidase activity has been recognized. Several comprehensive reviews of these subjects have appeared recently (1–7). This article will outline briefly some of the more significant advances in these fields. The major portion of the paper will be devoted to a discussion of milk intolerance on a lactase-deficiency and lactose-intolerance basis.

RECENT ADVANCES

These important advances include: 1) demonstration that the sugar-splitting activity of the succus entericus (intestinal fluid) was inadequate to explain the digestion of dietary disaccharides; 2) localization of the major portion of disaccharidase activity to the membranous portion of microvilli of the small intestinal mucosa; 3) characterization of several specific disaccharide-splitting enzymes in man (four maltases, two sucrases, one or two lactases, and one trehalase); 4) confirmation that hydrolysis of the disaccharides to absorbable monosaccharides is the rate-limiting step in the digestion and absorption of dietary disaccharides; 5) recognition that clinical conditions exist in which the levels of activity of a specific intestinal disaccharidase(s) are inadequate to digest the usual dietary intake of that disaccharide; 6) demonstration that undigested disaccharides cannot be absorbed, and subsequently an excess quantity of fluid enters the bowel lumen to dilute the sugar, motility is increased, and the subject may develop abdominal cramps, bloating, or diarrhea; 7) description of several diarrheal and abdominal pain syndromes that can result from disaccharidase deficiencies (Table 16-1); 8) identification of sucrase and isomaltase deficiency as an inherited defect with resultant intolerance of table sugar; 9) documentation of the occurrence of disaccharidase deficiencies as a secondary result of small intestinal mucosal damage; 10) recognition of milk intolerance in school-age children and adults on the basis of an isolated lactase deficiency with intolerance of lactose; 11) documentation that this isolated lactase deficiency occurs in otherwise healthy adults and is very common in nonwhite populations; and 12) the suggestion that intestinal lactase levels are probably under genetic control.

IS MILK INTOLERANCE ON A LACTASE-DEFICIENCY BASIS AN INTERNATIONAL NUTRITIONAL PROBLEM?

The advances in the understanding of disaccharide digestion and absorption have opened many new areas of investigation in the field of nutrition (7). One important question that has arisen is: does lactase deficiency and the resultant milk intolerance constitute a major international nutrition problem? The remainder of this paper will be devoted to exploring evidence on both sides of this question.

LACTASE DEFICIENCY AND LACTOSE INTOLERANCE

Lactase deficiency causing lactose and milk intolerance can occur either as an isolated, often genetically determined, defect or as an acquired sequel secondary to mucosal damage. As an example of acquired lactase deficiency, acute infectious diarrhea in infants may be followed by severe diarrhea and malnutrition aggravated by lactose intolerance. Kwashiorkor, celiac disease, and tropical sprue can also cause a secondary lactase deficiency as a result of intestinal damage.

TABLE 16-1

Classification of Disaccharide Intolerance

I. Lactose intolerance
  A) Congenital physiologic lactase deficiency in premature infants
  B) Congenital lactase deficiency, presumably genetically determined; Some with lactosuria (? related to D2)
  C) Acquired lactase deficiency in older children and adults, probably genetically determined
  D) Acquired, secondary to diffuse mucosal damage and part of generalized disaccharidase deficiency
      1) Mucosal damage (celiac disease, tropical sprue, Whipple's disease, intestinal lymphangiectasis, et cetera)
      2) Acute gastroenteritis (especially infants)
      3) Kwashiorkor
      4) Neomycin administration (only lactase studied)
      5) *Giardia lamblia* infestation (presumed mucosal damage)
      6) Post-bowel surgery in infants (presumed mucosal damage)
  E) Acquired, secondary to alteration in intestinal transit
      1) Small bowel resection (may have decreased lactase level also)
      2) Postgastrectomy or pyloroplasty (may unmask a preexisting deficiency)
      3) Lactose intolerance without lactase deficiency
  F) Suggested disease associations
      1) Ulcerative colitis
      2) Regional enteritis
      3) Irritable-colon syndrome
      4) Osteoporosis
II. Sucrose intolerance
  A) Congenital, sucrase-isomaltase deficiencies, genetically determined
  B) Seemingly acquired sucrose intolerance in adults, sucrase-isomaltase deficiency, presumably also genetically determined and similar to the congenital deficiencies
  C) Acquired, secondary to diffuse mucosal damage

Isolated lactase deficiency, presumably transmitted on a genetic basis, is the main subject under discussion in this report. Affected individuals include otherwise healthy children, adolescents, and adults. They are intolerant of as little as 12 g of lactose (equivalent to 1 glass of milk) and notice the development of bloating, cramps, or diarrhea after drinking 1 or 2 glasses of milk. There is no evidence of intestinal damage. Milk intolerance is often not suspected in these children complaining of abdominal pain, because they had been able to drink milk without difficulty as infants. Adults with lactase deficiency and milk intolerance usually limit their milk intake to amounts that will not cause a laxative effect. If they increase the amount of milk consumed, as in the treatment of peptic ulcer, symptoms appear (8).

PREVALENCE OF LACTOSE INTOLERANCE IN ADULTS

Lactose intolerance is defined as development of cramps, bloating, or diarrhea after ingestion of 50–100 g of lactose. This is the amount of lactose in 1–2 quarts of milk. The development of symptoms and the lack of a significant rise in blood sugar on a lactose-tolerance test correlates well with low jejunal lactase levels (3, 8).

Lactose intolerance has been demonstrated in 5–10% of the adult white population of the United States (8, 9). These figures were based on random samples of 20 males in Baltimore and 100 people in Rochester, Minnesota, respectively. The findings from selected groups from Chicago and Oklahoma supported these results (10, 11). The prevalence was much higher in groups of Negroes and groups of Orientals. Seventy per cent of 20 unselected healthy Negro adults in Baltimore were lactose intolerant and lactase deficient (8). In Uganda, Africa, 72% of 40 healthy Negroes were lactase deficient, as determined by jejunal biopsy and lactase assay. There was a marked tribal difference in prevalence. For example, 21 of 22 Bantus had lactase levels less than 2 units whereas only 1 of 7 Batutsi had these low enzyme levels (12). The results of surveys of Orientals have been quite consistent. Nineteen of twenty Filipinos, Taiwanese, and American-born Chinese in Baltimore were intolerant of 50 g of lactose (13). Nineteen of 20 Chinese and Indian students studied in Australia developed symptoms with a lactose load (14). A group of 11 Oriental subjects (2 Japanese, 4 Korean, and 5 Chinese) were studied in Rochester, Minnesota, and all 11 had an isolated lactase deficiency (15).

DOES INTOLERANCE OF A LACTOSE LOAD INDICATE CLINICAL MILK INTOLERANCE?

This depends on the quantity of milk consumed at one time and the individual's "sensitivity." Fifteen of twenty lactose-intolerant men gave a history of watery diarrhea, flatulence, or cramps after ingesting 1 or 2 glasses of milk. The other five noted symptoms only after rapidly drinking 1 quart of milk (8). Fourteen of nineteen lactose-intolerant Oriental medical personnel had noted intolerance of 1 or 2 glasses of milk. Some patients claim that as little as 100 cc of milk (5 g of lactose) will cause them to have cramps or abdominal distention.

Experimentally, ingestion of as little as 6 g of lactose by a lactase-deficient subject causes noticeable dilution of barium and an increased rate of intestinal transit on X-ray study (16). The subject's underlying bowel-motility pattern will also influence the degree of symptoms produced. Some persons with an "irritable colon" syndrome and lactase deficiency or patients who have had gastric surgery seem to have particularly severe cramps and diarrhea with lactose-tolerance tests.

Factors that increase small intestine transit, such as drinking the milk as a very cold beverage, might increase symptomatology. Conversely, milk might be better tolerated by a lactose-intolerant person, if it was consumed with a meal, which would delay gastric emptying. Some lactase-deficient persons can consume as much as a quart of milk daily if they drink small amounts at several times through the day (17).

IS LACTOSE INTOLERANCE A PROBLEM IN CHILDREN?

Since school-age children are the target of some nutritional programs utilizing milk as a source of protein and calories, it is important to be sure that lactose intolerance is not influencing the acceptance of milk supplements.

Milk intolerance, severe diarrhea, and malnutrition can occur during infancy on a lactase-deficiency basis, but this is not common. There are a few reported incidences of inherited congenital lactase deficiencies, but most lactase deficiency in infancy seems to be secondary to mucosal damage associated with acute "infectious" diarrhea. These latter problems should be self-limited and end in infancy or early childhood, unless there are complicating factors such as malnutrition or parasitosis. This aspect will be discussed in a subsequent paragraph.

Milk intolerance is usually not considered as a cause of abdominal cramps or diarrhea in children who had been able to drink milk as infants. Clinical evidence suggests that otherwise healthy teenagers and adults with an isolated lactase deficiency have decreasing levels of intestinal lactase after infancy (8, 18, 19). Many of these subjects, both white and Negro, give a clear history of the onset of milk intolerance as adolescents or young adults (8). Among Negroes in whom there is a very high incidence of milk and lactose intolerance in adults, there is a decreasing ability to tolerate a lactose load with increasing age after infancy (18, 19). In a group of 15 healthy Negro children, ages 6–11 years, 45% developed cramps or diarrhea, or both, with a lactose load equivalent to 1–2 pints of milk. In young Negro adults this rate exceeds 70% (18).

We are becoming increasingly aware that white and Negro children with lactose intolerance can present to the physician with recurrent abdominal pain that is subsequently eliminated by limiting the intake of lactose-containing products including milk, ice cream, powdered soft drinks, and chocolate drinks.

DOES THE AMOUNT OF MILK CONSUMED AFFECT IN-TESTINAL LACTASE LEVELS?

There is no evidence that lactase levels in man can be lowered or raised by alterations in lactose intake. A group of 10 children, ages 7–17 years, with galactosemia, who had carefully avoided all lactose-containing materials since early infancy had normal lactose-tolerance tests, suggesting that their lactase levels had not decreased during this prolonged period of lactose abstinence. The one exception was a 15-year-old Negro child who was lactase deficient, as determined by biopsy assay (20). Complete lactose deprivation for 42 days was not associated with a significant change in lactase activity or in lactose tolerance in adults (21).

Conversely, there are no reports of induction of higher lactase levels by milk or lactose feeding to either normals (22) or lactase-deficient persons (17). In addition, some of our lactase-deficient patients were consuming milk at the time of the study of their lactase levels (8, 12). Some milk-intolerant subjects have been able to gradually increase their milk intake while others have not (13, 17).

DOES THIS COEXISTENCE OF MALNUTRITION OR GASTRO-ENTERITIS FURTHER INCREASE THE INCIDENCE OF LAC-TASE DEFICIENCY IN POPULATIONS THAT ALREADY HAVE A HIGH INCIDENCE IN ADULTS?

Studies of children in Africa with kwashiorkor indicate that the rate of lactose maldigestion was as high as 65%. Generalized disaccharidase deficiencies were found by assay of jejunal mucosa and, presumably, they occur because of intestinal mucosal damage. After therapy, lactose maldigestion persisted (23).

Children with presumed infectious diarrhea in the United States also were lactose intolerant for several weeks after the onset of the acute episode (24). As further evidence of presumed lactase deficiency with mucosal damage, infants in Puerto Rico with acute gastroenteritis had a decreased ability to digest lactose (25). Giardiasis has been associated with a reversible decrease in lactase activity (26). Malnutrition in adults living in tropical climates can be associated with generalized disaccharidase deficiencies (27). Thus, residence in a tropical area with exposure to malnutrition, acute diarrhea, and parasitosis and with probable damage to the jejunal mucosa, would be expected to increase the incidence of lactase deficiency in children. It is not known if the presence of gut mucosal damage would increase the "secretory" response to osmotically active materials and thus cause a greater outpouring of fluid when undigested lactose is in the lumen.

ROLE OF MILK IN WORLD NUTRITION PRACTICES AND PLANS

Fluid milk and cream consumption averaged 279 lbs./person per year in the major milk-producing countries of the world in 1965 (28). These areas include Canada, the United States, the Scandinavian nations, Western Europe, Australia, and New Zealand. In some of these dairy-producing countries, there are well publicized programs to encourage the consumption of milk by preschool and school-age children. Some emphasis is also directed at teenagers and adults. The percentage of the population that consumes milk varies with age, sex, income, geographic location, urbanization, and number of children in the household (29). There are varying estimates of how many people consume liquid milk. In one English study, an average of 42% of a group of 4,637 school pupils, ages 11–19 years, both male and female, consumed 90 ml of milk per weekday when it was offered as a free dietary supplement. The range in nine different schools comprising the study was 23–60% (30). Among women in Iowa, age 30–59 years, less than 50% drank any milk and the majority of the "milk drinkers" consumed 1 cup or less per day, (29). Powdered milk is provided as a food supplement in many welfare programs in this country. Presumably a major portion of the intended recipients are Negroes.

In most countries that have a low rate of milk production, the amount of milk consumed per capita is much lower. The African countries produced only 4% of the total estimated world milk production in 1953. In the Gold Coast, the per capita production was 11 lbs./person per year plus imports of 7 lbs./person. In the Congo, the estimated per capita production was 6 lbs./year (31). Trypanosomiasis makes cattle raising very difficult in most parts of tropical Africa and is a major factor in the low rate of milk production. It was estimated that in 1965 milk was absent from the diet of more than three-fourths of the population of Nigeria (32). In the Belgian Congo in 1953, it was estimated that milk and dairy products contributed 15 cal/day to the total calorie intake per person (31). Some of these nations import dairy products, usually in the form of dried milk, canned milk, or cheese. Milk products are being exported as commercial material and as foreign aid donations to many countries in Asia and Africa.

Malnutrition is a major international problem and is thought to affect the lives of perhaps two-thirds of the world's children. Milk is being used to combat protein malnutrition in areas of Africa, Asia, and South America. Many of these programs have been directed at older infants, preschool and school-age children. School meal programs (½ pint of milk, equivalent to 12 g of lactose) have been a common method of distribution. Skim milk has frequently been used as a protein source in the therapy of kwashiorkor.

MILK TOLERANCE IN AFRICA AND IN ASIA

There is little information available in the medical literature or in nutrition surveys concerning the tolerance of milk products in children and adults in the developing nations in Africa and Asia. It is generally assumed that milk is well tolerated in all parts of the world. Powdered and canned milk are successful commercial products in many parts of Africa. The New Zealand whole-milk biscuit, which contains 6.82 g of lactose as skim-milk powder and whey powder has been used as a nutritional source for schoolchildren in areas of the Middle-East, Asia, and Africa. It was stated in a recent letter to *Nature* that "at no stage have any adverse effects or symptoms indicative of any degree of lactose intolerance been reported" (32). The director of a study in Iran reported that 40 schoolchildren seemed to be pleased to receive one milk biscuit each day for 5 weeks (34).

Cook and Kajubi studied 20 healthy African medical personnel in Kampala, Uganda. Eighteen had "flat" blood sugar curves after 50 g of lactose and 16 had symptoms indicating a high incidence of lactase deficiency. Only 3% of these intelligent individuals, mainly from upper-class Ugandan homes, have a history of milk intolerance beginning in early adult life, but the milk intake of the entire study group was estimated at only ⅛ to ¼ of a pint of milk per person per day (12). Thus, these lactase-deficient subjects were probably not consuming enough milk to develop symptoms.

There are a few anecdotal reports of milk intolerance in the general population in Asia and Africa. Jeffries reported that Wiya tribesmen in Nigeria, Africa, would not drink cow's milk because they claimed that milk upset their stomachs and produced diarrhea. At the time of the initial observation, the author assumed that the milk had become contaminated because of unsanitary conditions (35).

A newspaper article describing the factors that contribute to widespread malnutrition in Africa cited presumably incorrect local beliefs that block the efforts of UNICEF to improve nutrition. Among the "unscientific attitudes" was the belief that "diarrhea results from drinking milk and that many mothers condemn all milk, depriving their children of a basic food" (36).

One of the authors of this review (Huang) was aware of milk intolerance among his medical school classmates in Taiwan when milk supplements were provided by the United States as foreign aid. A nutritionist in the Phillipines said that she had to dilute powdered skim milk to avoid intolerance among schoolchildren (C. B. Sutter, personal communication).

In the course of a nutrition study on an Andean Indian population of

Vicos, Peru, it was found that reconstituted powdered milk caused stomach pain among the recipients. When the milk supplements were incorporated into sugar-sweetened chocolate, it seemed to make it acceptable to the schoolchildren. In this area, the local population converts a major portion of its locally produced milk into cheese (H. F. Dobyns, personal communication).

There are also suggestions that "soft curd" milks, in which lactose is fermented to lactic acid, are being consumed in parts of Africa. Preliminary studies among Mashona tribes in Salisbury, Rhodesia, demonstrated a 100% incidence of lactose intolerance and of lactase deficiency. These same people were able to tolerate large amounts of milk, but this milk was always soured by a traditional ritual (D. Clain, personal communication). It is also said that fermented mares' milk is a regular part of the diet of Mongolians. Yogurt, a fermented milk product, is also popular in the Arab Middle East.

CONCLUSION

Inadequate intestinal digestion of lactose and resultant milk intolerance is very common among Negroes and Asian adults as determined by small population samples. This high incidence of lactose intolerance based on lactase deficiency is presumably determined on a genetic basis. Most of these subjects were studied in temperate climates and were in good nutritional balance. Since milk is widely recommended as an important protein source in areas of widespread malnutrition, including Africa and Asia, it is essential to determine if liquid milk, as given in aid programs, is well tolerated by children, teenagers, and adults in these underdeveloped countries. If liquid milk is poorly tolerated it can be given in smaller amounts, as a warm beverage or with meals in an effort to lessen lactose intolerance. Alternatively, other forms of dairy products, such as cheese or yogurt, or other sources of protein and calcium could be utilized to combat protein malnutrition.

REFERENCES

1. Miller, D., and R. K. Crane. The digestion of carbohydrates in the small intestine. *Am. J. Clin. Nutr.* 12:220, 1963.
2. Prader, A., and S. Auricchio. Defects of intestinal disaccharide absorption. *Ann. Rev. Med.* 16:345, 1965.
3. Littman, A., and J. B. Hammond. Diarrhea in adults caused by deficiency in intestinal disaccharidases. *Gastroenterology* 48:237, 1965.
4. Haemmerli, U. P., and H. Kistler. Disaccharide malabsorption. In: *Disease-a-Month*. Chicago: Year Book, 1966, p. 1–59.

5. Townley, R. R. W. Disaccharidase deficiency in infancy and childhood. *Pediatrics* 38:127, 1966.
6. Davidson, M. Disaccharide intolerance. *Pediat. Clin. N. Am.* 14:93, 1967.
7. Bayless, T. M., and N. L. Christopher. Disaccharidase deficiency. *Am. J. Clin. Nutr.* 22:181, 1968.
8. Bayless, T. M., and N. S. Rosensweig. A racial difference in incidence of lactase deficiency. *J. Am. Med. Assoc.* 197:968, 1966.
9. Newcomer, A. D., and D. B. McGill. Disaccharidase activity in the small intestine: Prevalence of lactase deficiency in 100 healthy subjects. *Gastroenterology* 53:881, 1967.
10. Littman, A., A. B. Cady and J. Rhodes. Lactase and adult disaccharidase deficiencies in a hospital population. *Israel J. Med. Sci.* 4:110, 1968.
11. Welsh, J. D., V. Rohrer, K. B. Knudsen and F. F. Paustian. Isolated lactase deficiency. *Arch. Internal Med.* 120:261, 1967.
12. Cook, G. C., and S. K. Kajubi. Tribal incidence of lactase deficiency in Uganda. *Lancet* 1:725, 1966.
13. Huang, S. S., and T. M. Bayless. Milk and lactose intolerance in healthy Orientals. *Science* 160:83, 1968.
14. Davis, A. E., and T. Bolin. Lactose intolerance in Asians. *Nature* 216:1244, 1967.
15. Chung, M. H., and D. B. McGill. Lactase deficiency in Orientals. *Gastroenterology* 54:225, 1968.
16. Laws, J. W., J. Spencer and G. Neale. Radiology in the diagnosis of disaccharides deficiency. *Brit. J. Radiol.* 40:594, 1967.
17. Cautrecacas, P., D. H. Lockwood and J. R. Caldwell. Lactase deficiency in the adult. *Lancet* 1:14, 1965.
18. Huang, S. S., and T. M. Bayless. Lactose intolerance in healthy children. *New Engl. J. Med.* 276:1283, 1967.
19. Cook, G. C. Lactase activity in newborn and infant Baganda. *Brit. Med. J.* 1:527, 1967.
20. Kogut, M. D., G. N. Donnell and K. N. F. Shaw. Studies of lactose absorption in patients with galactosemia. *J. Pediat.* 71:75, 1967.
21. Knudson, K. B., E. M. Bradley, F. L. Lecocq, H. M. Bellancy and J. D. Welsh. Effect of fasting and refeeding on the histology and disaccharidase activity of the human intestine. *Gastroenterology* 55:46, 1968.
22. Rosensweig, N. S., and R. H. Herman. Control of jejunal sucrase and maltase activity by dietary sucrose or fructose in man. *J. Clin. Invest.* 47:2253, 1968.
23. Bowie, M. D., G. O. Barbezat and J. D. L. Hansen. Carbohydrate absorption in malnourished children. *Am. J. Clin. Nutr.* 20:89, 1967.
24. Sunshine, P., and N. Kretchmen. Studies of small intestine during development. III. Infantile diarrhea asociated with intolerance to disaccharides. *Pediatrics* 34:38, 1964.
25. Torres-Pinedo, R., and H. Rodriquez. Studies on infant diarrhea. IV. Composition of jejunal fluid after a single feeding. *Soc. Pediat. Res.* May 4, 1968.
26. Hoskins, L. C., S. J. Winawer, S. A. Broitman, L. S. Gottlieb and N.

Zamcheck. Clinical giardiasis and intestinal malabsorption. *Gastroenterology* 53:265, 1967.

27. Halstead, C. H., S. Sheir, N. Sourial and V. N. Patwardhan. Small intestine structure and absorption in Egypt: Influence of hookworm infestation and malnutrition. In press.

28. Foreign agriculture circular. *Per Capita Consumption of Dairy Products, 1964 and 1965.* Washington, D.C.: U.S. Dept. Agr., January, 1967.

29. Burk, M. C., and A. E. Hammill. Milk consumption patterns of a sample of Iowa women. *J. Am. Dietet. Assoc.* 49:319, 1966.

30. McKenzie, J. C., J. Mattinson and J. Yudkin. Milk in schools: an experiment in nutrition education. *Brit. J. Nutr.* 21:811, 1967.

31. McCabe, T. W. *Pattern of World Milk Production.* Washington, D. C.: U.S. Dept. Agr., 1955.

32. May, J. M. *Ecology of Malnutrition in Middle Africa.* New York: Hafner, 1965.

33. McGillivary, W. A. Lactose intolerance. *Nature* 219:615, 1968.

34. Gaughey, A. Quoted in: *New Zealand Dairy Exporter*, March, 1968, pp. 25–26.

35. Jeffries, M. D. W. The Wiya tribe. *African Studies* 21:83, 174, 1962.

36. Associated Press News Service, New York. As quoted by: *The Evening Sun*, Baltimore, July 16, 1968.

The anthropological implications of the differential distribution of the genotypes for lactose production among populations is discussed in the review article by Robert D. McCracken, "Lactase deficiency: An example of dietary evolution" in *Current Anthropology*, 12:479–517, October-December 1971.

See also Norman Kretchmer, "Lactose and lactase," *Scientific American*, 227:70–79, October 1972. [Editor's footnote]

17

HUMAN ADAPTATION
TO HIGH ALTITUDE

Paul T. Baker

Recently, anthropologists have been concentrating on in-depth studies of the biology and culture of populations. The following article is an example of this kind of anthropological biology, reporting on a study of the people of a small district in southern Peru. Biological and cultural factors in the adaptation of the Nuñoans are discussed, and significant aspects of the concept of adaptation are illuminated.

The idea of adaptation is crucial in understanding evolution, both in the microevolutionary sense, and in the origin of taxa, such as species and genera. The general definition of biological adaptation, as "any structural or behavioral trait which enhances the efficiency of an individual or a population in reproduction, nutrition and survival" is really too vague to be useful for more than a point of departure.

For a human population, we must identify and distinguish between cultural and biological factors. This is not as simple as it sounds, particularly since every human group adapts in both modes. Furthermore, there is the difficulty of distinguishing genetic traits that represent adaptiveness, from developmental traits that do not involve gene differences. Physiological traits can be either genetic and developmental, or they can be both. The intensive study of a population is an aid in understanding how an array of individuals in a specialized environment is adapted by both kinds of trait.

This paper by Dr. Paul T. Baker can help to define the problems in research terms, and serve to suggest investigation strategies in studying human populations.

Dr. Baker is Professor of Anthropology at Pennsylvania State University.

Reprinted from *Science, 163*: March 14, 1969, pages 1149–1156. Copyright 1969 by the American Association for the Advancement of Science. By permission of the publisher and the author.

Stretching along western South America from Colombia to Chile lies a large section of the Andean plateau, or *Altiplano*, which rises above 2500 meters (about 8250 feet) (Figure 17-1). This area is suitable for human habitation up to the permanent snow line, which is generally above 5300 meters (17,590 feet). There are now more than 10 million people living in this zone, and the historical and archeological records indicate that it has been densely populated for a long time. Indeed, before Europeans arrived, the Inca empire, which had its center in this zone, formed one of the two major civilizations of the Western Hemisphere and, in A.D. 1500, probably contained about 40 percent of the total population of the hemisphere.

With such a history, one would assume this to be an ideal environment for man and the development of his culture. Yet, in point of fact, modern man from a sea-level environment ("sea-level man") finds this one of the world's more uncomfortable and difficult environments. The historical records show that such was the case even in the 1500's, when the Spanish complained of the "thinness of the air," moved their capital from the highlands to the coast, and reported that, in the high mining areas, the production of a live child by Spanish parents was a rare, almost unique, phenomenon (1). Today this environmental zone remains the last major cultural and biological center for the American Indian. The population has an extremely low admixture of genes from European peoples and virtually none from African peoples. The few cities are Hispanicized, but the rural areas retain a culture which, in most aspects, antedates the arrival of the Spaniards.

It would be far too simplistic to suggest that this unique history is explicable entirely on the basis of the effects of altitude on sea-level man. Yet, there is sound scientific evidence that all sea-level men suffer characteristic discomfort at high altitudes, the degree of discomfort depending upon the altitude. There is evidence, also, of long-term or permanent reduction in their maximum work capacity if they remain at these altitudes, and evidence that they undergo a number of physiological changes, such as rises in hemoglobin concentrations and in pulmonary arterial pressure. In a few individuals the initial symptoms develop into acute pulmonary edema, which may be fatal if untreated. On the basis of less complete scientific evidence, other apparent changes are found for sea-level man at high altitudes: temporary reduction in fertility, reduction in the ability of the female to carry a fetus to term, and a high mortality of newborn infants (2, 3).

With these problems in mind, a group of scientists from Pennsylvania State University, in collaboration with members of the Instituto de Biología Andina of Peru, decided to investigate the biological and cultural characteristics of an ecologically stable Peruvian Quechua population living in traditional fashion at a high altitude. We chose for study the most stable population known to us at the highest location reasonably accessible. It was hoped that some insight could be gained into the nature of this quite obviously successful and unusual example of human adaptation. In this article I review some of the results available from this continuing study.

The general problem may be defined by three questions: (i) What are the unique environmental stresses to which the population has adapted? (ii) How has the population adapted culturally and biologically to these stresses? (iii) How did the adaptive structures become established in the population?

Our basic method of study, in attempting to answer these questions, was a combination of ecological comparisons and experimental analysis.

THE STUDY POPULATION

The population chosen for study lives in the political district of Nuñoa, in the department of Puno in southern Peru. In 1961 the district had a population of 7750 and an area of about 1600 square kilometers. Geographically, the district is formed of two major diverging river valleys, flat and several kilometers broad in the lower parts but branching and narrow above. These valleys are surrounded by steep-sided mountains. In the lower reaches of the valley the minimum altitude is 4000 meters; the higher parts of some valleys reach above 4800 meters. The intervening mountains rise, in some parts, to slightly above 5500 meters.

The climatic conditions of the district are being studied from weather stations on the valley floors and on the mountain sides at different altitudes. From present records, the pattern seems fairly clear. The lower valley floor appears to have an average annual temperature of about 8°C with a variation of only about 2°C from January, the warmest month, to June, the coldest. This is much less than the diurnal variation, which averages about 17°C. The seasonal variation in temperature is due almost entirely to cloud cover associated with the wet season. Some snow and rain fall in all 12 months, but significant precipitation begins around October, reaches its peak in January, and ends in April. Since the diurnal variation is high, some frost occurs even in the wet months. Mean temperatures fall in proportion to increases in altitude (by about 1° C per 100 meters), but, because of the sink effect in the valleys, minimum temperatures on the valley floors are usually somewhat lower than those on the lower mountain sides.

Except for two small areas of slow-growing conifers, almost all of the district is grassland. Because of the existing climatic and floral conditions, herding has become the dominant economic activity. Alpaca, llama, sheep, and cattle, in that order, are the major domestic animals. Agricul-

Fig. 17-1   The high-altitude areas of South America and the location of Nuñoa.
Shading indicates altitude of 2500 meters (8200 feet) or more.

ture is limited to the cultivation of frost-resistant subsistence crops, such
as "bitter" potatoes, *quinoa,* and *cañihua* (species of genus *Cheno-
podium*). Even these crops can be grown only on the lower mountain
sides and in limited areas on the lower valley floors. In recent years, crop
yields have been very low because of drought, but they are low even in
good years.

A single town, also called Nuñoa, lies within the district and contains
about one-fourth of the district's population; the other three-fourths live in
a few native-owned settlements called *allyus,* or on large ranches or
haciendas, which are frequently owned by absentee landlords. The social
structure may be loosely described as being made up of three social
classes: a small (less than 1 percent) upper class, whose members are
called *mestizos;* a larger intermediate class of individuals called *cholos;*
and the Indians, or *indígenas,* who constitute over 95 percent of the
population. Membership in a given social class is, of course, based on a
number of factors, but the primary ones are degree of westernization and

wealth. Race appears to be a rather secondary factor, despite the racial connotations of the class designations: *mestizos*, of mixed race; *cholos*, transitional; *indígenas*, indigenous inhabitants. Biologically the population is almost entirely of Indian derivation.

By Western economic and medical-service standards, this district would be considered very poor. If we exclude the *mestizo*, we find per capita income to be probably below $200 per year. The only medical treatment available in the district is that provided by a first-aid post. The upper class has access to a hospital in a neighboring district. Yet repeated surveys suggest that the diet is adequate, and that death rates are normal, or below normal, for peasant communities lacking modern medicine. Indeed, this population must be viewed as being one nearly in ecological balance with its technology and its physical environment.

Superficial archeological surveys reveal that the central town predated the Spanish conquest, and suggest that the district population has been fairly stable for at least 800 or 900 years.

From this general survey of the Nuñoa population and its environs we have concluded that the unusual environmental stresses experienced by this population are hypoxia and cold; other stresses, more common to peasant groups in general, which the Nuñoa population experiences are specific infectious diseases, the problems of living in an acculturating society, and, possibly, nutritional deficiencies.

HYPOXIA

At elevations such as those of the district of Nuñoa, the partial pressure of atmospheric oxygen is 40 percent or more below the values at sea level. As noted above, such a deficiency of oxygen produces a multitude of physiological changes in sea-level man, at all ages. We therefore attempted to evaluate the native Indian's responses to altitude at all stages of the life cycle.

DEMOGRAPHY

A survey of more than 10 percent of the population revealed an average completed fertility of about 6.7 children for each female. This is quite a high fertility by modern standards, but there was no evidence of voluntary birth control, and cultural practices appeared in many ways designed to provide maximum fertility; under these conditions, 6.7 children is no more than average. This same survey, partially summarized in Table 17-1, did not show an unusually high rate of miscarriage but did reveal two unusual features. (i) The earliest age at which any woman gave birth to a child was 18 in the low valleys and something over 18 at higher elevations. The

TABLE 17-1

Statistics on reproduction and viability of offspring for the district of Nuñoa, based on a sample of approximately 14 percent of the population of the district.

SAMPLE	Number in sample	Mean age (yr)	MARRIED WOMEN				SEX RATIO (MALES TO FEMALES) OF OFFSPRING		MORTALITY OF OFFSPRING (%)[2]	
			Mean age at first pregnancy (yr)	Offspring (No.)	Mean number of offspring per woman	Mean number of surviving[1] offspring per woman	At birth	Surviving[1]	Male	Female
Total	136	36.2	19.5+	608	4.5	3.2	124	129	30	33
Postmenopausal individuals	31	45+	20.1	207	6.7	4.4	113	146	27	44

[1] "Surviving" refers to time of the census.
[2] During the period of growth.

average age of first pregnancy was also higher for women at the higher altitudes. (ii) The sex ratio was highly unusual in that there was a large number of excess males. Furthermore, there was a higher mortality of females than of males throughout the period of growth. In an associated study of newborns it was found that, for Quechua mothers in Cuzco (altitude, 3300 meters), placenta weights at childbirth were higher and infant birth weights were lower than corresponding weights for comparable mothers near sea level (4). Finally, an analysis of the Peruvian census showed that, as in the United States, the mortality of newborn infants is higher at higher elevations; this does not appear to be primarily a socioeconomic correlation. From the results so far obtained, we conclude that fecundity and survival through the neonatal period is probably adversely affected by high altitude, even in the native populations of high-altitude regions. However, it is clear that the Nuñoans can still maintain a continuing population increase. Our data do not provide a basis for deciding whether, at high altitudes, fecundity and survival of offspring through the neonatal period are greater for natives than they are for immigrant lowlanders (5).

GROWTH

Intensive studies on growth were carried out on over 25 percent of the Nuñoa-district children, from newborn infants to young people up to the age of 21. A number of unusual growth features were apparent shortly after birth. Thus, as shown in Table 17-2, a slower rate of general body growth than is standard in the United States is apparent from a very early age. In addition, developmental events such as the eruption of deciduous teeth and the occurrence of motor behavior sequences occur late relative to U.S. standards. For example, the mean number of teeth erupted at 18 months was 11.5 for Nuñoa infants as compared with 13 for U.S. infants. The median age at which Nuñoa children briefly sat alone was 7 months, and the median age at which they walked alone was 16.2 months. These

TABLE 17-2

Stature and weights of Nuñoa infants and of infants in the United States

AGE (MONTHS)	STATURE, MALES (CM)		STATURE, FEMALES (CM)		WEIGHT, MALES (KG)		WEIGHT, FEMALES (KG)	
	Nuñoa	U.S.	Nuñoa	U.S.	Nuñoa	U.S.	Nuñoa	U.S.
6	62	66	61	65	6.9	7.6	6.6	7.3
12	71	75	69	74	7.9	10.1	7.3	9.7
24	76	87	75	87	9.9	12.6	9.0	12.3

data were collected by means of the technique developed by Bayley, who reported that the median ages at which U.S. children sat and walked alone were 5.7 months and 13.2 months, respectively (see 6).

Some of the growth characteristics in later development are shown in Figures 17-2 and 3. In these growth studies it was possible to compare our results with cross-sectional data for groups from lower elevations (Huánuco and Cajamarca, 2500 meters; Lima and Ica, 300 meters). We have also collected some semi-longitudinal data in order to evaluate growth rates. These combined data (7) showed, for Nuñoa children, (i) lack of a well-defined adolescent growth spurt for males, and a late and poorly defined spurt for females; (ii) a very long period of general body growth; and (iii) larger chest sizes, in all dimensions and at all ages, than those of children from lower elevations.

In explanation of the unusual growth aspects of the Nuñoa population, at least three hypotheses may be suggested: (i) all Quechua have an unusual growth pattern, genetically determined; (ii) malnutrition and disease are the prime causes; (iii) hypoxia is the major factor. Our present data are not adequate for testing these hypotheses. However, a number of observations suggest that hypoxia is a major factor. As discussed below, we have been unable to find any evidence of widespread malnutrition or of

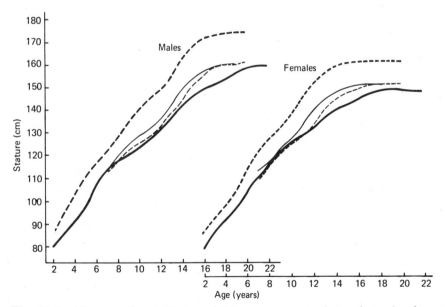

Fig. 17-2 The growth of Nuñoa children as compared to that of other Peruvian populations and of the U.S. population. (Heavy dashed lines) U.S. population; (light solid lines) Peruvian sea-level population; (light dashed lines) Peruvian moderate altitude 1990 to 2656 meters) population; (heavy solid lines) Nuñoa population (altitude 4268 meters).

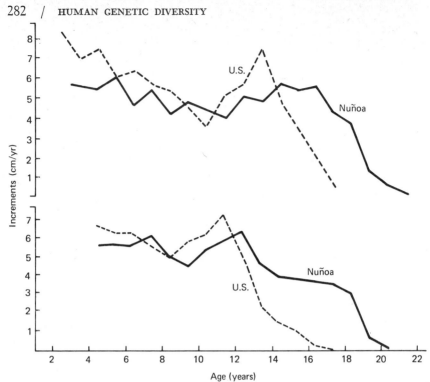

Fig. 17-3 Rates of general body growth for children in Nuñoa as compared with that for children in the United States. (Top) males; (bottom) females.

unusual disease patterns. What data are available on the growth of other Quechua show growth patterns different from those found for the Nuñoans (8). Finally, hypoxia has been shown to affect growth in a number of animals other than man (3).

WORK PHYSIOLOGY

The most striking effect of high altitude (4000 meters) on newcomers, apparent after the first few days of their stay, is a reduced capacity for sustained work. This reduction is best measured through measurement of the individual's maximum oxygen consumption. For young men from a sea-level habitat, the reduction is in the range of 20 to 29 percent; the men who had received physical training generally showed a greater reduction than the untrained (9). Some rise in maximum oxygen consumption occurs during a long stay at high altitudes, but studies extending over periods of as much as a year have failed to show a recovery to even near low-altitude values for adult men (10).

Maximum oxygen consumption for any individual or group is controlled by a large number of factors, among which the level of continuing exercise is of major importance. Among young men of European descent, mean values for maximum oxygen consumption range from below 40 milliliters of oxygen per kilogram of body weight for sedentary groups, through 45 milliliters per kilogram for laborers, up to more than 55 milliliters per kilogram for highly trained runners (11). The high degree of variability in this parameter makes it difficult to determine whether the native of a high-altitude region has a work capacity different from that of an individual from a low-altitude habitat, and makes it even more difficult to determine whether the complex physiological differences between groups from high and low altitudes have resulted in a better adaptation, with respect to work capacity, for the high-altitude Quechua.

In order to help clarify these questions, we determined maximum oxygen consumption for a number of carefully selected samples of contrasting populations; in all cases the method of determination was the same. Some of the results of these studies are presented in Table 17-3. It should be noted that the individuals referred to as White Peruvian students are so classified on the basis of morphology, and it is quite possible that they contain some admixture of native Quechua genes.

On the basis of results obtained for students alone, one investigator surmised that the differences in maximum oxygen consumption for high-altitude and low-altitude groups might be only a matter of life-long exposure, plus a high state of physical fitness in the highland Quechua (12). In an article based on partial results (13), some of the investigators, including myself, pointed out that trained athletes from lower altitudes could, at high altitudes, achieve the same oxygen consumption per unit of body weight as the Nuñoa native. While the data now available support the idea that physical training and life-long exposure to hypoxia act to increase maximum oxygen consumption, these two factors appear insufficient to explain the total results. The data, instead, suggest that a fairly random sample of Nuñoa males between the ages of 18 and 40 have a vastly greater maximum oxygen consumption than a group of reasonably physically fit researchers from the United States. The oxygen consumption of the Nuñoans also significantly exceeded that of young native students from lower altitudes, and equaled that of a group of highly trained U.S. athletes who had spent a month at high altitude. Furthermore, the heart rate and ventilation rate for the Nuñoans remained low. The Nuñoans do walk more than people from the United States, but nothing in the personal history of the Nuñoa subjects suggested that they had experienced physical training or selection comparable to that involved in becoming a college track athlete. To me, the data suggest that a high-altitude Quechua heritage confers a special capacity for oxygen consumption at 4000 meters. In the absence of more precise data, such a conclusion remains tentative. However, the data can certainly be interpreted as showing that the Nuñoa native in his high-altitude habitat has a maximum oxygen consumption equal to, or above, that of sea-level dwellers in their oxygen-rich environments. This conclusion is in agreement with the results obtained by Peruvian researchers (2).

COLD

THE MICROENVIRONMENT

By the standards of fuel-using societies, the temperatures in the Nuñoa district are not very low, and we would consider the typical daily weather equivalent to that of a pleasant fall day in the northern United States. However, the lack of any significant source of fuel made us suspect that at least some segments of the population might suffer from significant cold stress. As mentioned above, few trees grow in the district, and those that do are slow-growing. At present, these trees are used primarily as rafters for houses; only an occasional member of the upper class uses wood for cooking. The fuel used almost universally for cooking is dried llama dung or alpaca dung. This dung provides a hot but rapid fire, and is burned in a clay stove, which provides little external heat. Since cooking is done in a building separate from the other living quarters, only the women and children benefit from the fire. Bonfires are lit only on ceremonial occasions; the winter solstice is celebrated by a pre-Hispanic ceremony in which many bonfires are lit all over the district. The use of fires during solstice ceremonies throughout the world is often considered an act of sympathetic magic to recall the sun. For Nuñoans it may also recall the warmth.

The houses of the native pastoralists, with walls of stacked, dry stone and roofs of grass, provide no significant insulation. Measurements made within the dwelling units generally showed the temperature to be within 2° or 3°C of the outdoor temperature. This is in sharp contrast to the situation in the adobe houses used by the upper classes, by some agriculturalists in the Nuñoa district, and by all classes in slightly lower areas of the Altiplano. Adobe houses provide good insulation, and indoor temperatures are frequently 10°C above outdoor temperatures during the cold nights of the dry season.

From this analysis and other observations we concluded that the Nuñoa native depends upon his own calories for heat, and relies, for heat conservation, primarily upon his clothing and upon certain customs, such as spending the early evening in bed and having as many as four or five individuals sleep in the same bed. As shown in Figure 17-4, his clothing is layered and bulky, with windproof materials on the outside. Thus, it provides good insulation for his body. However, the Nuñoa native's wardrobe does not include gloves, and the only foot coverings are sandals occasion-

TABLE 17-3

Some data from tests of maximum oxygen consumption (max Vo₂) at high and low altitudes for contrasting populations.

GROUP OF SUBJECTS	Altitude at which tested (m)	Duration of exposure to high altitude	Number	Mean age (yr)	Mean height (cm)	Mean weight (kg)	Max Vo₂ (liter/ min)	Aerobic capacity[2] (ml/kg/ min)	Maximum ventilation (liter/ min)	Ventilation equivalent[3] (V/Vo₂ min)	Maximum heart rate (beat/ min)	Oxygen pulse (ml/ beat)
					SUBJECT			RESPONSE[1]				
Nuñoa Quechua	4000	Life	25	25	160	57	2.77	49.1	75	27.3	171	16.0
U.S. white researchers	300		6	30	183	79	3.92	50.4	131	33.7	185	21.2
U.S. white researchers	4000	4 weeks+	12	27	181	75	2.78	38.1	91	32.9	173	16.6
U.S. white athletes	300		6	20	179	71	4.58	64.2	131	28.8	175	26.5
U.S. white athletes	4000	4 weeks	6	20	179	71	3.14	46.6	105	33.7	172	19.4
Quechua from sea level	100		10	22	160	62	3.01	49.3	108	36.2	187	16.7
Quechua from sea level	4000	4 weeks	10	22	160	62	2.67	44.5	87	33.4	190	14.5
Peruvian students Quechua	3830	Life	10	23.8	162	60	2.79	46.8	72	25.8	188	15.1
White	3830	Life	13	23.5	169	61	2.62	42.8	74	28.2	186	14.2

[1] The measurements were made by means of a bicycle ergometer in 10-minute progressive exhaustion tests.
[2] Aerobic capacity is the maximum oxygen consumption per kilogram of body weight and, as such, is the most significant measure available of the success of the individual's (and, by inference, the group's) biological oxygen transport system. It is also assumed to be one of the best measures available for judging an individual's work capacity relative to his body size. The lower the value, the greater the relative efficiency in supplying oxygen.
[3] Ventilation equivalent is the ventilation volume per unit oxygen uptake per unit time.

Fig. 17-4 A Nuñoa native beside his house during a brief snow squall.

ally worn by men. The insulating effect of native clothing was tested under laboratory-controlled cold conditions. It was found that at 10°C the clothing increased mean body temperature by about 3°C and raised the temperature of hands and feet despite the lack of gloves and shoes.

From the total assessment we concluded that the Nuñoa native probably experiences two types of exposure to cold: (i) total-body cooling during the hours from sunset to dawn and (ii) severe cooling of the extremities, particularly in the daytime during periods of snow and rain. To assess the degree of stress due to cold we took measurements of rectal and skin temperatures of individuals in samples selected by sex, age, and altitude of habitat. These studies indicated that at night the adult women experience very little stress from cold, whereas adult men showed some evidence of such stress. During the day, women, because they are less active than men, may experience slight stress from cold, whereas men do not, except for their extremities. Active children show no evidence of such stress; however, during periods of inactivity, as at night, their skin temperature and rectal temperature are low. Indeed, at these times, all indices show an inverse relationship between age, size, and body temperatures (14).

Physiological Responses

In order to characterize the Nuñoa native's responses to cold, we used three types of laboratory exposures: (i) total-body cooling at 10°C with nude subjects, for 2 hours; (ii) cooling of the subject's hands and feet at 0°C, for 1 hour; (iii) cooling of the feet with cold water. The subjects were Nuñoa males and females, North American white males, and Quechua males from low-altitude habitats.

Some results of the studies of total-body cooling are summarized in Table 17-4. Since the Nuñoa Indians are smaller than either the coastal Indians or U.S. whites, the data are presented in terms of surface area (15). Viewed in this way, the data show that the native Quechua from a high-altitude habitat produced more body heat during the first hour than individuals from low-altitude habitats, but produced amounts similar to those for such individuals during the second hour. By contrast, heat loss was much greater for the Nuñoans than for members of the other groups during the first hour and similar during the second hour. The findings on heat loss are perhaps the more interesting, since they conform with results of two other studies of cooling responses in native Quechua from high-altitude habitats (16).

When the source of the greater heat loss was closely examined, it proved to be almost entirely the product of high temperatures of the extremities, and these temperatures, in turn, seem to be produced by a high flow of blood to the extremities during exposure to cold. When a comparison was made, as between Nuñoa males and females, of the temperatures of the extremities, the women were found to have warmer hands

TABLE 17-4

Exchange of body heat as exemplified in heat production and heat loss. The values are averages for two exposures at 10° C, expressed on the basis of body-surface area.

SUBJECTS	NUMBER IN SAMPLE	HEAT PRODUCTION (KCAL/M²)			HEAT LOSS[1] (KCAL/M²)		
		1st 60 minutes	2nd 60 minutes	Total time	1st 56 minutes[2]	2nd 60 minutes	Total time
Whites	19	51.5	62.2	113.7	29.6	11.8	41.4
Nuñoans	26	58.1	65.5	123.6	51.0	12.9	63.9
Lowland Indians	10	54.7	65.5	120.2	30.1	14.7	44.8

[1] Heat not replaced by metabolic activity.

[2] Because perfect equivalence in body temperature prior to exposure to cold was not achieved, heat loss in the first 4 minutes of exposure has been excluded from this calculation.

and somewhat colder feet. In both sexes the temperature for hands and feet were significantly above corresponding values for white male subjects. The specific studies of hand and foot cooling made with exposures of types ii and iii shed further light on the subject, showing that the maximum differences between populations occurred with moderate cold exposure; that the high average temperatures of hands and feet were the result of a slow decline in temperature, with less temperature cycling than is found in whites; and that the population differences were established by at least the age of 10 (the youngest group we could test) (17).

Since the oxygen exchange between hemoglobin and tissue bears a close positive relationship to temperature (18), it is clear that, when the temperature of the peripheral tissues remains high, more oxygen is available to these tissues. Therefore, the high temperatures of the extremities of the Nuñoans at low atmospheric temperatures may be considered not only an adaptation to cold but also a possible adaptation to hypoxia.

NUTRITION AND DISEASE

In the complex web of adaptations that are necessary if a peasant society is to survive, adequate responses to nutritional needs and prevalent diseases are always critical. For a people living at high altitude, these responses are important to an interpretation of the population's response to altitude and cold.

NUTRITION

The analysis of nutritional problems in the Nuñoa district has proceeded through a number of discrete studies, including a study of dietary balance in individuals, a similar study for households, an analysis of food intake by individuals, and a study of the metabolic cost, to the community, of food production. Of these studies, the first two have been completed, the third is in the analysis stage, and the fourth is in the data-collection stage. The results to date suggest that the Nuñoa population has a very delicate, but adequate, balance between nutritional resources and needs.

The dietary-balance study was carried out with six native adult males, chosen at random. Food requirements were predicted from U.N. Food and Agriculture Organization standards, on the basis of weight and temperature. The food used in the study consisted wholly of native foodstuffs and was prepared by a native cook. The results showed that, for these individuals, protein, caloric, and fluid balance remained good, and indicated that caloric and protein balance was good prior to the time of the study (19). The household survey suggested that nutrition for the population as a whole was generally adequate, although the method did not permit conclusions on the adequacy of nutrition for special subgroups, such as children and pregnant women (20).

The household survey also suggested that the diet might be somewhat deficient in vitamin A and ascorbic acid. Subsequent, more detailed surveys of individuals now cast doubt on the validity of this conclusion and suggest that, if malnutrition exists, it is probably no more common than in U.S. society. Indeed, in the light of the modern concept of "overnutrition," we might even say that the Nuñoans have a better dietary balance than the U.S. population. As noted above, the balance is delicate, and there must be years and times of the year when certain dietary deficiencies exist. Furthermore, the balance is subtle. To cite an example, the basic foods available are very low in calcium, yet adequate calcium is obtained, primarily by use of burned limestone as a spice in one type of porridge (21).

DISEASE

Our data on infectious disease are particularly inadequate. Health questionnaires are almost useless, since native concepts of health are only partially related to modern medicine. The *indígenas* attribute over 50 percent of all illness and death to *susto*, a word best translated as "fright." As noted above, no regular medical treatment is available, so records are lacking, even for a subsample. In our general survey we encountered the usual variety of infectious diseases and had the impression that respiratory ailments, such as tuberculosis and pneumonia, were common. On the other hand, we did not find evidence of deficiency diseases—not even goiters.

Perhaps the most striking results of the survey were those relating to cardiovascular disease. In the survey, heart murmurs were common among children, but no evidence of myocardial infarction or stroke was seen. Casual blood pressures of individuals from age 10 to 70+ were taken. They revealed a complete absence of hypertension; the highest pressure encountered was 150/90 mm-Hg. Other researchers have reported similar results for high-altitude populations, and it has been suggested that hypoxia may directly reduce the incidence of hypertension (22). In an attempt to trace the etiology of the low blood pressures, we subdivided our sample into a series of paired groups, first into lower- and higher-altitude groups, next into urban and rural groups, finally into more acculturated and less acculturated groups.

Significant differences in the effect of age on blood pressure appear when the sample is divided on the basis of any of these three criteria. However, it is not possible to assess the extent to which the environmental factors are independently related to blood pressure, since the total sample is too small to provide six independent subsamples large enough to give meaningful analytical results. It is our present belief that acculturation is the most significant of the factors, since altitude, urban residence, and acculturation are interrelated within the Nuñoa district population and the group differences are most striking when acculturation is taken as the criterion of subdivision. Children aged 10 to 20 years were classified as "acculturated" if they were in school, whereas children in the same age bracket who were not in school were classified as "unacculturated." Among adults, evidence of schooling, knowledge of Spanish, and use of modern clothing and specific material items, such as radios, were taken as signs of acculturation.

The results of these comparisons are shown in Figures 17-5 and 6. Certainly the regressions cannot be taken as evidence that altitude does not affect blood pressure, since none of the Nuñoa males, even those in the acculturated group, have high blood pressures in old age. However, the analysis does show that, within a native population living at high altitude, something associated with the process of acculturation into general Peruvian society leads to significant increases in systemic blood pressure with age. Similar results have been reported for peasant and "primitive" populations at low altitudes, and some researchers have attributed

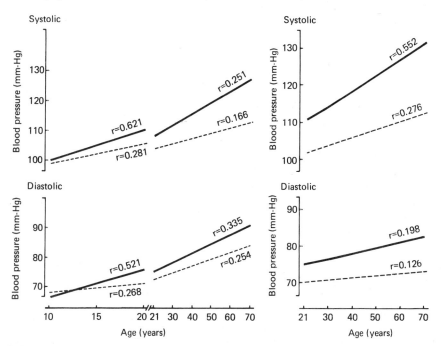

Fig. 17-5 (left)   Changes with age in the blood pressure of male Nuñoa natives, according to level of acculturation. (Solid lines) More acculturated; (dashed lines) less acculturated; r Pearson correlation coefficient. Fig. 17-6 (right)   The same data for female Nuñoa natives.

the increase to psychological stress associated with modern culture (23). Such an explanation might apply to the results of our study, but it does not appear safe to conclude that this is the case before carefully examining nutritional and disease correlates. The available nutritional data are being examined for evidence of possible nutritional differences between the groups.

DISCUSSION AND CONCLUSIONS

From earliest recorded history it has been recognized that men from different populations vary in physical (that is, anatomical) characteristics and in cultures, but it is only recently that the variety of *physiological* differences has been revealed. The physiological differences so far shown are of the same general magnitude as the anatomical variations. That is, the available information suggests a basic commonality with respect to functions such as temperature regulation, energy exchange, and response to disease, but comparison of different populations has revealed a number of specific variations in response to environmental stress.

Probably the most controversial aspect of these new findings concerns the mechanisms underlying the physiological differences. Because of the time and expense involved in studying the problem, population samples have often been small. Thus, with respect to specific findings, such as the high oxygen-consumption capacity reported by some workers for natives living at high altitudes, it has been suggested that biased sampling explains the difference. Other differences have been explained in terms of short-term acclimatization or of variations in diet or in body composition. Of course, genetic differences and long-term or developmental acclimatization have also been suggested, but the short-term processes have been more commonly accepted as explanations because they are based on known mechanisms.

We believe that a more general application of the extensive and intensive methods used in studying the Nuñoa population is a necessary next step in the search for the sources of differences in functional parameters in different populations. Thus, with respect to the Nuñoa population, a study of growth alone would probably have led to the conclusion that the observed differences in growth were the result of malnutrition.

On the other hand, the results of the growth study considered together with results of detailed studies of nutritional and other responses suggest hypoxia as a better explanation. Similar examples could be cited, from other aspects of the program, to show that a set of integrated studies of a single population can provide insights not obtainable with data pertaining to a single aspect of population biology.

In general, the data are still not adequate to treat the third question originally posed, on the sources of adaptation. Indeed we cannot even clearly differentiate genetic factors from long-term and developmental acclimatization. For this purpose one would require comparable data on several populations that vary in genetic structure, in altitude of habitat, and in other aspects of environmental background. Fortunately, collection of such data is contemplated as part of the Human Adaptability Project of the International Biological Programme. The importance of understanding the sources of population differences seems obvious in a world where the geographical and cultural mobility of peoples is greater than it has ever been throughout history.

It seems clear that the native of the high Andes is biologically different from the lowlander, and that some of the differences are the result of adaptation to the environment. How well and by what mechanisms a lowland population could adapt to this high-altitude environment has not yet been adequately explored. Moreover, almost nothing is known about the biological problems faced by the highlander who migrates to the lowlands.

SUMMARY

The high-altitude areas of South America are in many ways favorable for human habitation, and they have supported a large native population for millennia. Despite these facts, immigrant lowland populations have not become predominant in these areas as in other parts of the New World, and lowlanders experience a number of biological difficulties on going to this region.

In order to learn more about the adaptations which enable the native to survive at high altitudes, an intensive study of a native population is being carried out in the district of Nuñoa in the Peruvian Altiplano. In this area hypoxia and cold appear to be the most unusual environmental stresses. Results to date show a high birth rate and a high death rate, the death rate for females, both postnatal and prenatal (as inferred from the sex ratio at birth), being unusually high. Birth weights are low, while placenta weights are high. Postnatal growth is quite slow relative to the rate for other populations throughout the world, and the adolescent growth spurt is less than that for other groups. The maximum oxygen consumption (and thus the capacity for sustained work) of adult males is high despite the reduced atmospheric pressure at high altitude. All lowland groups brought to this altitude showed significant reductions in maximum oxygen consumption. The Nuñoa native's responses to cold exposure also differ from those of the lowlander, apparently because blood flow to his extremities is high during exposure to cold. The disease patterns are not well known; respiratory diseases appear common, whereas there seems to be almost no cardiovascular disease among adults. Systemic blood pressures are very low, particularly those of individuals living in traditional native fashion. Nutrition appears to be good, but analysis of the nutrition studies is continuing.

The results of these studies are interpreted as showing that some aspects of the natives' adaptation to high altitudes require lifelong exposure to the environmental conditions and may be based on a genetic structure different from that of lowlanders.

REFERENCES AND NOTES

1. C. Monge, *Acclimatization in the Andes* (Johns Hopkins Press, Baltimore, 1948).
2. *Life at High Altitudes; Proceedings of the Special Session held during the Fifth Meeting of the PAHO Advisory Committee* (Pan American Health Organization, Washington, D.C., 1966); *Arch. Inst. Biol. Andina* (this is a new journal devoted entirely to the subject of acclimatization at high altitude; address, Apartado 5073, Lima, Peru).
3. E. J. Van Liere and J. C. Stickney, *Hypoxia* (Univ. of Chicago Press, Chicago, 1963).
4. J. P. McClung, *Amer. J. Phys. Anthropol.* 27, 248 (1967).
5. C. Hoff, thesis, Pennsylvania State University (1968).
6. T. S. Baker, A. V. Little, P. T. Baker, "Infant Development at High Altitude," in preparation.
7. P. T. Baker, R. Frisancho, R. B. Thomas, in *Human Adaptability to Environments and Physical Fitness*, M. S. Malhotra, Ed. (Defense Institute of Physiology and Allied Sciences, Madras, India, 1966).
8. R. Frisancho, thesis, Pennsylvania State University (1966).
9. E. R. Buskirk, J. Kollias, R. F. Akers, E. K. Prokop, E. Picon-Reategui, *J. Appl. Physiol.* 23, 259 (1967).
10. From results of a study of adult men over a 1-year period and inferences from cardiovascular changes found only in the high-altitude newborn, Velasquez concludes that adults from lowland regions can never adapt to high altitudes as well as natives do [T. Velasquez, in *The Physiological Effects of High Altitude*, W. H. Weihe, Ed. (Macmillan, New York, 1964)].
11. K. L. Andersen, in *The Biology of Human Adaptability*, P. T. Baker and J. S. Weiner, Eds. (Clarendon, Oxford, 1966).
12. R. B. Mazess, thesis, University of Wisconsin (1967).
13. J. Kollias, E. R. Buskirk, R. F. Akers, E. K. Prokop, P. T. Baker, E. Picon-Reategui, *J. Appl. Physiol.* 24, 792 (1968).
14. J. M. Hanna, thesis, University of Arizona (1968); P. T. Baker, in *Human Adaptability and Its Methodology*, H. Yoshimura and J. S. Weiner, Eds. (Japan Society for the Promotion of Sciences, Tokyo, 1966).
15. P. T. Baker, E. R. Buskirk, J. Kollias, R. B. Mazess, *Human Biol.* 39, 155 (1967).

16. P. T. Baker, in *The Biology of Human Adaptability*, P. T. Baker and J. S. Weiner, Eds. (Clarendon, Oxford, 1966); R. W. Elsner and A. Bolstad, "Thermal and Metabolic Responses to Cold Exposure of Andean Indians at High Altitude," *Ladd Air Force Base Tech. Rep. AALTOR 62–64* (1963).
17. M. A. Little, thesis, Pennsylvania State University (1968).
18. The conclusion is based on the shift in the oxygen disassociation curve produced by lowering blood temperature. A standard graph, such as that presented by Consolazio and his associates [C. F. Consolazio, R. E. Johnson, L. J. Pecora, *Physiological Measurements of Metabolic Functions in Man* (McGraw-Hill, New York, 1963), p. 112] illustrates this effect.
19. E. Picon-Reategui, "Food Requirements of High Altitude Peruvian Natives," unpublished.
20. R. B. Mazess and P. T. Baker, *Amer. J. Clin. Nutr.* 15, 341 (1964).
21. P. T. Baker and R. B. Mazess, *Science* 142, 1466 (1963).
22. M. J. Fregly and A. B. Otis, in *The Physiological Effects of High Altitude*, W. H. Weihe, Ed. (Macmillan, New York, 1964).
23. N. A. Scotch, *Amer. J. Public Health* 53, 1205 (1963).

18

THE DISTRIBUTION OF GENETIC DIFFERENCES IN BEHAVIOR POTENTIAL IN THE HUMAN SPECIES

Benson E. Ginsburg and William S. Laughlin

In man as well as animals, observable behavior is the result of a complex and subtle set of interactions between the genetically determined potentialities of the organism, and the kinds of experiences presented by the environment. In man, cultural institutions play a major role in shaping the experiences through which the

Reprinted from *Science and the Concept of Race*, edited by Margaret Mead, Theodosius Dobzhansky, Ethel Tobach, and Robert E. Light. New York: Columbia University Press, 1968, pages 26–36. By permission of the authors and the publisher.

*individual passes. "Is heredity more important than environment?"
is the familiar question-begging format for the kinds of inquiry
required so that we can ultimately understand the precise ways in
which genotype and environment (culture) together produce the
human behavioral phenotypes which we can observe, test and
compare. Intelligence tests are thus means of describing and
comparing phenotypes.*

*Performance tests purporting to measure mental abilities have
been employed for decades. They have been utilized mainly as
devices for placement in educational systems, such as public schools,
and military training programs. It was unavoidable that the results
of such testing should be used also as a means of documenting
and supporting ideas of the relationships, social, political and
economic, between Black people and White, as structured by our
social system. The results of mental ability tests have been employed
for ideological warfare: On the one hand to pretend that they truly
establish the "innate intellectual inferiority" of submerged groups;
on the other hand, as evidence of the systematic suppression of
minorities by the "white establishment." The significance of such
tests thus has been drowned in waves of angry rhetoric. The question
whether it is ever possible to design genuine culture-free instruments
to isolate genetic potentials for valid genotype comparisons has
been changed to the question whether it should ever be attempted.*

*Against this background investigators have tried to formulate
the complex problems which must be solved before tentative answers
can be generated. How can behavioral genotypes be identified?
How can groups be compared?*

*It should be repeated here that most anthropologists reject the
use of any measures of intellectual abilities to prop up doctrines
of racial inferiority. Our species seems to have been the product
of phyletic evolution, with arrays of interbreeding hominid populations becoming transformed as a single unit into more progressive
hominid populations of* Homo sapiens. *There is no evidence that
any clusters of human population were isolated long enough for
significant differences in mental potential to result. With "intelligence" as a human adaptation to culture and culture building, it
is hardly likely that the genes for stupidity could ever have been
adaptive for any human populations.*

*The following paper, read at a symposium on the race concept
in the sciences, is a sober attempt to define the problems—and even,
in this context, such terms as race itself. The authors suggest ways
that study of laboratory animals can illuminate the relation between
genes and behavior, particularly mental abilities.*

Dr. Ginsburg is in the Department of Biology, and Dr. Laughlin is in the Department of Anthropology, at the University of Connecticut.

The topic of the symposium, "The Utility of the Construct of Race," appears to invite us to choose sides, or at least to vote *yes* or *no* regarding the usefulness of this concept. However, the topics of the three sessions of the symposium—"Behavior-Genetic Analyses and Their Relevance to the Construct of Race," "Biological Aspects of Race in Man," and "Social and Psychological Aspects of 'Race' "—tell us that the rubric "race" and the various concepts with which it has been identified are with us, have been with us, and will continue to be with us, whatever euphemisms we may attempt to substitute for the controversial four-letter word, and however much we may deplore the political, cultural, and biological misuses to which it has been subjected. We should remind ourselves that the term *race* does not have a merely human connotation. There are races of fruit flies, mice, and plants. As Professor Dobzhansky has pointed out elsewhere in this symposium, if there were no such construct, we should have had to invent it in order to account for local genetic differences between population groups which are only partially isolated from each other and continue to exchange genes but also to maintain some obvious differences. It is our concern to demonstrate that current evidence must be interpreted as indicating that behavioral equipotentiality exists for all such populations of reasonable size and that populations can direct their areas of biological achievement by internal genetic restructuring. Behavioral and social forces that direct such restructuring constitute the major selective agencies determining the further evolution of the human species.

The human species is not panmictic and was undoubtedly less so in the past than it is today. Its subpopulations are separated by barriers of distance, geography, language, religion, and a host of other cultural customs, ensuring that now, as in the past, the total species remains a breeding reticulum composed of partial isolates within which there is significantly more genetic exchange than there is between component groups, some of which are as recognizably different from each other as are races, breeds, and varieties of other species. Physically we come packaged in a range of colors and a variety of body types. Particular clusters of these attributes are not randomly distributed, but, rather, differentiate Eskimos from American Indians from Indian Indians from a variety of Mongolian types, a myriad of Negro types, and a diversity of Caucasoids. We do not dismiss or minimize these physical differentiators—and could not if we would. They serve primarily as convenient external markers by which we identify individuals and classify populations.

The identifying hallmarks differentiating component populations within the human species are more than skin deep. They extend to skeletal structures, blood groups, the prevalence of various abnormalities such as sickle-cell anemia, Tay-Sachs disease, and favism, and if we are to be quantitative rather than qualitative, to a variety of physiological, clinical, and behavioral measures.

Man is not unique in these respects. Physical, physiological, and behavioral differences have been described for subpopulations within many other well-studied animal species. In fact, it would be nothing short of remarkable if we were to find that the Ainu and the Zulu were alike in their genetic capacities and therefore in their morphological, physiological, and behavioral characteristics. There is neither a moral nor a scientific dilemma, so far as the physical attributes are concerned, as no superiority or inferiority inheres in skin color, hair texture, or body conformation, *per se*, throughout the normal range of variation. Nor do we pretend that there is genetic egalitarianism with respect to the way these properties are distributed among the various "races" within the human species. Despite demonstrations that there are dietary effects on growth and hormonal effects on pigment, we find no need to place exaggerated emphasis upon these environmental contributions to differences in stature or skin color, or to denigrate the biological components of these aspects of phenotype.

When it comes to physiological, biochemical, or immunological differences, we are similarly unconcerned from a moral or sociological point of view. When it comes to differences in behavioral potential, however, the issues suddenly become explosive to the point where the mere admission of the possibility of comparable biological differences in this realm becomes, not an object for discussion and scientific study, but one for moral censure and apologistic adventurism.

We suspect, therefore, that there are two avenues open to those working in behavior genetics and biological anthropology. One avenue is to maintain (as some anthropologists have) that biologically based differences in behavioral capacities are superficial, and that all important human behavioral capacities are so fundamental to the entire species that all subgroups share these as a common evolutionary heritage. On this view, individual differences in behavioral capacities within races of moderate to large population size constitute an adequate sample of the totality of human behavioral variation, and significant differences in the behavioral realm between human ethnic groups are therefore culturally determined. Genetic differences could have been obtained only by sustained differential selection, and proponents of the view that there are no important biological differences in behavioral capacities between diverse human population groups generally argue that selection for behavioral differences would have had to be intense, consistent, and of long duration to accomplish such a result. They maintain that this has not been the case; they are critical of the possibility that important sampling differences may have occurred; and they "explain" any seemingly contradictory data on the basis of cultural determination.

A second avenue is provided by the argument that animal data, for one reason or another, may not have relevance to man, since human data rest on such a complex of nature-nurture interactions in which *nature* is represented by an unanalyzable polygenic system, that no analyses on human populations in this area are presently meaningful. Many social anthropologists, sociologists, and educators make the further argument that nature is something we can do nothing about, while we have only begun to scratch the surface of what can be accomplished by manipulating cultural, social, and psychological variables, and that these, therefore, represent the hopeful frontiers of research. Behavioral genetic research on the human species, they say, is premature and, moreover, dangerous, since it lends itself to a racist point of view and is, therefore, inimical to any civilized concept of human rights and elementary political or personal decency.

We are here to argue that, if anything, the reverse is true. The pretenders to having cultural equalizing devices go far beyond their data in offering these as panaceas. In fact, recognition of genetic inequalities in behavior potential within the different breeding partial isolates of the human species and characterization of these differences permit us to both maximize the behavioral endowments we possess and offer the major scientific hope for upgrading our biological condition. We are not here considering the yet unrealized possibility of rewriting the DNA code script in such a way as to correct for genetic deficiencies at the molecular source. Neither are we arguing that groups need to share their genetic endowments by interbreeding in order to achieve behavioral parity—a notion which assumes the inherent superiority of one group over another and seeks to upgrade the "inferior" groups by introducing more favorable genes from a better endowed population. The gist of the genetic argument is quite different and entails no need for looking at things other than as they are in order to point the way clearly toward making them more nearly as we wish they were.

Behavior genetic analyses have demonstrated the following relevant facts:

1. A measure of central tendency with respect to a behavioral attribute of a genetically variable group provides very little useful information. The isomorphism of the attribute being measured with the way in which the organism's behavioral capacities are organized is not to be inferred from the fact that members within a given population can be measured in this particular respect. Individuals graded alike on such a measure may yet be genetically different with respect to the behavioral potential the test purports to measure. In genetic terms, the attribute may be only one aspect of a natural phenotype, or may confound overlapping regions between a variety of such phenotypes as well as confound a phenotype with its

phenocopy. Moreover, even if the attribute tested corresponded to the natural organization of behavioral capacities, there is more than one genetic way to a given phenotype, as well as a variety of nongenetic routes to a phenocopy (1, 2).

2. While it is true that behavior is complicated, and that behavioral capacities are likely to be affected by many genes, it does not follow that the genetic substratum for such capacities are unanalyzable or that particular major genes do not exert major determining influences on these capacities or potentials.

3. Many seemingly highly correlated phenomena have no necessary causal connections so that data demonstrating correlations among physical, physiological, and behavioral attributes in particular populations must be further interpreted in order to evaluate the significance of the correlations obtained (3).

In order to demonstrate the relevance of these three desiderata for the topic in question, we should like to cite several examples from the behavioral realm.

Strain differences in susceptibility to sound-induced seizures in mice have been heavily researched by a number of laboratories for more than three decades (4, 5, 6, 7, 8, 9). A variety of genetic hypotheses have been offered, ranging from monogenic determination to complex polygenic models for this easily recognizable phenotype. Every important neural transmitter and several hormones have been correlated with this behavior. A number of pigment genes also have been shown to be associated with it. The periods of maximum susceptibility for a given strain or genotype have been authoritatively but variously described—all this with exactly the same strains (members of which are very nearly genetic replicates) under comparable conditions of rearing and testing. Strain differences in aggressiveness of male mice have also been reliably differentiated, with similar remarkable differences in results (1, 10). In both these cases, and in many others, further analytic work has demonstrated that there are individual major genic differences contributing to the characteristics in question and that these can be analyzed one at a time in terms of some of their intermediary metabolic and physiological effects, which may then, in turn, be related to the behavioral capacities (1, 2). Each genetic substitution can be put on a variety of different inbred strain backgrounds, so that interaction effects may also be studied. These interactions account for differences in scaling and timing, and for some of the other seeming discrepancies in the earlier literature. It is possible also to demonstrate, by this precise genetic methodology, that some of the correlations with pigment genes, neural transmitter levels, and endocrine factors result from linkage or other noncausal associations, while others are not dissociable from the genetic substitution and must be considered to be causally related to the differences in behavior under study. Through the identification of some of the gene-controlled mechanisms, the "natural phenotype" can be inferred. In the examples analyzed by this method, the phenotype is invariably broader than the original attribute discovered from either observation or testing.

Differences in results on mouse strain aggressiveness are particularly instructive because some can be resolved by instituting different treatments in the very early pre-weaning period. How animals respond to a variety of handling and other regimes, and when they are maximally responsive to such treatments, is a function of strain and, therefore, of genotype (1). Similar differences in aggressiveness and emotionality in relation to early handling paradigms have been reported for various breeds of dogs (11). Selection experiments with both mice and dogs have demonstrated that it is possible to affect behavioral capacities profoundly within 7 to 10 generations, and that this can be achieved from an initial population that has been selected to quite a different set of behavioral criteria, if this base population is genetically variable (3, 10). In these instances, phenotypic selection for the first set of criteria included a variety of genetic ways of arriving at the phenotype in a normative spectrum of environmental conditions. Selection not only can move the population to a new phenotypic norm without resorting to crosses with outside individuals but it also can repackage the behavioral and morphological attributes so that a particular color type and body conformation that was previously associated with a given constellation of behavioral characters, either by accident or selection, can later become associated with quite a different behavioral profile (12).

If these arguments are extended to the human species, similar inferences may be drawn despite the broader behavioral lability and the obviously greater effects of experiential factors. Differences in sensitive periods for response to stress as well as the effects of such early stimulation on the direction and magnitude of later emotional behavior rest on homologies of mechanisms that can be investigated, as well as on analogies of behavioral effects. Differences in intelligence (however we may parse its constituent capacities), in perceptual abilities, musical ability, mathematical ability, and many other aspects of human achievement and performance show high heritability (13). While it is obvious that individuals placed in evironments where these capacities are undetected and unnurtured may never exhibit their potential, the converse is also true.

The search for meaningful behavior phenotypes in human populations, which has leaned heavily on sibship and twin data, must be expanded to include a comparative study of the entire species and especially of small population isolates in which there is appreciable consanguinity. Such isolates may be expected to provide a variety of samples from the total gene pool and to favor the expression of various recessive and otherwise buffered genes. Most such studies have thus far included relatively narrow samples of behavioral attributes.

It would seem, by analogy with animal experiments, including homologies of structure and mechanism, that human behavioral phenotypic variability is affected by many genetic factors and that the human genome, no less than the nonhuman, has ensured itself of a variety of pathways to similar behavioral capacities. The inference that follows from the genetic dissimilarities in behavior potential among the diverse populations comprising the human species is not that these differences, which inhere in the populations, cannot be bridged without an exchange of genes, but rather that a tremendous amount of genetic variability with respect to behavioral potential has been assimilated to a much narrower range of phenotypic expression. Many of these differences can be detected and magnified by cultural devices, which also affect the structuring of the gene pool, whether or not genetic exchange between population groups is involved.

Teleologically speaking, it is advantageous for a population to contain a high amount of genetic variability, as this represents the future evolutionary potential of the species. Wright has estimated that such genotypic variability may amount to an average of ten alleles for each genetic locus (14). The process of biparental reproduction magnifies this variability exponentially by recombining the genetic constellations that may occur in a large, random-breeding population in proportion to the frequency of occurrence of each allele, restricted by the phenomenon of linkage, expanded by the process of mutation, and guided by natural selection (14). Natural selection can act only indirectly to modify the frequency of a given gene through the average fitness of the phenotypes that result from its occurrence over a range of combinations with other genes. If each unique genetic combination expressed itself equivalently on a scale of equal increments, the phenotypic variability for each genetically affected attribute of an organism would be almost infinite, and a population would consist of a collection of highly disparate individuals in appearance, behavior, and physiology. We should, all of us, be freaks, by comparison with one another. Despite our individual genetic uniqueness, except for monozygotic twins and members of clones and of highly inbred strains, we are relatively minor variants of a modal phenotypic theme that is more narrowly defined by the race or variety representing the pool of unrestricted genetic exchange to which our immediate genealogies may be traced, and more broadly characterized by the subspecies or species to which the narrower taxonomic categories (which exchange genes much more freely within the populations by which they are defined than between them) are genetically allied.

Clearly, the large amount of genetic variability in a population is severely buffered in its very limited phenotypic expression. These modal expressions represent the adaptive phenotypes resulting from the evolutionary history of each taxonomic group (15). The buffering devices include dominance, epistatic interactions, and a variety of limitations on penetrance and expressivity, many of which are now more completely understood as a result of recent findings in physiological and developmental genetics. Population groups may, therefore, share similar phenotypes but differ significantly genotypically. Correlatively, severe phenotypic selection within a noninbred population can produce the appearance of uniformity by means of effects on marker genes and on genes that modify the expression of other genes, while the population remains genetically variable to a degree that cannot be inferred from the range of phenotypic variability exhibited under natural conditions. Such evolved normative phenotypes are, in this sense, highly buffered and can assimilate relatively high amounts of genetic infusion from neighboring populations having somewhat different phenotypic profiles to their own normative phenotypes. This is evidently the case with white and Negro populations in the United States, where the amount of genetic exchange has been estimated to be on the order of 30 per cent (16).

Given the common evolutionary history of the human species, the multiplicity of genetic routes to a phenotypic potential, the assimilation of genetic variability to a comparatively narrow range of phenotypic expression, the relative lack of inbreeding, and the sharing of genes between various human populations both now and in the past, the existence of recognizable subdivisions of the human species that differ from each other genetically and phenotypically cannot be taken to mean that abilities more commonly encountered in one group than in another under present conditions of ascertainment and culture necessarily reflect the true biological differences between the groups under comparison. One does not expect such groups to be genetically or phenotypically identical. Dissimilarities, however, are not in themselves measures of biological fitness. Moreover, under most systems of equal opportunity and equivalent selection, any numerically significant segment of the human species could, by virtue of its genetic variability, probably replace any other with respect to behavioral capacities.

Mating within human population groups is, as previously mentioned, assortative rather than panmictic. Society and culture set the styles in mate selection. Where ethnic groups are differentially treated with respect to the premium that a prevailing culture places on the recognition and nurturing of the full spectrum of their abilities, including economic opportunities which ensure the possessor of certain talents of a chance to identify, develop, and use them, a differential selection system is set up with respect to the partitioning of the gene pool among the groups whose opportunities differ in these respects. Fortunately, this does not create a negative selection in a disadvantaged group. It simply fails to exploit fully the genetic potential available. The democratization of our educational system, the equalization of economic opportunities, and the availability of specialized training to persons possessing special attributes, without regard to racial origin, would be expected to set up assortative mating tracks with respect to such attributes. Even on the extreme assumption that each ethnic group would be reproductively isolated from every other, this way of detecting genetically based abilities and of providing opportunities for persons of comparable endowment to meet and marry would provide for relatively rapid phenotypic equivalence where human abilities are concerned. Social and educational factors are thus viewed as providing tests for the detection of phenotypic variability in behavioral capacities, methods for maximizing the potential of each phenotype, and opportunities for positive phenotypic assortative mating through the selection exerted by colleges, universities, conservatories, art schools, institutes for mechanical training, and many others in which persons of comparable endowment and interest often meet and marry on the basis of these abilities and interests. This has been true of certain segments of the population for a long time. The present social revolution may be expected to apply it even more throughly to these segments and to extend it to other segments as well.

There is, therefore, a reciprocity or feedback between the genetic potential of a population and its social structure, such that not only does the former determine what the latter can be, but the latter exerts an important biological effect on the former. It is fortunate that the potential behavioral adaptations do not in any way atrophy from disuse. The deuces remain in the deck, whether the rules of the game are that the deuce is wild or not. While the present assortment of genetic abilities in various population groups may not be numerically equal, they are, in our view, equipotential, and it is our contention that any genetically diverse population existing in reasonable numbers could replace any other on the face of the earth with respect to behavioral abilities, given the prerequisite opportunities for the detection and nurturing of these abilities and for assortative mating to occur on these bases.

REFERENCES

1. Benson E. Ginsburg, "All Mice Are Not Created Equal: Recent Findings on Genes and Behavior," *Social Service Review, 40* (1966), 121–34.
2. Benson E. Ginsburg, "Genetics as a Tool in the Study of Behavior," *Perspectives in Biology and Medicine, 1* (1958), 397–424.
3. Benson E. Ginsburg and William S. Laughlin, "The Multiple Bases of Human Adaptability and Achievement: A Species Point of View," *Eugenics Quarterly, 13* (1966), 240–57.

4. M. L. Watson, "The Inheritance of Epilepsy and Waltzing in *Peromyscus*," *Contributions from the Laboratory of Vertebrate Genetics* (University of Michigan), No. 11 (1939), 1–24.

5. Governor Witt and Calvin S. Hall, "The Genetics of Audiogenic Seizures in the House Mouse," *Journal of Comparative and Physiological Psychology*, 42 (1949), 58–63.

6. John L. Fuller, Clarice Easler, and Mary E. Smith, "Inheritance of Audiogenic Seizure Susceptibility in the Mouse," *Genetics*, 35 (1950), 622–32.

7. Benson E. Ginsburg and D. S. Miller, "Genetic Factors in Audiogenic Seizures," in *Psychophysiologie, Neuropharmacologie et Biochimie de la Crise Audiogene* (Colloques Internationaux du Centre de la Recherche Scientifique), No. 112 (1963), 217–28.

8. Benson E. Ginsburg, "Causal Mechanisms in Audiogenic Seizures," in *Psychophysiologie, Neuropharmacologie et Biochimie de la Crise Audiogene* (Colloques Internationaux du Centre de la Recherche Scientifique), No. 112 (1936), 229–40.

9. K. Schlesinger, W. Boggan, and D. X. Freedman, "Genetics of Audiogenic Seizures: I. Relation to Brain Serotonin and Norepinephrin in Mice," *Life Sciences*, 4 (1965), 2345–51.

10. Benson E. Ginsburg, "Genetic Parameters in Behavior Research," in Jerry Hirsch, ed., *Behavior-Genetic Analysis* (New York, McGraw-Hill Book Co., 1967).

11. D. G. Freedman, "Constitutional and Environmental Interactions in Rearing of Four Breeds of Dogs," *Science*, 127 (1958), 585–86.

12. William S. Laughlin and Benson E. Ginsburg, "Repackaging People," Kinescope tape, Elizabeth Bailey, producer (University of Wisconsin education television station WHA, 1966).

13. Curt Stern, *Principles of Human Genetics* (2nd ed., San Francisco, W. H. Freeman and Co., 1960).

14. S. Wright, "Statistical Genetics in Relation to Evolution," *Actualités Scientifiques et Industrielles* (Paris, Hermann, 1939).

15. H. C. Bumpus, "The Elimination of the Unfit as Illustrated by the Introduced Sparrow, *Passer domesticus*," *Biological Lectures Delivered at the Marine Biological Laboratory of Wood's Hole* (Boston, Ginn and Co., 1899).

16. H. Bentley Glass and C. C. Li, "The Dynamics of Racial Intermixture: An Analysis Based on the American Negro," *American Journal of Human Genetics*, 5 (1953), 1–20.

PART FIVE

IS MAN
STILL EVOLVING?

We are fairly certain that man, as we know him today, has been essentially the same kind of animal for over thirty thousand years. He has been at a plateau of evolution for at least a thousand generations. At this point in his history, it is not speculative to assert that there is a greater likelihood of man's next evolutionary step being his extinction rather than morphologic change.

If man does not destroy himself, and if any major change were to occur in the course of time, he would no longer be *Homo sapiens sapiens,* so closely have anthropologists defined the species. As the object, then, of large-scale evolutionary change, man probably is finished. But evolutionary factors continue to operate on human populations. The conditions of life within which humans operate, the human ecological niche, include not only the physical environment, but, as we have seen, the requirements imposed by culture-building, culture maintenance, and culture transmission. This means that the conditions under which people work together, live together, or fail to live together have implications for the direction in which future human evolution will alter.

Studies of other mammals under laboratory conditions may give clues to the role that population density and distribution may play in governing the development of group behavior patterns. Such research may ultimately assist us in understanding such questions in relation to the human population crisis. But first we must, as Aristide Esser says, be able to explain why living in groups may be advantageous to the evolution of man, before we can deal with the contemporary problems of the distribution of human populations.

In the following papers, some lessons are suggested for biological research on human groups. Lenneberg discusses the biological basis for that species-specific human character, language. McClintock reports on physiological alterations of individuals in a group setting. Neel describes research on small homogeneous isolated human populations and the indications for the improvement of life for populations of the developed world.

The question implicit in this section is whether man can evolve or invent the systems which will enable us to live together in a post-Pleistocene world.

The reader is invited to construct his own predictions concerning the future course of human evolution.

19

ON EXPLAINING LANGUAGE

Eric H. Lenneberg

It has often been remarked that man produces culture, at the same time that man as a species has been produced by culture. The phrase used by the archeologist Vere Gordon Childe as the title of his book on culture history, Man Makes Himself, *expresses this idea succinctly.*

Language is a good example of what is meant by the phrase. Every normal human has the biological capacity to learn a language. At the same time, what language he learns depends on the nature of the language community he has participated in during infancy and early childhood. That linguistic behavior (or cognition, as Lenneberg uses the word in his article) is a species-specific biological capability there is no doubt; that language is a fact of culture at the same time is equally obvious.

The capacity for language is the product of evolution, just as the capacity for communication by gesture and signal is the result of evolution in other animal species. Whatever the set of experiences were that acted selectively on protohominid populations and their gene pools to develop the capacity for language, the impact of language use itself must have been one of these kinds of experiences. The nature of human life with even rudimentary patterns of cognition and symbolizing must have been among the selective agents that shaped and oriented the direction of human biological evolution. Thus man has been caught in a loop of shaping his evolution by means of the product of that evolution.

In the following article, Dr. Eric H. Lenneberg reviews the problems of investigating the biology of language. Although he is careful to point out at the outset that his concern is with the development of language in the individual rather than the evolution of language ability in the species, Dr. Lenneberg does deal with the relation of human language capacity to behavior phylogenies of

Reprinted from *Science*, Vol. 164, May 9, 1969, pages 635–643. Copyright 1969 by the American Association for the Advancement of Science. By permission of the author and the publisher.

other kinds of animals. He remarks on the dangers of viewing
human language capability simply as an added genetic component
superimposed on a prehuman primate base.

The reader should also take note of how the author integrates
data from research in the psychology of speech and language with
concepts from linguistics, cultural anthropology and human evolu-
tionary biology.

Dr. Lenneberg, Professor of Psychology at Cornell University,
is the author of Biological Foundations of Language.

Many explanations have been offered for many aspects of language; there is little agreement, however, on how to explain various problems or even on what there is to be explained. Of course, explanations differ with the personal inclinations and interests of the investigator. My interests are in man as a biological species, and I believe that the study of language is relevant to these interests because language has the following six characteristics. (i) It is a form of behavior present in all cultures of the world. (ii) In all cultures its onset is age correlated. (iii) There is only one acquisition strategy—it is the same for all babies everywhere in the world. (iv) It is based intrinsically upon the same formal operating characteristics whatever its outward form (1). (v) Throughout man's recorded history these operating characteristics have been constant. (vi) It is a form of behavior that may be impaired specifically by circumscribed brain lesions which may leave other mental and motor skills relatively unaffected.

Any form of human behavior that has all of these six characteristics may likewise be assumed to have a rather specific biological foundation. This, of course, does not mean that language cannot be studied from different points of view; it can, for example, be investigated for its cultural or social variations, its capacity to reflect individual differences, or its applications. The purpose of this article, however, is to discuss the aspects of language to which biological concepts are applied most appropriately (2). Further, my concern is with the development of language in children —not with its origin in the species.

PREDICTABILITY OF LANGUAGE DEVELOPMENT

A little boy starts washing his hands before dinner no sooner than when his parents decide that training in cleanliness should begin. However, children begin to speak no sooner and no later than when they reach a given stage of physical maturation (Table 19-1). There are individual variations in development, particularly with respect to age correlation. It

TABLE 19-1

Correlation of motor and language development (3, pp. 128–130)

AGE (YEARS)	MOTOR MILESTONES	LANGUAGE MILESTONES
0.5	Sits using hands for support; unilateral reaching	Cooing sounds change to babbling by introduction of consonantal sounds
1	Stands; walks when held by one hand	Syllabic reduplication; signs of understanding some words; applies some sounds regularly to signify persons or objects, that is, the first words
1.5	Prehension and release fully developed; gait propulsive; creeps downstairs backward	Repertoire of 3 to 50 words not joined in phrases; trains of sounds and intonation patterns resembling discourse; good progress in understanding
2	Runs (with falls); walks stairs with one foot forward only	More than 50 words; two-word phrases most common; more interest in verbal communication; no more babbling
2.5	Jumps with both feet; stands on one foot for 1 second; builds tower of six cubes	Every day new words; utterances of three and more words; seems to understand almost everything said to him; still many grammatical deviations
3	Tiptoes 3 yards (2.7 meters); walks stairs with alternating feet; jumps 0.9 meter	Vocabulary of some 1000 words; about 80 percent intelligibility; grammar of utterances close approximation to colloquial adult; syntactic mistakes fewer in variety, systematic, predictable
4.5	Jumps over rope; hops on one foot; walks on line	Language well established; grammatical anomalies restricted either to unusual constructions or to the more literate aspects of discourse

is interesting that language development correlates better with motor development than it does with chronological age. If we take these two variables (motor and language development) and make ordinal scales out of the stages shown in Table 19-1 and then use them for a correlation matrix, the result is a remarkably small degree of scatter. Since motor development is one of the most important indices of maturation, it is not unreasonable to propose that language development, too, is related to physical growth and development. This impression is further corroborated by examination of retarded children. Here the age correlation is very poor, whereas the correlation between motor and language development continues to be high (3). Nevertheless, there is evidence that the statistical relation between motor and language development is not due to any immediate, causal relation; peripheral motor disabilities can occur that do not delay language acquisition.

Just as it is possible to correlate the variable language development with the variables chronological age or motor development, it is possible to relate it to the physical indications of brain maturation, such as the gross weight of the brain, neurodensity in the cerebral cortex, or the changing weight proportions of given substances in either gray or white matter. On almost all counts, language begins when such maturational indices have attained at least 65 percent of their mature values. (Inversely, language acquisition becomes more difficult when the physical maturation of the brain is complete.) These correlations do not prove causal connections, although they suggest some interesting questions for further research.

EFFECT OF CERTAIN VARIATIONS IN SOCIAL ENVIRONMENT

In most of the studies on this topic the language development of children in orphanages or socially deprived households has been compared with that of children in so-called normal, middle-class environments. Statistically significant differences are usually reported, which is sometimes taken as a demonstration that language development is contingent on specific language training. That certain aspects of the environment are absolutely essential for language development is undeniable, but it is important to distinguish between what the children actually do, and what they can do.

There is nothing particularly surprising or revealing in the demonstration that language deficits occur in children who hear no language, very little language, or only the discourse of uneducated persons. But what interests us is the underlying capacity for language. This is not a spurious question; for instance, some children have the capacity for language but do not use it, either because of peripheral handicaps such as congenital deafness or because of psychiatric disturbances such as childhood schizophrenia; other children may not speak because they do not have a sufficient capacity for language, on account of certain severely retarding diseases.

There is a simple technique for ascertaining the degree of development of the capacity for speech and language. Instead of assessing it by means of an inventory of the vocabulary, the grammatical complexity of the utterances, the clarity of pronunciation, and the like, and computing a score derived from several subtests of this kind, it is preferable to describe the children's ability in terms of a few broad and general developmental stages, such as those shown in Table 19-1. Tests which are essentially inventories of vocabulary and syntactic constructions are likely to reflect simply the deficiencies of the environment; they obscure the child's potentialities and capabilities.

I have used the schema described to compare the speech development of children in many different societies, some of them much more primitive

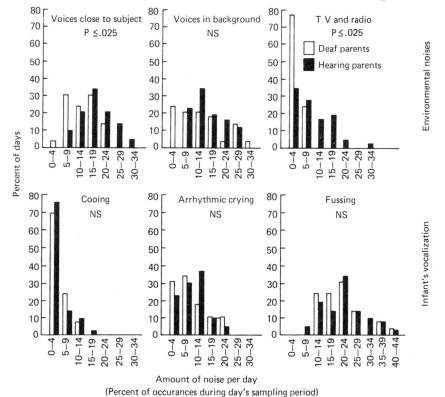

Amount of noise per day
(Percent of occurances during day's sampling period)

Fig. 19-1   Frequency of various noises. The basic counting unit is individual recording days.

than our own. In none of these studies could I find evidence of variation in developmental rate, despite the enormous differences in social environment.

I have also had an opportunity to study the effect of a dramatically different speech environment upon the development of vocalizations during the first 3 months of life (4). It is very common in our culture for congenitally deaf individuals to marry one another, creating households in which all vocal sounds are decidedly different from those normally heard and in which the sounds of babies cannot be attended to directly. Six deaf mothers and ten hearing mothers were asked, during their last month of pregnancy, to participate in our study. The babies were visited at home when they were no more than 10 days old and were seen biweekly thereafter for at least 3 months. Each visit consisted of 3 hours of observation and 24 hours of mechanical recording of all sounds made and heard by the baby. Data were analyzed quantitatively and qualitatively. Figure 19-1 shows that although the environment was quantitatively quite different in the experimental and the control groups, the frequency distributions of various baby noises did not differ significantly; as seen in Figure 19-2, the developmental histories of cooing noises are also remarkably alike in the two groups. Figure 19-3 demonstrates that the babies of deaf parents tend to fuss an equal amount, even though the hearing parents are much more likely to come to the child when it fusses. Thus the earliest development of human sounds appears to be relatively independent of the amount, nature, or timing of the sounds made by parents.

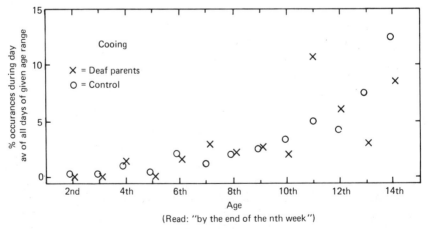

(Read: "by the end of the nth week")

Fig. 19-2   Each baby's day was divided into 6-minute periods; the presence or absence of cooing was noted for each period; this yielded a percentage for each baby's day; days of all babies were ordered by their ages, and the average was taken for all days of identical age. Nonaveraged data were published in (4).

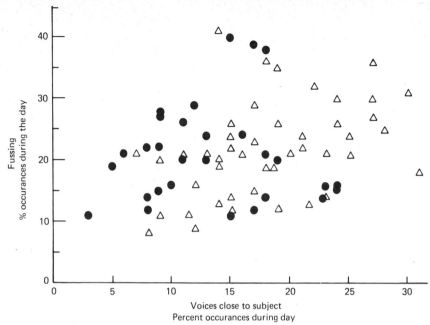

Fig. 19-3 Relation between the amount of parents' noises heard by the baby and the amount of fussing noises made by the baby. Each symbol is one baby's day; (solid circles) deaf parents; (triangles) hearing parents.

I have observed this type of childrearing through later stages, as well. The hearing children of deaf parents eventually learn two languages and sound systems: those of their deaf parents and those of the rest of the community. In some instances, communication between children and parents is predominantly by gestures. In no case have I found any adverse effects upon the language development of standard English in these children. Although the mothers made sounds different from the children's, and although the children's vocalizations had no significant effect upon attaining what they wanted during early infancy, language in these children invariably began at the usual time and went through the same stages as is normally encountered.

Also of interest may be the following observations on fairly retarded children growing up in state institutions that are badly understaffed. During the day the children play in large, bare rooms, attended by only one person, often an older retardate who herself lacks a perfect command of language. The children's only entertainment is provided by a large television set, playing all day at full strength. Although most of these retarded children have only primitive beginnings of language, there are always some among them who manage, even under these extremely deprived circumstances, to pick up an amazing degree of language skill. Apparently they learn language partly through the television programs, whose level is often quite adequate for them!

From these instances we see that language capacity follows its own natural history. The child can avail himself of this capacity if the environment provides a minimum of stimulation and opportunity. His engagement in language activity can be limited by his environmental circumstances, but the underlying capacity is not easily arrested. Impoverished environments are not conducive to good language development, but good language development is not contingent on specific training measures (5); a wide variety of rather haphazard factors seems to be sufficient.

EFFECT OF VARIATIONS IN GENETIC BACKGROUND

Man is an unsatisfactory subject for the study of genetic influences; we cannot do breeding experiments on him and can use only statistical controls. Practically any evidence adduced is susceptible to a variety of interpretations. Nevertheless, there are indications that inheritance is at least partially responsible for deviations in verbal skills, as in the familial occurrence of a deficit termed congenital language disability (2, chapter 6). Studies, with complete pedigrees, have been published on the occurrence and distribution of stuttering, of hyperfluencies, of voice qualities, and of many other traits, which constitute supporting though not conclusive evidence that inheritance plays a role in language acquisition. In addition to such family studies, much research has been carried out on twins. Particularly notable are the studies of Luchsinger, who reported on the concordance of developmental histories and of many aspects of speech and language. Zygosity was established in these cases by serology (Figure 19-4). Developmental data of this kind are, in my opinion, of greater relevance to our speculations on genetic background than are pedigrees.

The nonbiologist frequently and mistakenly thinks of genes as being directly responsible for one property or another; this leads him to the fallacy, especially when behavior is concerned, of dichotomizing everything as being dependent on either genes or environment. Genes act merely on intracellular biochemical processes, although these processes have indirect effects on events in the individual's developmental history. Many alterations in structure and function indirectly attributable to genes are more immediately the consequence of alterations in the schedule of developmental events. Therefore, the studies on twins are important in that they show that homozygotes reach milestones in language development at the same age, in contrast to heterozygotes, in whom divergences are relatively common. It is also interesting that the nature of the deviations—the symptoms, if you wish—are, in the vast majority, identical in homozygotes but not in heterozygotes.

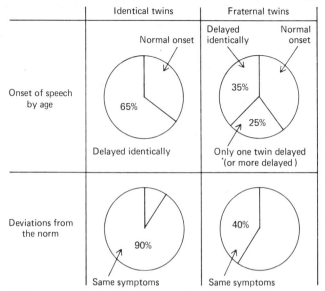

Fig. 19-4   The onset of speech and its subsequent development tend to be more uniform among identical twins than fraternal twins.

Such evidence indicates that man's biological heritage endows him with sensitivities and propensities that lead to language development in children who are spoken to (in contrast to chimpanzee infants, who do not automatically develop language—either receptive or productive—under identical treatment). The endowment has a genetic foundation, but this is not to say that there are "genes for language," or that the environment is of no importance.

ATTEMPTS TO MODIFY LANGUAGE DEVELOPMENT

Let us now consider children who have the capacity for language acquisition but fail to develop it for lack of exposure. This is the case with the congenitally deaf, who are allowed to grow up without either language or speech until school age, when suddenly language is brought to them in very unnatural ways. Before this time they may have half a dozen words they can utter, read, write, or finger-spell, but I have known of no profoundly deaf child (in New England, where my investigations were conducted) with whom one could communicate by use of the English language before school age.

When deaf children enter an oralist school, lipreading and speech become the major preoccupation of training. However, in most children these activities remain poor for many more years, and in some, throughout life. Their knowledge of language comes through learning to read and write. However, teachers in the oral tradition restrict expression in the graphic medium on the hypothesis that it interferes with lipreading and speech skills. Thus, exposure to language (i) comes much later in these children's lives than is normal, (ii) is dramatically reduced in quantity, (iii) is presented through a different medium and sensory modality, and (iv) is taught to the children rather as a second language is taught, instead of through the simple immersion into a sea of language that most children experience. The deaf children are immediately required to use grammatically correct sentences, and every mistake is discussed and explained to them.

The results of this procedure are interesting but not very encouraging from the educational point of view. During the early years of schooling, the children's spontaneous writings have a very unusual pattern; there is little evidence that the teachers' instruction in "how to compose correct sentences" is of any avail. Yet, careful analysis of their compositions shows that some subtleties of English syntax that are usually not part of the grammar taught in the school do make their appearance, sometimes quite early. There can be no question that the children do not simply imitate what they see; some of the teachings fall by the wayside, whereas a number of aspects of language are automatically absorbed from the written material given to the children.

There are other instances in which efforts are made to change a child's language skills by special training, as in the mildly retarded, for example. Many parents believe that their retarded child would function quite normally if somebody could just teach him to speak. At Children's Hospital in Boston I undertook a pilot study in which a speech therapist saw a small number of children with Down's syndrome (mongolism) for several hours each week, in an effort to speed up language development. Later, two graduate students in linguistics investigated the children's phonetic skills and tried to assess the capacities of each child for clearer enunciation. Throughout these attempts, it was found that if a child had a small repertoire of single words, it was always possible to teach him yet another word, but if he was not joining these words spontaneously into phrases, there was nothing that could be done to induce him to do so. The articulatory skills were somewhat different. It was often possible to make a child who had always had slurred speech say a specific word more clearly. However, the moment the child returned to spontaneous utterances, he would fall back to the style that was usual for him. The most interesting results were obtained when the retarded children were required simply to repeat well-formed sentences. A child who had not developed to a stage in which he used certain grammatical rules spontaneously, who was still missing the syntactic foundations and prerequisites, could not be taught to repeat a sentence that was formed by such higher rules. This was true even in sentences of very few words. Similar observations have since been made on normal children (6), with uniformly similar results; normal children, too, can repeat correctly only that which is formed by rules they have already mastered. This is the best indication that language does not come about by simple imitation, but that the child abstracts regularities or relations from the language he hears, which he then applies to building up language for himself as an apparatus of principles.

WHAT SETS THE PACE OF LANGUAGE DEVELOPMENT?

There is a widespread belief that the development of language is dependent on the motor skills of the articulating organs. Some psychologists believe that species other than man fail to develop language only because of anatomical differences in their oral structures. However, we have evidence that this is not so.

It is important that we are clear about the essential nature of language. Since my interests are in language capacities, I am concerned with the development of the child's knowledge of how language works. This is not the same as the acquisition of "the first word." The best test for the presence and development of this knowledge is the manner in which discourse is understood. In most instances, it is true that there is a relation between speech and understanding, but this relation is not a necessary one (7).

By understanding, I mean something quite specific. In the realm of phonology, understanding involves a process that roughly corresponds to the linquists' phonematization (in contrast, for example, to a "pictographic" understanding: phonematization results in seeing similarities between speech sounds, whereas pictographic understanding would treat a word as an indivisible sound pattern). In the realm of semantics, understanding involves seeing the basis on which objects are categorized, thus enabling a child to name an object correctly that he has never seen before. (The child does not start out with a hypothesis that "table" is the proper name of a unique object or that it refers to all things that have four appendages.) In the realm of grammar, understanding involves the extraction of relations between word classes; an example is the understanding of predication. By application of these tests, it can be shown empirically that Aunt Pauline's favorite lapdog does not have a little language knowledge, but, in fact, fails the test of understanding on all counts.

A survey of children with a variety of handicaps shows that their grasp of how language works is intimately related to their general cognitive growth, which, in turn, is partly dependent on physical maturation and partly on opportunities to interact with a stimulus-rich environment. In many retarding diseases, for example, language development is predicted best by the rate of advancement in mental age (using tests of nonverbal intelligence). In an investigation of congenitally blind children (8), we are again finding that major milestones for language development are highly correlated with physical development. A naive conception of language development as an accumulation of associations between visual and auditory patterns would be hard put to explain this.

In adults, language functions take place predominantly in the left hemisphere. A number of cortical fields have been related to specific aspects of language. The details are still somewhat controversial and need not concern us here. It is certain, however, that precentral areas of the frontal lobe are principally involved in the production of language, whereas the postcentral parietal and superior temporal fields are involved in sensory functions. These cortical specializations are not present at birth, but become only gradually established during childhood, in a process very similar to that of embryological history; there is evidence of differentiation and regulation of function. In the adult, traumata causing large leftsided central cortical lesions carry a highly predictable prognosis; in 70 percent of all cases, aphasia occurs, and in about half of these, the condition is irreversible (I am basing these figures on our experience with penetrating head injuries incurred in war).

Comparable traumatic lesions in childhood have quite different consequences, the prognosis being directly related to the age at which the insult is incurred. Lesions of the left hemisphere in children under age 2 are no more injurious to future language development than are lesions of the right hemisphere. Children whose brain is traumatized after the onset of language but before the age of 4 usually have transient aphasias; language is quickly reestablished, however, if the right hemisphere remains intact. Often these children regain language by going through stages of language development similar to those of the 2-year-old, but they traverse each stage at greater speed. Lesions incurred before the very early teens also carry an excellent prognosis, permanent residues of symptoms being extremely rare.

The prognosis becomes rapidly worse for lesions that occur after this period; the young men who become casualties of war have symptoms virtually identical with those of stroke patients of advanced age. Experience with the surgical removal of an entire cerebral hemisphere closely parallels this picture. The basis for prognosticating operative success is, again, the age at which the disease has been contracted for which the operation is performed.

If a disturbance in the left hemisphere occurs early enough in life, the right hemisphere remains competent for language throughout life. Apparently this process is comparable to regulation, as we know it from morphogenesis. If the disease occurs after a certain critical period of life, namely, the early teens, this regulative capacity is lost and language is interfered with permanently. Thus the time at which the hemispherectomy is performed is less important than the time of the lesion.

CRITICAL AGE FOR LANGUAGE ACQUISITION

The most reasonable interpretation of this picture of recovery from aphasia in childhood is not that there is vicarious functioning, or taking over, by the right hemisphere because of need, but rather that language functions are not yet confined to the left hemisphere during early life. Apparently both hemispheres are involved at the beginning, and a specialization takes place later (which is the characteristic of differentiation), resulting in a kind of left-right polarization of functions. Therefore, the recovery from aphasia during preteen years may partly be regarded as a reinstatement of activities that had never been lost. There is evidence that children at this age are capable of developing language in the same natural way as do very young children. Not only do symptoms subside, but active language development continues to occur. Similarly, we see that healthy children have a quite different propensity for acquiring foreign languages before the early teens than after the late teens, the period in between being transitional. For the young adult, second-language learning is an academic exercise, and there is a vast variety in degree of proficiency. It rapidly becomes more and more difficult to overcome the accent and interfering influences of the mother tongue.

Neurological material strongly suggests that something happens in the brain during the early teens that changes the propensity for language acquisition. We do not know the factors involved, but it is interesting that the critical period coincides with the time at which the human brain attains its final state of maturity in terms of structure, function, and biochemistry (electroencephalographic patterns slightly lag behind, but become stabilized by about 16 years). Apparently the maturation of the brain marks the end of regulation and locks certain functions into place.

There is further evidence that corroborates the notion of a critical period for primary language acquisition, most importantly, the developmental histories of retarded children. It is dangerous to make sweeping generalizations about all retarded children, because so much depends on the specific disease that causes the retardation. But if we concentrate on diseases in which the pathological condition is essentially stationary, such as microcephaly vera or mongolism, it is possible to make fairly general predictions about language development. If the child's mental developmental age is 2 when he is 4 years old (that is, his I.Q. is 50), one may safely predict that some small progress will be made in language development. He will slowly move through the usual stages of infant language, although the rate of development will gradually slow down. In virtually all of these cases, language development comes to a complete standstill in the early teens, so that these individuals are arrested in primitive stages of language development that are perpetuated for the rest of their lives. Training and motivation are of little help.

Development in the congenitally deaf is also revealing. When they first enter school, their language acquisition is usually quite spectacular, considering the enormous odds against them. However, children who by their early teens have still not mastered all of the principles that underlie the production of sentences appear to encounter almost unsurmountable difficulties in perfecting verbal skills.

There is also evidence of the converse. Children who suddenly lose their hearing (usually a consequence of meningitis) show very different degrees of language skill, depending on whether the disease strikes before the onset of language or after. If it occurs before they are 18 months old, such children encounter difficulties with language development that are very much the same as those encountered by the congenitally deaf. Children who lose their hearing after they have acquired language, however, at age 3 to 4, have a different prospect. Their speech deteriorates rapidly; usually within weeks they stop using language, and so far it has proved impossible to maintain the skill by educational procedures [although new techniques developed in England and described by Fry (9) give promise of great improvement]. Many such children then live without language for a relatively long time, often 2 to 3 years, and when they enter the schools for the deaf, must be trained in the same way that other deaf children are trained. However, training is much more successful, and their language habits stand out dramatically against those of their less fortunate colleagues. There appears to be a direct relation between the length of time during which a child has been exposed to language and the proficiency seen at the time of retraining.

BIOLOGICAL APPROACH: DEFINING LANGUAGE FURTHER

Some investigators propose that language is an artifact—a tool that man has shaped for himself to serve a purpose. This assumption induces the view that language consists of many individual traits, each independent of the other. However, the panorama of observations presented above suggests a biological predisposition for the development of language that is anchored in the operating characteristics of the human brain (10). Man's cognitive apparatus apparently becomes a language receiver and transmitter, provided the growing organism is exposed to minimum and haphazard environmental events.

However, this assumption leads to a view different from that suggested by the artifact assumption. Instead of thinking of language as a collection of separate and mutually independent traits, one comes to see it as a profoundly integrated activity. Language is to be understood as an operation rather than a static product of the mind. Its *modus operandi* reflects that of human cognition, because language is an intimate part of cognition. Thus the biological view denies that language is the cause of cognition, or even its effect, since language is not an object (like a tool) that exists apart from a living human brain.

As biologists, we are interested in the operating principles of language because we hope that this will give us some clues about the operating principles of the human brain. We know there is just one species *Homo sapiens*, and it is therefore reasonable to assume that individuals who speak Turkish, English, or Basque (or who spoke Sanskrit some millennia ago) all have (or had) the same kind of brain, that is, a computer with the same operating principles and the same sensorium. Therefore, in a biological investigation one must try to disregard the differences between the languages of the world and to discover the general principles of operation that are common to all of them. This is not an easy matter; in fact, there are social scientists who doubt the existence of language universals. As students of language we cannot fail to be impressed with the enormous differences among languages. Yet every normal child learns the language to which he is exposed. Perhaps we are simply claiming that common denominators must exist; can we prove their existence? If we discovered a totally isolated tribe with a language unknown to any outsider, how could we find out whether this language is generated by a computer that has the same biological characteristics as do our brains, and how could we prove that it shares the universal features of all languages?

As a start, we could exchange children between our two cultures to discover whether the same language developmental history would occur in those exchanged. Our data would be gross developmental stages, correlated with the emergence of motor milestones. A bioassay of this kind (already performed many times, always with positive results) gives only part of the answer.

In theory, one may also adduce more rigorous proof of similarity among languages. The conception of language universals is difficult to grasp intuitively, because we find it so hard to translate from one language to another and because the grammars appear, on the surface, to be so different. But it is entirely possible that underneath the structural difference that makes it so difficult for the adult speaker to learn a second language (particularly one that is not a cognate of his own) there are significant formal identities.

Virtually every aspect of language is the expression of relations. This is true of phonology (as stressed by Roman Jakobson and his school), semantics, and syntax. For instance, in all languages of the world words label a set of relational principles instead of being labels of specific objects. Knowing a word is never a simple association between an object and an acoustic pattern, but the successful operation of those principles, or application of those rules, that lead to using the word "table" or "house" for objects never before encountered. The language universal in this instance is not the type of object that comes to have a word, nor the particular relations involved; the universal is the generality that words stand for relations instead of being unique names for one object.

Further, no language has ever been described that does not have a second order of relational principles, namely, principles in which relations are being related, that is, syntax in which relations between words are being specified. Once again, the universal is not a particular relation that occurs in all languages (though there are several such relations) but that all languages have relations of relations.

Mathematics may be used as a highly abstract form of description, not of scattered facts but of the dynamic interrelations—the operating principles—found in nature. Chomsky and his students have done this. Their aim has been to develop algorithms for specific languages, primarily English, that make explicit the series of computations that may account for the structure of sentences. The fact that these attempts have only been partially successful is irrelevant to the argument here. (Since every native speaker of English *can* tell a well-formed sentence from an ill-formed one, it is evident that some principles must exist; the question is merely whether the Chomskyites have discovered the correct ones.) The development of algorithms is only one province of mathematics, and in the eyes of many mathematicians a relatively limited one. There is a more exciting prospect; once we know something about the basic relational operating principles underlying a few languages, it should be possible to characterize formally the abstract system *language* as a whole. If our assumption of the existence of basic, structural language universals is correct, one ought to be able to adduce rigorous proof for the existence of homeomorphisms between any natural languages, that is, any of the systems characterized formally. If a category calculus were developed for this sort of thing, there would be one level of generality on which a common denominator could be found; this may be done trivially (for instance by using the product of all systems). However, our present knowledge of the relations, and the relations of relations, found in the languages so far investigated in depth encourages us to expect a significant solution.

ENVIRONMENT AND MATURATION

Everything in life, including behavior and language, is interaction of the individual with its milieu. But the milieu is not constant. The organism itself helps to shape it (this is true of cells and organs as much as of animals and man). Thus, the organism and its environment is a dynamic system and, phylogenetically, developed as such.

The development of language in the child may be elucidated by applying to it the conceptual framework of developmental biology. Maturation may be characterized as a sequence of states. At each state, the growing organism is capable of accepting some specific input; this it breaks down and resynthesizes in such a way that it makes itself develop into a new state. This new state makes the organism sensitive to new and different types of input, whose acceptance transforms it to yet a further state, which opens the way to still different input, and so on. This is called epigenesis. It is the story of embryological development observable in the formation of the body, as well as in certain aspects of behavior.

At various epigenetic states, the organism may be susceptible to more than one sort of input—it may be susceptible to two or more distinct kinds or even to an infinite variety of inputs, as long as they are within determined limits—and the developmental history varies with the nature of the input accepted. In other words, the organism, during development, comes to crossroads; if condition A is present, it goes one way; if condition B is present, it goes another. We speak of states here, but this is, of course, an abstraction. Every stage of maturation is unstable. It is prone to change into specific directions, but requires a trigger from the environment.

When language acquisition in the child is studied from the point of view of developmental biology, one makes an effort to describe developmental stages together with their tendencies for change and the conditions that bring about that change. I believe that the schema of physical maturation is applicable to the study of language development because children appear to be sensitive to successively different aspects of the language environment. The child first reacts only to intonation patterns. With continued exposure to these patterns as they occur in a given language, mechanisms develop that allow him to process the patterns, and in most instances to reproduce them (although the latter is not a necessary condition for further development). This changes him so that he reaches a new state, a new potential for language development. Now he becomes aware of certain articulatory aspects, can process them and possibly also reproduce them, and so on. A similar sequence of acceptance, synthesis, and state of new acceptance can be demonstrated on the level of semantics and syntax.

That the embryological concepts of differentiation, as well as of determination and regulation, are applicable to the brain processes associated with language development is best illustrated by the material discussed above under the headings "brain correlates" and "critical age for language acquisition." Furthermore, the correlation between language development and other maturational indices suggests that there are anatomical and physiological processes whose maturation sets the pace for both cognitive and language development; it is to these maturational processes that the concept differentiation refers. We often transfer the meaning of the word to the verbal behavior itself, which is not unreasonable, although, strictly speaking, it is the physical correlates only that differentiate.

PSEUDO-HOMOLOGIES AND NAIVE "EVOLUTIONIZING"

The relation between species is established on the basis of structural, physiological, biochemical, and often behavioral correspondences, called homologies. The identification of homologies frequently poses heuristic problems. Common sense may be very misleading in this matter. Unless there is cogent evidence that the correspondences noted are due to a common phylogenetic origin, one must entertain the possibility that resemblances are spurious (though perhaps due to convergence). In other words, not all criteria are equally reliable for the discovery of true homologies. The criteria must pass the following two tests if they are to reveal common biological origins. (i) They must be applicable to traits that have a demonstrable (or at least conceivable) genetic basis; and (ii) the traits to which they apply must not have a sporadic and seemingly random distribution over the taxa of the entire animal kingdom. Homologies cannot be established by relying on similarity that rests on superficial inspection (a whale is not a fish); on logical rather than biological aspects (animals that move at 14 miles per hour are not necessarily related to one another); and on anthropocentric imputation of motives (a squirrel's hoarding of nuts may have nothing in common with man's provisions for his future).

Comparisons of language with animal communication that purport to throw light on the problem of its phylogenetic origins infringe on every one of these guidelines. Attempts to write generative grammars for the language of the bees in order to discover in what respect that language is similar to and different from man's language fail to pass test (i). Syntax does not have a genetic basis any more than do arithmetic or algebra; these are calculi used to describe relations. It may be that the activities or circumstances to which the calculi are applied are in some way related to genetically determined capacities. However, merely the fact that the calculus may or may not be applied obviously does not settle that issue.

The common practice of searching the entire animal kingdom for communication behavior that resembles man's in one aspect or another fails test (ii). The fact that some bird species and perhaps two or three cetaceans can make noises that sound like words, that some insects use discrete signals when they communicate, or that recombination of signals has been observed to occur in communication systems of a dozen totally unrelated species are not signs of a common phylogeny or genetically based relationship to language. Furthermore, the similarities noted between human language and animal communication all rest on superficial intuition. The resemblances that exist between human language and the language of the bees and the birds are spurious. The comparative criteria are usually logical (12) instead of biological; and the very idea that there must be a common denominator underlying all communication systems of animals and man is based on an anthropocentric imputation.

Everything in biology has a history, and so every communication system is the result of evolution. But traits or skills do not have an evolutionary history of their own, that is, a history that is independent of the history of the species. Contemporary species are discontinuous groups (except for those in the process of branching) with discontinuous communication behavior. Therefore, historical continuity need not lead to continuity between contemporary communication systems, many of which (including man's) constitute unique developments.

Another recent practice is to give speculative accounts of just how, why, and when human language developed. This is a somewhat futile undertaking. The knowledge that we have gained about the mechanisms of evolution does not enable us to give specific accounts of every event of the past. Paleontological evidence points to the nature of its fauna, flora, and climate. The precursors of modern man have left for us their bones, teeth, and primitive tools. None of these bears any necessary or assured relation to any type of communication system. Most speculations on the nature of the most primitive sounds, on the first discovery of their usefulness, on the reasons for the hypertrophy of the brain, or the consequences of a narrow pelvis are in vain. We can no longer reconstruct what the selection pressures were or in what order they came, because we know too little that is securely established by hard evidence about the ecological and social conditions of fossil man. Moreover, we do not even know what the targets of actual selection were. This is particularly troublesome because every genetic alteration beings about several changes at once, some of which must be quite incidental to the selective process.

SPECIES SPECIFICITIES AND COGNITIVE SPECIALIZATION

In the 19th century it was demonstrated that man is not in a category apart from that of animals. Today it seems to be necessary to defend the view (before many psychologists) that man is not identical with all other animals—in fact, that every animal species is unique, and that most of the commonalities that exist are, at best, homologies. It is frequently claimed that the principles of behavioral function are identical—in all vertebrates, for example—and that the differences between species are differences of magnitude, rather than quality. At other times, it is assumed that cognitive functions are alike in two species except that one of the two may have additionally acquired a capacity for a specific activity. I find fault with both views.

Since behavioral capacities (I prefer the term cognition) are the product of brain function, my point can well be illustrated by considering some aspects of brain evolution. Every mammalian species has an anatomically distinct brain. Homologies are common, but innovations can also be demonstrated. When man's brain is compared with the brain of other primates, extensive correspondences can be found, but there are major problems when it comes to the identification of homologies. Dramatic differences exist not only in size but also in details of the developmental histories; together with differences in cerebrocortical histology, topography, and extent, there are differences in subcortical fiber-connections, as pointed out by Geschwind (13) most recently and by others before him. The problem is, what do we make of the innovations? Is it possible that each innovation (usually an innovation is not a clear-cut anatomical entity) is like an independent component that is simply added to the components common to all the more old-fashioned brains? And if so, is it likely that the new component is simply adding a routine to the computational facilities already available? Both presumptions are naive. A brain is an integrated organ, and cognition results from the integrated operation of all its tissues and suborgans. Man's brain is not a chimpanzee's brain plus added "association facilities." Its functions have undergone reintegration at the same pace as its evolutionary developments.

The identical argument applies to cognitive functions. Cognition is not made up of isolated processes such as perception, storing, and retrieval. Animals do not all have an identical memory mechanism except that some have a larger storage capacity. As the structure of most proteins, the morphology of most cells, and the gross anatomy of most animals show certain species specificities (as do details of behavioral repertoires), so we may expect that cognition, too, in all of its aspects, has its species specificities. My assumption, therefore, is that man's cognition is not essentially that of every other primate with merely the addition of the capacity for language; instead, I propose that his entire cognitive function, of which his capacity for language is an integral part, is species-specific. I repeat once more that I make this assumption not because I think man is in a category all of his own, but because every animal species must be assumed to have cognitive specificities.

CONCLUSION

The human brain is a biochemical machine; it computes the relations expressed in sentences and their components. It has a print-out consisting of acoustic patterns that are capable of similar relational computation by machines of the same constitution using the same program. Linguists, biologists, and psychologists have all discussed certain aspects of the machine.

Linguists, particularly those developing generative grammar, aim at a formal description of the machine's behavior; they search mathematics for a calculus to describe it adequately. Different calculations are matched against the behavior to test their descriptive adequacy. This is an empirical procedure. The raw data are the way a speaker of a language understands collections of words or the relationships he sees. A totally adequate calculus has not yet been discovered. Once available, it will merely describe, in formal terms, the process of relational interpretation in the realm of verbal behavior. It will describe a set of operations; however, it will not make any claims of isomorphism between the formal operations and the biological operations they describe.

Biologists try to understand the nature, growth, and function of the machine (the human brain) itself. They make little inroads here and there, and generally play catch-as-catch-can; everything about the machine interests them (including the descriptions furnished by linguists).

Traditionally, learning theory has been involved neither in a specific description of this particular machine's behavior nor in its physical constitution. Its concern has been with the use of the machine: What makes it go? Can one make it operate more or less often? What purposes does it serve?

Answers provided by each of these inquiries into language are not intrinsically anatagonistic, as has often been claimed. It is only certain overgeneralizations that come into conflict. This is especially so when claims are made that any one of these approaches provides answers to all the questions that matter.

REFERENCES AND NOTES

1. E. H. Lenneberg, in *The Structure of Language, Readings in the Philosophy of Language*. J. A. Fodor and J. J. Katz, Eds. (Prentice-Hall, Englewood Cliffs, N.J., 1964).
2. For complete treatment, see E. H. Lenneberg, *Biological Foundations of Language* (Wiley, New York, 1967).
3. E. H. Lenneberg, I. A. Nichols, E. F. Rosenberger, in *Disorders of Communication* D. Rioch, Ed. (Research Publications of Association for Research in Nervous and Mental Disorders, New York, 1964), vol. 42.
4. E. H. Lenneberg, F. G. Rebelsky, I. A. Nichols, *Hum. Develop.* 8, 23 (1965).
5. R. Brown, C. Cazden, U. Bellugi, in *The 1967 Minnesota Symposium on Child Psychology*, J. P. Hill, Ed. (Univ. of Minnesota Press, Minneapolis, in press).

6. D. Slobin, personal communication.
7. E. H. Lenneberg, *J. Abnorm. Soc. Psychol.* 65, 419 (1962).
8. ———, S. Fraiberg, N. Stein, research in progress.
9. D. B. Fry, in *The Genesis of Language: A Psycholinguistic Approach*, F. Smith and G. A. Miller, Eds. (MIT Press, Cambridge, 1966).
10. For details, see E. H. Lenneberg, *Perception and Language*, in preparation.
11. N. Chomsky, "The formal nature of language" (in 2, appendix A).
12. See, for instance, C. F. Hockett, in *Animal Communication*, W. E. Lanyon and W. N. Tavolga, Eds. (American Institute of Biological Sciences, Washington, D.C., 1960); and in *Sci. Amer.* 203, 89 (1960).
13. N. Geschwind, *Brain* 88, 237, 585 (1965).
14. I thank H. Levin and M. Seligman for comments and criticisms.

20

MENSTRUAL SYNCHRONY AND SUPPRESSION

Martha K. McClintock

To say that behavior has biological and cultural substrates is to state a truism; to untangle the biological from the cultural elements is another matter. For some behavioral scientists, this is the major task: to specify the precise nature of the interaction between experiences and genetic capabilities, inventing a whole new way of describing behavior and the elements which comprise it.

The ecological approach in cultural anthropology is one valuable contribution to this kind of understanding. Ecological anthropologists seek to analyze culture systems as adaptive behavioral systems, much as the interrelationships among other populations of animal and plant life are studied as adaptations.

Are there physiological elements which are integrated within social systems? To what extent are physiological processes limited and triggered, altered and accelerated by interaction between individuals? We have evidence from studies of smaller mammals, such as mice, that the presence of other individuals influences

Reprinted from *Nature*, Vol. 229, January 22, 1971, pages 244–245. By permission of the author and the publisher.

internal body functions, even biological rhythms, such as menstrua-
tion. In the following article, Dr. McClintock reports on an
analogous study of human behavior. Note the scientific way this
study was conducted, even though her subjects could not be
observed in a laboratory setting.

Remember that physiological processes are based on genes, as are
the structural traits of organisms. Thus, one question with a set of
potentially interesting answers is the way that human interpersonal
behavior, and the genotypes underlying physiological processes
may have interacted in the evolution of man.

Dr. McClintock is in the Department of Psychology at the
University of Pennsylvania.

Studies of the influence of pheromones on the estrous cycles of mice,[1-4] and of crowding on variables such as adrenalin production in mice and other species[5] have suggested that social grouping can influence the balance of the endocrine system. Although there has been little direct investigation with humans, anecdotal and indirect observations have indicated that social groupings influence some aspects of the menstrual cycle. Menstrual synchrony is often reported by all-female living groups and by mothers, daughters and sisters who are living together. For example, the distribution of onsets of seven female lifeguards was scattered at the beginning of the summer, but after 3 months spent together the onset of all seven cycles fell within a 4-day period.

Indirect support is given by the investigation of Collet *et al.*[6] on the effect of age on menstrual cycle patterning. A higher percentage of anovulatory cycles were reported for college age women than for older women. Although Collet *et al.* attributed this to a maturational factor, it is interesting that most of the college aged women attended all female schools. Considering the parallel with the Lee–Boot effect in mice [1] (groups consisting only of females become pseudopregnant or anestrous), it seems possible that an interpersonal factor is operating together with the maturational factor.

Subjects were 135 females aged 17–22 yr—all residents of a dormitory in a suburban women's college. The dormitory in which they resided has four main corridors each with approximately twenty-five girls living in single and double rooms. Six smaller living areas, separated from the main corridors by at least one door, each house approximately eight girls in single rooms.

Three times during the academic year, each subject was asked when her last and second to last menstrual periods had begun; thus the date of onset was determined for all cycles between late September and early April. The average duration of menstruation and presence of dysmenorrhoea were noted. In addition, subjects estimated how many times each week they were in the company of males and listed by room number the girls ($N \leq 10$) with whom they spent the most time, indicating which two of these they saw most often.

The date of menstrual onset was compared for room mates and closest friends, for close friend groups and for living groups. Two people qualified as "closest friends" only if both had indicated that they saw each other most often. While menstrual cycle timing in women using birth control pills is individually invariant, these women were still included in the analysis, because their influence on the menstrual cycles of the others was unknown. For room mates and closest friends, the difference between the date of onset in October for one arbitrarily chosen member of the pair and the closest date of onset for the other was calculated. This difference was compared with a difference for March calculated in a similar way, but with one change: instead of choosing the closest onset dates for the pair, both onsets for March were chosen to follow the initial October onset by an equal number of cycles. For example, if onset 6 occurred on March 10 for the first member of the pair, and onsets 5 and 6 for the other member occurred on March 1 and March 29 respectively, then the March 10 and March 29 dates were used to calculate the difference in onset. This procedure was used to minimize chance coincidences that did not result from a trend towards synchrony.

The Wilcoxon matched-pairs signed-ranks test[7] was used to test for a significant decrease in the difference between onset dates of room mates and closest friends. This test utilizes both the direction and magnitude of change in differences and is therefore a relatively powerful test.

There was a significant increase in synchronization (that is, a decrease in the difference between onset dates) among room mates ($P \leq 0.0007$), among closest friends ($P \leq 0.003$) and among room mates and closest friends combined ($P \leq 0.0003$). The increase in synchrony for room mates did not differ significantly from the increase for closest friends. The increase in synchrony was further substantiated by non-overlapping confidence intervals, calculated for the median difference on onset dates[8] (Table 20-1).

TABLE 20-1

Confidence intervals (>0.99, in days) for the median difference in onset date between members of the pair

	OCTOBER	MARCH
Close friends and room mates		
N = 66	$7 < M < 10$	$3 < M < 7$
Random pairs		
N = 33	$6 < M < 14$	$5 < M < 15$

This synchrony might be due to some factor other than time spent with an individual; Koford[9] has attributed synchrony of the breeding season in *Macaca mulata* on Cayo Santiago to common seasonal changes in available food. The fact that the subjects generally eat as a dormitory group in a common dining room might be a significant factor in creating synchrony. A similar life pattern and common, repeated stress periods might also effect synchrony. Subjects were therefore randomly paired and tested for synchrony within the dormitory as a whole, but no significant trend (N.S., $P \leqslant 0.8$) was found, and the confidence intervals for the median difference in onset date overlapped completely.

Group synchrony was also investigated and the data were analyzed to verify that the decrease in difference between onset dates was a true measure of synchrony. All subjects were divided into fifteen groups of close friends ($5 \leqslant N \leqslant 10$), using the lists of close friends made by each subject. During the interview, it was stressed to each subject that her list of "close friends" should include the people she saw most often and with whom she spent the most time, not necessarily those with whom she felt the closest. But because there is usually some overlap, the term "close friends" was adopted. Only subjects who mutually listed each other were included in a group.

A mean onset date (μ_t) was determined for each group in October, late November, January, late February and April. As before, the onset dates (X_t), being compared, each followed the October onset (X_1) by an equal number of cycles. The mean individual difference from the group onset mean

$$\frac{\overset{n}{\underset{}{\Sigma}}(X_t - \mu_t)}{n}$$

was determined for each group and compared across time in two ways. First, a linear rank method, designed by Page[10] to test ordered hypotheses for multiple treatments, showed a significant decrease in individual differences from the group onset mean for close friend groups ($P \leqslant 0.001$). Second, a graph of this decrease as a function of time (Figure 20-1) indicated that the greatest decrease occurred in the first 4 months with little subsequent change. This asymptotic relation indicated that the decrease in difference between onset dates was indeed an increase in synchrony for close friend groups.

Usually those who considered themselves close friends lived together. Because this was not always the case, however, subjects were divided into thirteen living groups ($5 \leqslant N \leqslant 12$), solely on the basis of arrangement of rooms, to test the importance of geographic location. When grouped in this way, there was no significant increase in synchrony within groups.

Dewan[11] has suggested that the menstrual cycles of monkeys around the equator are synchronized because each cycle is locked in phase with the Moon. As the production by the pineal gland of a substance which inhibits the action of luteinizing hormone is suppressed by light, the continuous light of nights with a full moon would facilitate ovulation across a group of monkeys and induce synchrony. This suggests that the synchrony in close friend groups and among room mates comes from a common light-dark pattern, perhaps with common stress periods in which the subjects may stay up for a large part of the night. It would be expected that if synchrony arose from common light-dark cycles, room mates would exhibit a more significant amount of synchrony than do closest friends. The opposite trend was found, however, although it was not significant (room mates, $P \leq 0.007$; closest friends, $P \leq 0.003$). It does not seem likely therefore that a photoperiodic effect is a significant cause of synchrony. This is further supported by the lack of significant synchrony in random pairings in the dormitory.

Paralleling the Whitten effect in mice[3] (in which suppression of estrus in groups of females can be released by the introduction of a male pheromone) synchrony may result from a pheromonal interaction of suppression among close friend groups, followed by a periodic release due to

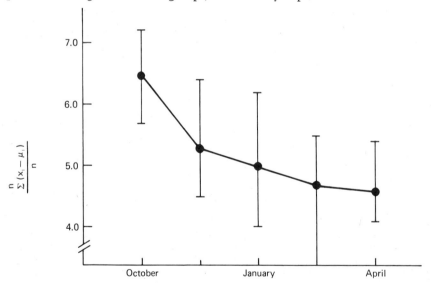

Figure 20-1 The median individual mean difference from the group onset mean $\dfrac{\sum\limits^{n}(X_t - \mu_t)}{n}$ as a function of time. The asymptotic relation and non-overlapping confidence intervals[8] for the medians in October and late February, and October and April (>0.99), indicate an increase in synchrony for close friend groups.

the presence of males on the weekend. However, this would be insufficient to explain the synchrony which occurred among room mates and close friends, but did not occur throughout the dormitory. Some additional pheromonal effect among individuals of the group of females would be necessary. Perhaps at least one female pheromone affects the timing of other female menstrual cycles.

Another possible source of synchrony might be the awareness of menstrual cycles among friends. A sample taken from the dormitory, however, indicated that 47% were not conscious of their friends' menstrual cycles, and, of the 53% who were, 48% (25% of the total) were only vaguely aware.

The significant factor in synchrony, then, is that the individuals of the group spend time together. Whether the mechanism underlying this phenomenon is pheromonal, mediated by awareness or some other process is a question which still remains open for speculation and investigation.

Subjects were divided into two groups: those who estimated that they spent time with males, once, twice or no times per week (N=42), and those who estimated that they spent time with males three or more times per week (N=33). Borderline cases and those taking birth control pills were discarded. After testing for homogeneity of variance, the mean cycle length and duration of menstruation was compared using Student's *t* test. Those who estimated seeing males less than three times per week experienced significantly $(P \leq 0.03)$ longer cycles than those of the other group whose mean cycle length corresponded with national norms (approximately 28 days)[12]. There was no significant difference in duration of menstruation itself $(P \geq 0.2$ Table 20-2).

The possibility that the results were confounded by a maturational factor was tested, as subjects included members of the freshman, sophomore, junior and senior classes. The subjects were regrouped and compared according to class: underclassmen were compared with upperclassmen. There was no significant difference in cycle length (underclassmen 29.6 ± 5.6 days; upperclassmen 29.9 ± 5.7 days).

TABLE 20-2
Mean cycle lengths and duration of menstruation

ESTIMATED EXPOSURE TO MALE (DAYS/WEEK)	LENGTH OF CYCLE (DAYS)	DURATION (DAYS)
0–2	30.0 ± 3.9	5.0 ± 1.1
N = 56		
3–7	28.5 ± 2.9	4.8 ± 1.2
N = 31		
P	≤ 0.03	N.S. ≤ 0.2

Exposure to males may not be the significant factor. It may be, for example, that those with longer cycles are less likely to spend time with males. However, many subjects spontaneously indicated that they became more regular and had shorter cycles when they dated more often. For example, one subject reported that she had a cycle length of 6 months until she began to see males more frequently. Her cycle length then shortened to 4.5 weeks. Then, when she stopped seeing males as often, her cycle lengthened again. Whether this is due to a pheromone mechanism similar to the Lee–Boot effect in mice[1] has yet to be determined.

Although this is a preliminary study, the evidence for synchrony and suppression of the menstrual cycle is quite strong, indicating that in humans there is some interpersonal physiological process which affects the menstrual cycle.

REFERENCES

1. Van der Lee, S., and Boot, L. M., *Acta Physiol. Pharmacol. Neerl.*, 5, 213 (1956).
2. Whitten, W. K., *J. Endocrinol.*, 18, 102 (1959).
3. Whitten, W. K., *Science*, 16, 584 (1968).
4. Parkes, A. S., and Bruce, H. M., *J. Reprod. Fertil.*, 4, 303 (1962).
5. Thiessen, D., *Texas Rep. Biol. Med.*, 22, 266 (1964); Leiderman, P. H., and Shapiro, D., *Psychobiological Approaches to Social Behavior* (Stanford University Press, 1964).
6. Collet, M. E., Wertenberger, G. E., and Fiske, V. M., *Fertil. Steril.*, 5, 437 (1954).
7. Siegal, S., *Nonparametric Statistics for the Behavioral Sciences* (McGraw-Hill, New York, 1956).
8. Nair, K. R., *Indian J. Statistics*, 4, 551 (1940).
9. Koford, C. B., in *Primate Behavior; Field Studies of Monkeys and Apes* (edit. by DeVore, I.) (Holt, Rinehart and Winston, New York, 1965).
10. Page, E. B., *Amer. Stat. Assoc. J.*, 58, 216 (1963).
11. Dewan, E. M., *Science Tech.*, 20 (1969).
12. Turner, C. D., *General Endocrinology* (Saunders, Philadelphia, 1965).

For some interesting implications of this kind of research see Alex Comfort's article "Communication May Be Odorous," in *New Scientist and Science Journal*, 49: pages 412–414, February 1971. [Editor's footnote]

21

LESSONS FROM
A "PRIMITIVE" PEOPLE

James V. Neel

The traditional laboratories for ethnographers and other cultural anthropologists have been the remnants of tribal peoples: small-scale, unstratified communities practicing more or less localized cultures, whose life ways have not yet been overwhelmed by contact with the agencies of the industrialized national cultures. There are not many such tribal units left. This century may be the setting for the final absorption of tribal cultures into national systems. And cultural anthropologists, feeling deeply the need to study and record such cultures before they disappear, have been mobilizing the resources of their profession for what may be termed "salvage anthropology" in those parts of the world where tribal cultures are threatened with imminent extinction: the highlands of New Guinea, and the Amazon basin tropical forest of South America.

In such studies they are being joined by physical anthropologists and population geneticists. One kind of problem to which biologically-oriented scientists address themselves is the extent to which the genetic data on tribal peoples—who are characterized by low population density and small-scale settlements—may illuminate earlier stages of human demographic history, when evolution was acting upon smaller, more discrete population units of Homo sapiens.

What have such studies revealed? Have they enabled scientists to understand more clearly how evolution might have proceeded in prehistoric times?

The following article by James Neel discusses some of these questions. Even more revealing is the extent to which the author sees implications of tribal or "primitive" population dynamics for the problems which face the developed technological societies of Europe and North America. A modern ecological approach to

Reprinted from *Science*, Vol. 170, November 20, 1970, pages 805–822. Copyright 1970 by the American Association for the Advancement of Science. By permission of the author and the publisher.

*such problems as the "population explosion" and a balanced
relation of people to the resources of their natural setting are
exemplified, in Neel's view, in the integration of today's remnants
of tribal peoples with their environment. The reader may well
ask himself in reading this article whether the author has succumbed to the anthropologist's Rousseauistic noble-savage biases,
or whether there really are lessons for us to learn in studying the
population biology of tribal groups.*

*Dr. Neel is Lee Dice Professor of Human Genetics at the Univer-
sity of Michigan Medical School.*

The field of population genetics is in a state of exciting intellectual
turmoil and flux. The biochemical techniques that are now so freely available have revealed a profusion of previously hidden genetic variability.
The way in which this variability arose and is maintained in populations—
to what extent by selection, past and present, and to what extent by simple
mutation pressure—is currently a topic of intensive discussion and debate,
and there is little agreement among investigators as to which are the most
promising approaches to the questions (see 1). At the same time, it is
becoming increasingly clear that the breeding structure of real populations
—especially those that approximate the conditions under which man
evolved—departs so very far from the structure subsumed by the classical
formulations of population genetics that new formulations may be necessary before the significance of this variation can be appraised by mathematical means.

Some 8 years ago, as the new population genetics began to emerge, my
co-workers and I began the formulation of a multidisciplinary study of
some of the most primitive Indians of South America among whom it is
possible to work (2). Scientists in the program ranged from the cultural
anthropologist to the mathematical geneticist. The general thesis behind
the program was that, on the assumption that these people represented the
best approximation available to the conditions under which human variability arose, a systems type of analysis oriented toward a number of
specific questions might provide valuable insights into problems of human
evolution and variability. We recognize, of course, that the groups under
study depart in many ways from the strict hunter-gatherer way of life that
obtained during much of human evolution. Unfortunately, the remaining
true hunter-gatherers are either all greatly disturbed or are so reduced in
numbers and withdrawn to such inaccessible areas that it appears to be
impossible to obtain the sample size necessary for tests of hypothesis. We
assume that the groups under study are certainly much closer in their
breeding structure to hunter-gatherers than to modern man; thus they

permit cautious inferences about human breeding structure prior to large-scale and complex agriculture.

I will present here four of our findings to date and will consider briefly some possible implications of these findings for contemporary human affairs. You will appreciate that I am the spokesman for a group of more than a dozen investigators, whose individual contributions are recognized in the appropriate source papers. The article is tendered with no great sense of accomplishment—we the participants in the endeavor realize how far we are from the solid formulations we seek. On the other hand, some of the data are already clearly germane to contemporary human problems. Thus, it will be maintained on the basis of evidence to be presented that, within the context of his culture and resources, primitive man was characterized by a genetic structure incorporating somewhat more wisdom and prudence than our own. How he arrived at this structure—to what extent by conscious thought, to what extent through lack of technology and in unconscious response to instinct and environmental pressures—is outside the purview of this presentation.

The studies to be described have primarily been directed toward three tribes: the Xavante of the Brazilian Mato Grosso, the Makiritare of southern Venezuela, and the Yanomama of southern Venezuela and northern Brazil. At the time of our studies, these were among the least acculturated tribes of the requisite size (>1000) in South America (see 3–8).

The four salient points about the Indian populations that we studied to be emphasized in this presentation are (i) microdifferentiation and the strategy of evolution, (ii) population control and population size, (iii) polygyny and the genetic significance of differential fertility, and (iv) the balance with disease (9).

MICRODIFFERENTIATION AND THE STRATEGY OF EVOLUTION

The term "tribe" conjures to most an image of a more or less homogeneous population as the biological unit of primitive human organization. We have now typed blood specimens from some 37 Yanomama, 7 Makiritare, and 3 Xavante villages with respect to 27 different genetic systems for which serum proteins and erythrocytes can be classified. A remarkable degree of intratribal genetic differentiation between Indian villages emerges from these typings (8, 10). One convenient way to express this differentiation in quantitative terms is to employ the distance function developed by Cavalli-Sforza and Edwards (11). On the basis of gene frequencies at the Rh, MNSs, Kidd, Duffy, Diego, and haptoglobin loci, the distances between seven of the central Yanomama villages and the distances between seven Makiritare villages are shown in Tables 21-1 and 21-2. The mean distance between the seven Yanomama villages is 0.330 unit, and between the Makiritare, 0.356 unit. Table 21-3 gives the distances between 12 Indian tribes of Central and South America, selected for consideration solely because 200 or more members have been studied for these same characteristics and the tribes were relatively free of non-Indian genes (admixture estimated at less than 5 percent) (12, 13). The mean distance is 0.385 unit. Thus the average distance between Indian villages is 85.7 (Yanomama) to 92.5 (Makiritare) percent of the distance between tribes. To some extent—an extent whose precise specification presents some difficult statistical problems—these distances result from stochastic events such as the founder effect of E. Mayr, sampling error, and genetic drift. But we have also begun to recognize structured factors in the origin of the differences. One is the "fission-fusion" pattern of village propagation, in consequence of which new villages are often formed by cleavages of established villages along lineal lines (fission), and migrants to established villages often consist of groups of related individuals (fusion) (14). A second such factor is a markedly nonisotopic (that is, a nonrandom and "unbalanced") pattern of intervillage migration (8). A third factor will be discussed below (see "Polygyny and genetic significance of differential fertility").

This situation, of subdivision of a population into genetically differentiated and competing demes, is one repeatedly visualized by Wright (15), beginning in 1931, as being most conducive to rapid evolution. Competition between these demes can only be termed intense (4, 6). On the basis of the genetic distance between these Indian tribes and an estimated arrival of the Indian in Central and South America some 15,000 years ago (16), we have with all due reservations calculated a *maximum* rate of gene substitution in the American Indian of 130,000 years per gene substitution per locus (13). This rate is approximately 100 times greater than an estimate of *average* rate based on amino acid substitutions in the polypeptides of a wide variety of animal species (17). For the present, we equate allele substitutions to amino acid substitutions on the assumption that the basic event in both cases is the partial or complete substitution of one codon for another. There is, of course, no logical discrepancy between these estimates, since one of them is a maximum estimate (18). On the other hand, part of the apparent difference may be valid. Thus, it seems a reasonable postulate that, all over the world in man's tribal days, the single most important step in the formation of a new tribe probably consisted of a village or a collection of related villages breaking away from its tribe and moving off into relative isolation. These villages, perhaps more than the break-away units of other animal populations, tended to consist of related individuals, thus providing unusual scope in man for what we have termed the "lineal effect" in establishing subpopulations whose gene frequencies are quite different from those of the parent population (19).

There is a current tendency to regard much of evolution, as measured by gene substitution, as non-Darwinian—that is, not determined by systematic pressures (17). We have recently argued that no matter whether one assumes a predominantly deterministic or indeterministic stance, the above-mentioned aspect of the social structure of primitive man resulted in genetic experiments of a type conducive to rapid evolution (18). Conversely, the current expansion and amalgamation of human populations into vast interbreeding complexes must introduce a great deal of inertia into the system.

POPULATION CONTROL AND POPULATION SIZE

The total human population apparently increased very slowly up to 10,000 years ago (20). If we may extrapolate from our Indian experience, the slowness of this increase was probably not primarily due to high infant and childhood mortality rates from infectious and parasitic diseases (see "The balance with disease"). We find that relatively uncontacted primitive man under conditions of low population density enjoys "intermediate" infant mortality and relatively good health, although not the equal of ours today (21–23). However, most primitive populations practiced spacing of children. Our data on how this spacing was accomplished are best for the Yanomama, where intercourse taboos, prolonged lactation, abortion, and infanticide reduce the average *effective* live birth rate to approximately one child every 4 to 5 years during the childbearing period (24, 25). The infanticide is directed primarily at infants whose older sibling is not thought ready for weaning, which usually occurs at about 3 years of age (25). Deformed infants and those thought to result from extramarital relationships are also especially liable to infanticide. Female infants are killed more often than male infants, which results in a sex ratio of 100–128 during the age interval 0 to 14 years (24). An accurate estimate of the frequency of infanticide still eludes us, but, from the sex-ratio imbalance plus other fragmentary information, we calculate that it involves perhaps 15 to 20 percent of all live births.

There have been numerous attempts to define the development in human evolution that clearly separated man from the prehominids. The phenomena of speech and of toolmaking have had strong proponents, whose advocacy has faltered in the face of growing evidence of the complexity of signaling and the ingenuity in utilizing materials that are manifested by higher primates. Population control may be such a key development. Among 309 skeletons of (adult) fossil man classified as to sex by

H. V. Vallois, 172 were thought to be males and 137 were classified as females, which gives a sex ratio of 125.6–100 (26). These finds were made over a wide area, and they extend in time depth from Pithecanthropus to Mesolithic man. I am aware of the controversy that surrounds the sexing of fossil skeletons, as well as the question of whether both sexes were equally subject to burial. Nevertheless, one interpretation is that preferential female infanticide is an old practice. In contrast to man, it appears that most higher primates must utilize their natural fecundity rather fully to maintain poulation numbers. I conclude, as has been suggested elsewhere (25), that perhaps the most significant of the many milestones in the transition from higher primate to man—on a par with speech and toolmaking—occurred when human social organization and parental care permitted the survival of a higher proportion of infants than the culture and economy could absorb in each generation and when population control, including abortion and infanticide, was therefore adopted as the only practical recourse available.

The deliberate killing of a grossly defective child (who cannot hope for a full participation in the society he has just entered) or of the child who follows too soon the birth of an older sibling (and thereby endangers the latter's nutritional status) is morally repugnant to us. I am clearly not obliquely endorsing a return to this or a comparable practice. However, I am suggesting that we see ourselves in proper perspective. The relationship between rapid reproduction and high infant mortality has been apparent for centuries. During this time we have condoned in ourselves a reproductive pattern which (through weaning diarrhea and malnutrition) has contributed, for large numbers of children, to a much more agonizing "natural" demise than that resulting from infanticide. Moreover, this reproductive pattern has condemned many of the surviving children to a marginal diet inconsistent with full physical and mental development.

We obviously cannot countenance infanticide. However, accepting the general harshness of the milieu in which primitive man functioned, I find

TABLE 21-1

Matrices of genetic distances between paired villages of the Maki- ritare indians (seven villages, six loci) [after (18)]

	DISTANCE MATRICES					
VILLAGE	BD	C	E	F	G	HI
A	.362	.558	.353	.345	.268	.336
BD		.250	.221	.432	.314	.296
C			.393	.588	.485	.444
E				.379	.249	.273
F					.394	.383
G						.158

TABLE 21-2

Matrices of genetic distances between paired villages of the
Yanomama indians (seven villages, six loci) [after (18)]

VILLAGE	DISTANCE MATRICES					
	B	C	D	E	H	I
A	.227	.228	.385	.157	.416	.243
B		.367	.506	.144	.537	.360
C			.298	.295	.346	.297
D				.464	.154	.364
E					.486	.296
H						.350

it increasingly difficult to see in the recent reproductive history of the civilized world a greater respect for the quality of human existence than was manifested by our remote "primitive" ancestors. Firth (27), in protesting the disturbance in population balance in the Pacific island of Tikopia when Christianity was substituted for ancient mores, expressed it thus:

> It might be thought that the so-called sanctity of human life is not an end in itself, but the means to an end, to the preservation of society. And just as in a civilized community in time of war, civil disturbance or action against crime, life is taken to preserve life, so in Tikopia infants just born might be allowed to have their faces turned down [see (28)], and to be debarred from the world they have merely glimpsed, in order that the economic equilibrium might be preserved, and the society maintain its balanced existence.

POLYGYNY AND GENETIC SIGNIFICANCE OF DIFFERENTIAL FERTILITY

The three Indian tribes among whom we have worked are polygynous (4, 5, 21), the reward in these nonmaterial cultures for male achievement (however judged) and longevity being additional wives. This pattern is found in many primitive cultures. As brought out in the preceding section, women seem to be committed to a pattern of child spacing, which in the Yanomama results, for women living to the age of 40 years, in a lower variance in number of reported live births than in the contemporary United States (24). By contrast with our culture, then, the mores of these primitive societies tend to minimize the variance of number of live births per female but maximize the variance of number of children per male.

One of our objectives is to understand the genetic consequences of polygyny. The translation of generalizations such as the above into the kind of hard data that can be employed in either deterministic formula-

TABLE 21-3

Matrix of pair-wise genetic distances for 12 South American tribes [after (18)]

TRIBE	CAKCHIQUEL	CAYAPA	CUNA	GUAYAMI	JIVARO	PEMON	QUECHUA	SHIPIBO	XAVANTE	YANO-MAMA	YUPA
Aymara	.260	.301	.355	.485	.370	.381	.288	.393	.374	.514	.450
Cakchiquel		.297	.224	.364	.342	.302	.278	.363	.250	.439	.326
Cayapa			.283	.446	.289	.346	.224	.486	.343	.473	.328
Cuna				.327	.381	.283	.331	.466	.227	.479	.239
Guayami					.444	.469	.398	.645	.410	.437	.433
Jivaro						.402	.270	.521	.375	.536	.433
Pemon							.319	.460	.371	.510	.354
Quechua								.433	.336	.479	.392
Shipibo									.335	.660	.479
Xavante										.549	.249
Yanomama											.452

tions or stochastic procedures based on population simultation designed to explore the genetic consequences of polygyny has proved quite difficult. A single example will suffice. We have earlier directed attention toward the unusual reproductive performance of certain headmen among the Xavante (21), and N. A. Chagnon has similar unpublished data for the Yanomama. During the past 2 years, J. MacCluer, in collaboration with Chagnon and myself, has been trying to develop a computer model that simulates the genetic and demographic structure of the Yanomama. The basic input has consisted of Chagnon's detailed demographic data from four villages. One of the several objectives of this simulation is to derive a better estimate of the amount of inbreeding than is possible from pedigree information in which the genealogical depth is so shallow, with particular reference to the complications introduced by polygyny. For some time there were great difficulties in reconciling certain aspects of the inbreeding results after 150 simulated years with the results of the first 20 simulated years (during which the real, input population dominated the findings), even though in many other respects—age at marriage, mean number of children, and distribution of polygyny—there was a good accord with the facts. During the first 20 simulated years, the members of the (real) population were, on the average, more closely related to each other than those in the simulated population at 150 years, so that, even when the model specified the maximum opportunity for consanguineous marriages consistent with field data, the level of inbreeding declined with time.

The reason finally became clear when the distribution of number of grandchildren per male was. considered. Table 21-4 contrasts the actual

TABLE 21-4

The number of grandchildren (who reach adult life) per male, for all males with at least one grandchild, in the input (real) popoulation and the artificial popoulation after 150 years of simulation. The simulation is based on the assumption of no correlation in fertility between generations.

NO. OF GRAND- CHILDREN PER MALE	INITIAL POPULATION	ARTIFICIAL POPULATION AFTER 150 YEARS
1	28	17
2	11	9
3	11	10
4	7	7
5	10	9
6	9	6
7	5	9
8	5	5
9	1	5

TABLE 21-4 *continued*

NO. OF GRAND- CHILDREN PER MALE	INITIAL POPULATION	ARTIFICIAL POPULATION AFTER 150 YEARS
10	7	5
11	3	5
12	2	3
13		7
14		5
15	1	2
16	2	3
17		3
18		1
19	1	1
20	2	
21	1	2
22	1	2
23		1
24	1	2
25	1	
26		
27		
28	1	
29		
30		1
31		
32		
33		
34		
35		
36		
37		
38		1
39		
40		
41	1	
42	1	
43		
44		
45		
46	1	
—	—	—
—	—	—
—	—	—
62	1	
Total	114	121

distribution in the input data with that predicted by the simulation program after 150 years on the assumption of no correlation in fertility between successive generations. It is clear that the assumption is incorrect. Note the disporportionate number of grandchildren born to some few males, a situation that greatly increases the possibilities for marriages between first cousins or half-first cousins. Added genetic significance is lent to this phenomenon by the fact that the four males whose living grandchildren outnumber those of any male in the computer population represent two father-son combinations. One reason for this "familial" fertility seems to be that, if because of the polygyny of his father a young man possesses many sisters or half-sisters (who can be "traded"), that young man has an advantage in forming alliances and obtaining extra wives; thus, polygyny begets polygyny. The phenomenon does not appear to be primarily genetic. Clearly here is the explanation of why, when MacCluer programmed for the maximum rate of marriage between relatives (but did not incorporate in the model this aspect of fertility), there was an apparent decline in consanguinity as the real world of input was replaced by the simulated world. To be sure, we might have recognized this aspect of polygyny in Indian culture before we programmed the model, but I prefer to view the development as an example of how simulation forces one to look at the real population more carefully. This degree of differential fertility must be another factor in the marked genetic micro-differentiation between villages. Since the number of wives a man obtains and holds also depends on personal attributes, unquestionably determined genetically to some extent, here is an example of an interaction between the genetic system and the social system. Such interactions occur at all cultural levels, but it was unexpected to find this one emerging so strongly in these circumstances.

The *possible* genetic implications of polygyny are clear, but some of the facts necessary to a meaningful treatment are still lacking. Thus, one of our projected future investigations is an attempt to contrast certain mental attributes of polygynous with nonpolygynous males. In many respects, Indian culture is much more egalitarian than our own. The children of a village have the same diet and, by our standards, a remarkably similar environment. There are minimal occupational differences, and we do not find the differentiation into fishing villages, mining villages, or farming communities encountered in many cultures. Even with allowance for the happy accident of a large sibship, the open competition for leadership in an Indian community probably results in leadership being based far less on accidents of birth and far more on innate characteristics than in our culture. Our field impression is that the polygynous Indians, especially the headmen, tend to be more intelligent than the nonpolygynous. They also tend to have more surviving offspring. Polygyny in these tribes thus appears to provide an effective device for certain types of natural selection. Would that we had quantitative results to support that statement!

THE BALANCE WITH DISEASE

Inasmuch as viral, bacterial, and parasitic diseases are commonly regarded as among the important agents of natural selection, a particular effort has been directed toward assaying the health of primitive man and the characteristics of his interaction with these disease agents. We have reported that the Xavante are, in general, in excellent physical condition (21, 22), and we have similar unpublished data on the Yanomama and Makiritare. In terms of morbidity, perhaps the most important disease is falciparum malaria, which is probably a post-Columbian introduction (29). Fortunately for our view of the health of the pre-Columbian American Indian, we can find villages in which malaria does not seem to be a problem. Figure 1 presents our working concept of the Yanomama life expectancy curve, to be improved as more data become available (23, 24). Although infant and childhood mortality rates are high by the standards of a civilized country such as present-day Japan, they are low in comparison with India at the turn of the century, especially since there was probably gross underreporting in the data from India. Note the relatively high Yanomama death rate during the third, fourth, and fifth decades, a substantial fraction of which is due to warfare. One way to view the differences between these three curves is that the advent of civilization dealt a blow to man's health from which he is only now recovering.

Dunn (30) has properly emphasized the degree to which the ecological setting influenced disease patterns in primitive man and the difficulty in reaching generalizations. Even so, certain common denominators may be emerging. For instance, the pattern of acquisition of immunity to endemic diseases in the Indian and possibly other primitives can already be seen to differ in a number of respects from the pattern in most civilized communities (23). Among the Xavante and Yanomama, for example, we find gamma globulin levels approximately two times those in civilized areas (31). Newborn infants presumably possess a high measure of maternal antibody acquired transplacentally. From the first, these infants are in an intimate contact with their environment that would horrify a modern mother—or physician. They nurse at sticky breasts, at which the young mammalian pets of the village have also suckled, and soon are crawling on the feces-contaminated soil and chewing on an unbelievable variety of objects. Our thesis is that the high level of maternally derived antibody, early exposure to pathogens, the prolonged period of lactation, and the generally excellent nutritional status of the child make it possible for him to achieve a *relatively* smooth transition from passive to active immunity to many of the agents of disease to which he is exposed. The situation is well illustrated by the manner in which concomitantly administered gamma globulin reduces the impact of a rubeola vaccination while still permitting the development of effective immunity. To be sure, civilized tropical populations also have relatively high globulin levels (see references in 21), so that there should be high placental transfer of passive immunity; however, because of the higher effective birth rate, the child of the civilizado is seldom nursed as long as the Indian child and thus falls prey to weanling diarrhea and malnutrition.

By his vaccination programs, then, modern man is developing a relatively painless immunity to his diseases, similar in some ways to the manner in which the Indian seems to have developed immunity to some of

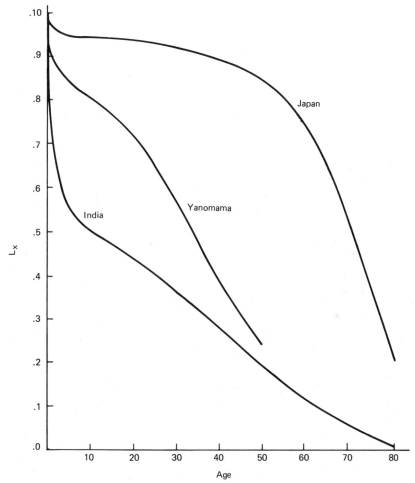

Fig. 21-1 "Life curves" for three types of populations: a highly urbanized and industrialized country (Japan in 1960); a densely populated, primarily agricultural country (India in 1901); and an unacculturated tribal population (the Yanomama) [after (23)]. Probability of survival is indicated by L_x.

his diseases. A danger for both groups is the sudden appearance of a "new" disease. Burnet (32) has described some of the possible consequences for civilized societies in the appearance in the laboratory of strains of pathogens with new combinations of antigenic and virulence properties, and Lederberg (33) has labeled this threat as one of the hidden dangers in experiments related to biological warfare. At the other extreme, we have recently witnessed at first hand the consequences of a measles epidemic among the Yanomama, known from antibody studies to be a "virgin-soil" population with respect to this virus (34). Although the symptomatic response of the Indian to the disease may be somewhat (but not markedly) greater than our own, much of the well-recognized enhanced morbidity and mortality in such epidemics is due to the secondary features of the epidemic—the collapse of village life when almost everyone is highly febrile, when mothers cannot nurse their infants, and when there is no one to provide for the needs of the community. After witnessing this spectacle, I find it unpleasant to contemplate its possible modern counterpart—when, in some densely populated area, a new pathogen, or an old one such as smallpox or malaria, appears and escapes control, and a serious breakdown of local services follows.

This relative balance with his endemic diseases is only one aspect of the generally harmonious relationship with his ecosystem that characterizes primitive man. There is an identification with and respect for the natural world, beautifully described by Radcliffe-Brown (35), Redfield (36), and Lévi-Strauss (37), among others, which we, who have walled it off so successfully while penetrating its secrets, find difficult to understand. In general, the religion of the tribes among whom we have worked is a pantheism in which both the heavens and the immediate environment are peopled by ubiquitous spirits, of human and nonhuman origin, whom it is vitally important to propitiate at every turn. To some extent their apparent respect for their ecosystem probably stems from ignorance and technical incompetence, but, in common with White (38), I believe that it also reflects the difference between a religion that regards man as a part of a system and one in which he is the divinely appointed master of the system.

A PROGRAM

In a world in which our heads are spinning under the impact of information overload, studies of primitive man provide, above everything else, perspective. Civilized man is a creature who each year is departing farther and farther from the population structure that obtained throughout most of human evolution and that was presumably of some importance to the evolutionary process. At the same time he is not only living far beyond a reasonable energy balance but is despoiling the resources for primary production so as to narrow increasingly the options available to redress the imbalances. The true dimensions of the dilemma that our present course has created are only now emerging (see especially 39). The intellectual arrogance created by our small scientific successes must be replaced by a profound humility based on the new knowledge of how complex is the system of which we are a part. To some of us, this realization carries with it the need for a philosophical readjustment which has the impact of a religious conversion.

We various members of the scientific community are all deeply engaged these days in speculating on the role that we will play in the next major cultural cycle. I find much of relevance to contemporary problems in my field, human genetics, in our studies of primitive man. It is clear that our primary objective—to understand the origin and significance of polymorphic variability—still eludes us. But there are other insights. In the light of our recent experiences among the Indian tribes, I shall now briefly consider some possible emphases in human genetics in the immediate future. In keeping with the new humility incumbent upon us all, it is not surprising that my suggestions are rather conservative; they are designed to preserve what we have rather than to promote unreal hopes of spectacular advances. They constitute in many respects an attempt to recreate, within limits, certain conditions that we have observed. These suggestions do not stem from any romanticism concerning the noble savage: Indian life is harsh and cruel, and it countenances an overt aggressiveness that is unthinkable today. Obviously the world should not return to a state of subdivision into demes of 50 to 200 persons constantly involved in a pattern of shifting loyalties and brutal conflict vis-à-vis neighboring demes. Nor are we likely to return to polygyny, with number of wives in part a function of one's "fierceness"—demonstrated by a series of duels with clubs or stylized bouts of chest poundings. Clearly we do not wish to abandon modern medical care to permit natural selection to have a better opportunity to work. But there are other, less disruptive aspects of primitive society for which there is a modern counterpart. These are enumerated below as a series of principles.

STABILIZATION OF THE GENE POOL

First principle: Stabilize the gene pool numerically. Throughout the world, primitive man seems to have curbed his intrinsic fertility to a greater extent than has the civilized world in recent centuries. Exactly how those curbs were relaxed with the advent of civilization is unclear, but the agricultural revolution undoubtedly played a part. Although it is currently fashionable to indict the great religions, on the basis of the Old Testament injunction to "be fruitful and multiply," their precise role (until recent times) is in my opinion unclear. The remaining pockets of dissent with the principle of population limitation are rapidly disappearing; the next 5 years will convince even the most reluctant. But by what precise formula should population limitation be accomplished? I have previously urged a simple quota system, set at three living children per couple on the thesis that failure to marry, infertility, and voluntary limitation to less than three would result in a realized average of approximately two children who reach the age of reproduction (40). I now wonder whether failure of contraception will not result in so many well-intentioned persons exceeding their quota that these guidelines are not sufficiently stringent; I would therefore amend the earlier suggestion to include provision for voluntary sterilization after the third child. You will recognize that this proposal implies relative stabilization of the present gene pool, a move that will tend to conserve all our present bewildering diversity but hinder evolution. It makes no value judgments about any specific group. There would be less opportunity for changes in gene frequency than with present patterns of differential fertility.

Such a policy cannot succeed if some religions and governments simply continue their present half-hearted admonitions and leave the rest to science, while other religions and governments actually oppose effective population control. What has been signally lacking thus far is a clear statement at every possible level of responsibility of the implications of continuing the present rate of population increase. Also lacking has been an administrative framework within which all peoples move toward population control simultaneously, thus dispelling deep-rooted fears that some sectors are being subjected to a subtle form of genocide.

PROTECTION OF THE GENE POOL

Second principle: Protect the gene against damage. If, as we have implied, polygyny among the Indian has eugenic overtones, there is no acceptable modern counterpart in view. However, we can at least protect the gene pool from obvious damage. The world of primitive man is remarkably uncontaminated. This fact, plus his lower mean age at reproduction, probably results in lower mutation rates than our own (41), but we have no direct evidence.

Until recently, the principal concomitant of civilization that appeared capable of damaging the gene pool was an increasing exposure to radiation. Now concern is shifting to the many potentially mutagenic chemicals being introduced into the environment as pesticides, industrial by-products, air contaminants, and so forth. The magnitude of this problem is currently undefined. About 6 percent of all newborn infants have been found to have defects partially or wholly of genetic origin (42). Let us assume that half of these defects (3 percent) result from recurrent mutation. Doubling the mutation rate would *eventually* double that 3 percent.

For all the work that has been done on the genetic effects of radiation, involving both man and experimental organisms, there still remain large areas of ambiguity, especially as regards the effects of low-level, intermittent, or chronic-type doses, such as characterize most human exposures. The current working estimate of the "doubling dose" of radiation of this type is 100 to 200 roentgens (42). In general, current man-made human exposures in the United States [probably less than 3 rem to the gonads in 30 years (42)] appear to be of the order of perhaps 1/30 to 1/60 of the "doubling dose," a price society thus far seems prepared to accept for the benefits of the medical uses of radiation and the development of nuclear sources of energy. It may be that genetically effective exposures to the chemical mutagens are as low or even lower, but we cannot be certain.

The technical advances of the past 20 years now render it possible and feasible to screen a representative 20 proteins in newborn infants for evidence of mutational damage (see 43); hence we need no longer rely, as in the studies at Hiroshima and Nagasaki, on the potential genetic effects of the atomic bombs, on such imprecise indicators of genetic damage as congenital malformations, survival rates, and sex ratio. A society that can afford to send man to the moon surely has the resources and the intelligence to monitor itself properly for increased mutation rates. If a significant increase is detected, however, the task of identifying the responsible agent or agents will, because of the many possibilities, be extremely difficult, and that agent, when identified, may be so relevant to the welfare of society that, as with radiation, the goal will be to minimize rather than to eliminate exposures. Despite these difficulties in detection and control, immediate steps to determine the facts are needed.

GENETIC COUNSELING AND PRENATAL DIAGNOSIS

Third principle: Improve the quality of life through parental choice based on genetic counseling and prenatal diagnosis. Both the pressures on the social system and its services and the increasing demands of society on the individual render it imperative that full advantage be taken of all morally acceptable developments that promise to minimize the number of unfortunate individuals incapable of full participation in this complex society. We will not return to infanticide, but there are ethical alternatives. Genetic counseling, which defines the high-risk family, represents one such development. In the past, once the identification had been made, the individuals who wished to limit the entry of defective children into the population had only two alternatives: to practice birth control or to apply for voluntary sterilization. Recently the possibilities inherent in prenatal diagnosis based on fetal cells obtained through amniocentesis during the first trimester of pregnancy have been receiving active attention (44). Where accurate diagnosis is possible and the presence of a defective fetus is established, the parents can be offered an abortion, usually with reasonable prospects of a normal child in the next pregnancy. Thus far the conditions that can be accurately diagnosed in the very early stages of pregnancy and the numerical impact of these entities are relatively small.

The moral issues that are involved cannot be evaded, and it is better in this time of reappraisal for society to face them forthrightly. At what point is the artificial termination of a pregnancy no longer ethical, even when the fetus concerned is incapable of marginal participation in society? Just what defects are of such gravity as to justify intervention? To what extent should persuasion be employed in implementing these new possibilities? In my opinion, once the principle of parental choice of a normal child is established, it seems probable that in large measure the parental desire for normal children can be relied on to result in the purely voluntary elimination of affected fetuses.

REALIZING THE GENETIC POTENTIAL OF THE INDIVIDUAL

Fourth principle: Improve the phenotypic expression of the individual genotype. It is a sobering thought that the relatively egalitarian structure of most primitive societies, plus the absence of large individual differences in material wealth, seems to ensure that, within the culturally imposed boundaries, each individual in primitive society leads a life (and enjoys reproductive success) more in accord with his innate capabilities than in our present democracy. In the difficult times ahead, society clearly needs the fullest possible participation of all its members. In the past, a very major effort has gone into the provision of special services for the physically and mentally handicapped. A retreat from such compassion is unthinkable, but it is apparent that a similar effort directed toward realizing the genetic potential of the underprivileged or the gifted would have far more impact on the solution of our problems.

Much of the thrust of the geneticist and those with allied interests has been directed toward the treatment of specific genetic diseases. Obviously these efforts need not only to be continued but to be greatly expanded. And equally obviously, the Indian contributes no insight into a program in this field of endeavor. Others are speaking eloquently to the needs and potentialities of this type of investigation.

But an even greater effort should be directed toward what I have elsewhere termed "culture engineering" (45), which merges at one extreme with the euphenics of Lederberg (46). There is presumably an environment (or group of environments) in which the still poorly understood potentialities of the human animal find the fullest and most harmonious expression. Although our present environment-culture reduces the impact of a number of previously important causes of mortality and morbidity, it creates a host of other "casualties of our times" (47). The challenge to culture engineering is, of course, greatest in the realm of the mind. It is not enough to think in terms of better schools and more attractive housing; the subtle and lasting influence of prenatal and early postnatal influences is becoming increasingly apparent (see 48). Experimental mammalian models are yielding fascinating evidence on the complexity of these interactions (49). It is doubtful whether our precipitous and helter-skelter attacks on our present world will yield an optimum environment. We cannot escape the consequences of the peculiar position in which we have placed ourselves; we must now cautiously and reverently accept the full responsibility for shaping our own world.

SUMMING UP

The foregoing principles constitute an extremely conservative program in human genetics, which advocates for the present a return to as many of the features of the population structure under which we evolved as is consistent with our present culture. The urgent need to understand the biomedical and social significance of human genetic variability as a basis for an eventual, more definite program should be clear, and yet we seem to be retreating from support of the necessary research while we are squandering billions in pursuit of dubious military goals.

There has been no mention in this presentation of that brand of genetic engineering concerned with controlled changes in transmissible genetic material. This omission is not due to oversight or limitations of time. My thesis is clearly a plea for a profound respect for ourselves and the system in which we function. It would be inconsistent with that thesis to suggest that, with our present limited knowledge of the human genome, we should in the near future think of intervening to alter it in ways we cannot completely understand. Research along these lines with experimental organisms is inevitable and desirable—but I question the wisdom of attempting, in the foreseeable future, to apply the results of that research to man.

The past decade has witnessed spectacular triumphs in the "inner space" of the cell and the "outer space" of the cosmos. Perhaps this decade will in retrospect be seen as the first of many decades of spectacular advances in our understanding of *"intermediate* space"—the biosphere —the space defined as that narrow life-supporting zone wherein occur the interactions between intact humans and other organisms and their environment which by definition are an ecosystem. As we realize the full complexity of intermediate space, it seems very probable that the scientific challenge to produce new knowledge will be equaled by the challenge of integrating the applications of that new knowledge smoothly into the ecosystem. In the most sophisticated way we can summon, we must return to the awe, and even fear, in which primitive man held the mysterious world about him, and like him we must strive to live in harmony with the only biosphere that we can be certain will be occupied by our descendants.

REFERENCES AND NOTES

1. J. V. Neel and W. J. Schull, *Perspect. Biol. Med.* 11, 565 (1968).
2. The term "primitive" is employed in the usual sense: preliterate; relatively untouched by civilization; with a very simple technology and with subsistence based on hunting, gathering, and elementary agricultural practices; and with a social structure in which concepts of kinship play a dominant organizational role.
3. O. Zerries, *Waika: Die kulturgeschichtliche Stellung der Waika-Indiener des oberen Orinoco in Rahmen der Völkerkunde Südamerikas*, vol. 1 of *Ergebnisse der Frobenius-Expedition 1954/1955 nach Südost Venezuela* (Klaus Renner, Munich, 1964).
4. D. Maybury-Lewis, *Akwẽ-Shavante Society* (Oxford Univ. Press, London, 1967), pp. xxxii and 356.
5. N. A. Chagnon, *Yanomamö: The Fierce People* (Holt, Rinehart & Winston, New York, 1968), pp. xiv and 142.
6. ———, in *War: The Anthropology of Armed Conflict and Aggression*, M. Fried, M. Harris, R. Murphy, Eds. (Natural History Press, Garden City, N.Y., 1968), p. 109.
7. T. Asch and N. A. Chagnon, *The Feast* [30-minute, 16-mm color film with synchronous sound (National Audiovisiual Center, Washington, D.C., 1970)]; H. Valero (as told to E. Biocca), *Yanoàma* (Dutton, New York, 1970), p. 382.
8. R. H. Ward and J. V. Neel, *Amer. J. Hum. Genet.*, in press.
9. Since this is not a formal review, I shall draw primarily on our own work in discussing our concrete findings. The reader will recognize, however, that there is considerable background literature, although much of it unfortunately lacks the quantitative detail necessary to precise formulations and the construction of population models.
10. J. V. Neel and F. M. Salzano, *Amer. J. Hum. Genet.* 19, 554 (1967); T. Arends, G. Brewer, N. Chagnon, M. Gallango, H. Gershowitz, M. Layrisse, J. Neel, D. Shreffler, R. Tashian, L. Weitkamp, *Proc. Nat. Acad. Sci. U.S.* 57, 1252 (1967).
11. L. L. Cavalli-Sforza and A. W. F. Edwards, *Amer. J. Hum. Genet.* 19, 233 (1967).
12. R. H. Post, J. V. Neel, W. J. Schull, in *Biomedical Challenges Presented by the American Indian, Scientific Publication 165* (Pan American Health Organization, Washington, D.C., 1968), pp. 141–185.
13. W. Fitch and J. V. Neel, *Amer. J. Hum. Genet.* 21, 384 (1969).
14. J. V. Neel and F. M. Salzano, *Cold Spring Harbor Symp. Quant. Biol.* 29, 85 (1964).
15. S. Wright, *Genetics* 16, 97 (1931).

16. J. M. Cruxent, in *Biomedical Challenges Presented by the American Indian, Scientific Publication 165* (Pan American Health Organization, Washington, D.C., 1968), pp. 11–16.
17. M. Kimura, *Nature 217*, 624 (1968); *Proc. Nat. Acad. Sci. U.S. 63*, 1181 (1969); J. L. King and T. H. Jukes, *Science 164*, 788 (1969).
18. J. V. Neel and R. H. Ward, *Proc. Nat. Acad. Sci. U.S. 65*, 323 (1970).
19. J. V. Neel, *Jap. J. Hum. Genet. 12*, 1 (1967).
20. E. S. Deevey, *Sci. Amer. 203* (3), 195 (1960).
21. J. V. Neel, F. M. Salzano, P. C. Junqueira, F. Keiter, D. Maybury-Lewis, *Amer. J. Hum. Genet. 16*, 52 (1964).
22. E. D. Weinstein, J. V. Neel, F. M. Salzano, *ibid. 19*, 532 (1967).
23. J. V. Neel, in *The Ongoing Evolution of Latin American Populations*, F. M. Salzano, Ed. (Thomas, Springfield, Ill., 1971).
24. —— and N. A. Chagnon, *Proc. Nat. Acad. Sci. U.S. 59*, 680 (1968); N. A. Chagnon, unpublished data.
25. J. V. Neel, in *Proceedings, VII International Congress of Anthropological and Ethnological Science, 1968* (Science Council of Japan, Tokyo, 1969), vol. 1, pp. 356–361.
26. H. V. Vallois, in *Social Life of Early Man*, S. L. Washburn, Ed. (Aldine, Chicago, 1961), p. 214.
27. R. Firth, *We, the Tikopia* (Macmillan, New York, ed. 2, 1958), pp. xxvi and 605.
28. A reference to infanticide by suffocation.
29. F. L. Dunn, *Hum. Biol. 37*, 385 (1965).
30. ——, in *Man the Hunter*, R. B. Lee and I. DeVore, Eds. (Aldine, Chicago, 1968), p. 221.
31. J. V. Neel, W. M. Mikkelsen, D. L. Rucknagel, E. D. Weinstein, R. A. Goyer, S. H. Abadie, *Amer. J. Trop. Med. Hyg. 17*, 474 (1968); T. Arends, unpublished data.
32. F. M. Burnet, *Lancet 1966-I*, 37 (1966).
33. J. Lederberg, "Biological warfare and the extinction of man," statement before the Subcommittee on National Security Policy and Scientific Developments, House Committee on Foreign Affairs (2 December 1969).
34. J. V. Neel, W. R. Centerwall, N. A. Chagnon, H. L. Casey, *Amer. J. Epidemiol. 91*, 418 (1970).
35. A. R. Radcliffe-Brown, *Structure and Function in Primitive Society* (Cohen & West, London, 1952), pp. 219.
36. R. Redfield, *The Primitive World and Its Transformations* (Cornell Univ. Press, Ithaca, N.Y., 1953), pp. xiii and 185.
37. C. Lévi-Strauss, *The Savage Mind* (Univ. of Chicago Press, Chicago, 1966), pp. xii and 290.
38. L. White, Jr., *Science 155*, 1203 (1967).
39. P. R. Erhlich and J. P. Holdran, *BioScience 19*, 1065 (1969); H. R. Hullett, *ibid. 20*, 160 (1970).
40. J. V. Neel, *Med. Clin. N. Amer. 33*, 1001 (1969).
41. ——, in *Proceedings, XII International Congress of Genetics, 1968* (Science Council of Japan, Tokyo, 1969), vol. 3, pp. 389–403.

42. United Nations, "Report of the United Nations Scientific Committee on the Effects of Atomic Radiation" (United Nations, New York, 1962), supplement No. 16 (A/5216).

43. J. V. Neel and A. D. Bloom, *Med. Clin. N. Amer.* 53, 1243 (1969).

44. J. Dancis, *J. Pediat.* 72, 301 (1968); J. W. Littlefield, *N. Engl. J. Med.* 280, 722 (1969); H. L. Nadler and A. B. Gerbie, *Amer. J. Obstet. Gynec.* 103, 710 (1969); R. DeMars, G. Sarto, J. S. Felix, P. Benke, *Science* 164, 1303 (1969).

45. J. V. Neel, *Harvey Lect.* 56, 127 (1961); see also R. Dubos, *WHO Chron.* 23, 499 (1969).

46. Euphenics is defined as the reprogramming of somatic cells and the modification of development. See J. Lederberg, in *Man and His Future*, G. Wolstenholme, Ed. (Little, Brown, Boston, 1963), p. 263.

47. A. B. Ford, *Science* 167, 256 (1970).

48. Pan American Health Organization, *Perinatal Factors Affecting Human Development, Scientific Publication 185* (Pan American Health Organization, Washington, D.C., 1969), p. 253; N. S. Scrimshaw and J. E. Gordon, *Malnutrition, Learning, and Behavior, Proceedings of an International Conference*, N. S. Scrimshaw and J. E. Gordon, Eds. (M.I.T. Press, Cambridge, 1968), pp. xiii and 566.

49. R. Dubos, *Perspect. Biol. Med.* 12, 479 (1969).

50. This article is adapted from a paper delivered as the Second Annual Lasker Lecture of the Salk Institute for Biological Studies on 21 May 1970. The investigations described have been supported in large measure by the U.S. Atomic Energy Commission. Principal colleagues in these studies include Dr. Miguel Layrisse and Tulio Arends of Venezuela, Dr. Francisco Salzano and Manuel Ayres of Brazil, and Drs. Napoleon Chagnon, Jean MacCluer, Lowell Weitkamp, Richard Ward, and Walter Fitch of the United States.

GLOSSARY

ABERRATION—Anomaly, defect, or abnormality.

ADAPIDAE—Extinct family of prosimians, believed to be similar to modern lemurs.

AEGYPTOPITHECUS—Oligocene fossil from the Fayum, in Egypt, believed to be ancestral to the modern pongids.

AEOLOPITHECUS—Oligocene fossil from the Fayum, considered to represent populations ancestral to modern pongids.

AGONISTIC—Aggressive or competitive.

AGRONOMIST—Specialist in scientific agriculture.

ALBINISM—Trait determined by a recessive gene in which the individual's system is unable to produce melanin, or pigment.

ALCAPTONURIA—Disorder determined by a recessive gene, in which the patient's system is unable to oxidize a certain chemical product of metabolism, which accordingly shows up in the urine.

ALGORITHM—Any special method of making computations.

ALLELE—When a gene occurs in two or more chemical forms, the variants are said to be allelic to one another.

ALTRUISTIC—Generous; concerned about other people's welfare.

ATAVISTIC—Having the character of a reversion to an earlier stage or type.

AMENORRHEA—Absence of menstrual activity.

AMNIOCENTESIS—Removal of fluid from the amniotic sac by suction.

AMPHIPLOIDY—Cellular anomaly in which the cell possesses two sets of chromosomes.

ANAPTOMORPHIDAE—Extinct Eocene prosimian family.

ANEUPLOIDY—Deficiency in chromosome number. The result of nondisjunction when the chromosomes are unequally divided between daughter cells in mitosis.

ANOMALY—Abnormality.

ANOPHTHALMIA—Abnormality characterized by failure of eyes to develop.

ANTHROPOID—Pertaining to the suborder Anthropoidea, made up of the monkeys, the apes, and human beings.

ANTIGEN—Foreign protein that, introduced into the body, elicits the production of a specific antibody that will neutralize or destroy it. The antigen-antibody response occurs in transfusions when bloods are incompatible for one or another of the hereditary blood factors.

ANTIRACHITIC—Pertaining to the prevention of the softening of the bones due to vitamin D deficiency (rickets).

ARBOREAL—Tree-living.

ATELES—Spider monkeys, South American cebids.

ARRHINENCEPHALY—Developmental abnormality characterized by absence of nose tissue.

AUSTRALOPITHECINES—Early Pleistocene populations of bipedal tool-using "man-ape" forms of Africa, some of which have been generally considered to be ancestral to later *Homo*.

AUTOSOME—Any of the nonsex chromosomes of the chromosome set.

BARIUM—Whitish metallic chemical element, used extensively in diagnosing and measuring digestive function.

BASI-OCCIPUT—Basion is a point on the forward margin of the foramen magnum, and the occiput is the rear bony plate forming the base of the brain case.

BIOTA—Animal and plant life characteristic of place or a period.

BIOTOPE—Smallest ecological area possessing uniform environmental conditions.

BIPEDALISM—Two-legged; erect posture and bipedal locomotion are characteristic of the hominids.

BRACHIATION—The characteristic mode of arboreal locomotion of such forms as chimpanzees, gibbons, orang-utans, and some monkeys. Brachiation consists of any of a number of variants of arm-over-arm swinging from branch to branch. Some anthropologists are convinced that prehominid forms went through a brachiating stage before adapting to terrestial life.

BRACHYDACTYLY—Genetic affliction characterized by abnormally short fingers or toes.

CARNIVOROUS—Meat-eating.

CAROTENOIDS—Yellow to deep red pigment substances found in animal and plant tissues.

CARPO-METACARPAL JOINT—Site of the joining of the wrist bone with a basal finger bone.

CARTILAGE-HAIR HYPOPLASIA SYNDROME—Genetic defect in the development of cartilage tissue and hair.

CATARRHINE—Pertaining to the monkeys of Africa and Asia, the apes (gorilla, chimpanzee, orang-utan, gibbon, siamang), and the hominidae. Catarrhines show, among other features, a dental formula that includes only *two* premolars in each jaw quadrant. New World platyrrhine forms show *three* premolars in each quadrant.

CEBIDAE—The taxon that includes the New World monkeys.

CEBOIDEA—Platyrrhine monkeys of the New World.

CEBUS—The capuchin monkey of South America.

CELIAC DISEASE—Disease of the intestinal system.

CERCOCEBUS—The mangabey monkeys of Africa. Some species are arboreal, others terrestial. They run on the ground with the tail arched over the back.

CERCOPITHECIDAE—Family of monkeys within the Primate order that includes the macaques, the baboons, mandrills, and drills, as well as colobus, or leafeating, monkeys. Many of these are largely terrestrial in habits.

CHALICOTHERES—Extinct group of odd-toed ungulates, with claws on their feet.

CHROMATOGRAPHY—Process that permits the physical separation of chemical substances for analysis.

CHOUKOUTIEN—Locality near Peking in China where important hominid fossil discoveries have been made of *Homo erectus* and Upper Pleistocene *sapiens* man.

CLADOGENESIS—Evolution by divergence and adaptive radiation from common ancestral populations.

CODON—A particular unit of DNA, consisting of three nucleotides, that directs the formation of an amino acid and thus is responsible for a particular phenotypic trait. The number of codon differences, as reflected in protein differences between two populations, is an index of phylogenetic distance between the two.

COLCHICINE—Chemical substance used in chromosome study (karyotyping) which aids in fixing and separating the chromosomes.

COLOBINAE—Monkey subfamily, within the Cercopithecidae, members of which are leaf eaters and possess sacculated stomachs.

COLOBOMATA—Eye defects due to abnormality of embryo development.

COLOBUS—Leaf-eating monkey with stomach pouches belonging, along with the baboons, macaques, drills, and mandrills, to the family Cercopithecidae.

CONSANGUINEAL—Related through direct descent from common parents or grandparents.

CONVERGENCE—When two populations or taxa show similarity of structure, and such likeness cannot be assigned to common ancestry, it is sometimes hypothesized that the two have adapted separately to similar environmental conditions.

CORPUS—Body, mass, major portion, as of a bone.

CRI DU CHAT SYNDROME—Genetic disorder in infants, characterized by mental retardation and a peculiar "cat's cry" as the child's crying pattern.

CYTOLOGY—The scientific study of cell structure and cell activity.

DELETERIOUS—Harmful or maladaptive. A deleterious gene is one that detracts from the efficiency of its possessor in carrying on life functions. Deleteriousness is not absolute, however; a gene that may be maladaptive, even lethal, for the individual, may be adaptive for a population. The sickle cell gene, for instance, is sublethal for the homozygote, yet by providing a number of heterozygotic individuals in each generation who are resistant to malaria, is adaptive for the population.

DEME—A Mendelian population; the smallest effective breeding population within a species.

DERMATOGLYPHIC—Pertaining to the study of palm-, foot-, and fingerprints.

DIASTEMA (plural "diastemata")—Gap between teeth, as between the incisor and canine teeth, within which the tooth from the opposing jaw rests.

DIK-DIK—Tiny antelope, found in Eastern and Southwest Africa.

DIMORPHISM—"Two-formed-ness," as in sexual dimorphism; the characteristic of pronounced differences in size and in structural proportions between males and females of the same species.

DIPLOID—Full set of chromosomes in pairs as they appear in somatic cells and immature sex cells. For man the diploid chromosome number is 46.

DISACCHARIDES—Any carbohydrates, such as sugars, that break down into simple sugars or monosaccharides by taking in water.

DIURNAL—Day-living; relating to daylight behavior, in contrast to nocturnal, night-living.

DIVERGENT STRABISMUS—Eye disorder in which the eyes cannot be properly focused because they diverge outward.

DRILL—Species of baboonlike, cheek-pouched cercopithecine monkeys found in Africa.

DUIKER—Small African antelope.

ECOLOGICAL NICHE—The sum total of environmental factors that act on a population as it maintains itself in a particular habitat. The presence of other populations of plants and animals, and characteristics of the physical environment, are features of a habitat to which the morphology and the behavior of the population must be adapted if it is to survive.

ECOLOGY—Study of the interrelationships of animal and plant populations with each other and with their physical environment.

ECONICHE—Coined word for "ecological niche."

ECOTONES—Areas where two different kinds of populations compete for dominance; transitional zones.

EDAPHIC—Pertaining to topography or the soil.

ELECTROPHORESIS—Separating different kinds of body-protein molecules electrically.

ELLIS-VAN CREVELD SYNDROME (CHONDROECTODERMAL DYSPLASIA)— Genetic form of dwarfism, inherited as a recessive trait. Its features include extra digits, malformation of fingernails, shortening of extremities, and premature eruption of the teeth.

ELONGATED—Lengthened.

ENDEMICITY—Property of a disease whereby it is a constant factor in a human environment, in contrast to a pattern of recurrence (epidemicity). Falciparum malaria is *endemic* in West Africa; rubella is *epidemic* in urban U.S.

ENIGMATIC—Puzzling, mysterious.

EOCENE—Epoch of the Cenozoic Era, from about 60 to 35 million years ago, during which early primates, prosimians, were already distributed in Europe, Asia, and North Africa.

EPHEMERAL—Transitory, passing.

EPIPHYSEAL UNION—Epiphyses are the sites of growth and fusion as cartilage is replaced by bony material during maturation.

ERYTHROCYTES—Red corpuscles of the blood.

ETIOLOGY—Study of the causes underlying, or the conditions leading to, the appearance of a disease or its symptoms.

EUNACHOID—Resembling a eunuch, or castrated male.

EXOGAMY—Practice of selecting a mate from a social group different from one's own.

FALCIPARUM MALARIA—Disease characterized by fever, anemia, and enlargement of the spleen, often fatal if untreated. It is carried by infection by the protozoan *Plasmodium,* species *falciparum.* The parasite is transmitted from person to person by the bite of the *Anopheles* mosquito.

FAUNA—Animal life characteristic of a region, locality, or sequence of geological time.

FAVISM—Hereditary enzyme deficiency causing severe reactions to certain chemicals. The sex-linked gene that causes it may provide resistance to malaria.

FAYUM—Site near Cairo, Egypt, where significant anthropoid fossils have been recovered.

FLATULENCE—Excessive gassiness in the digestive system.

FORAGERS—Food gatherers.

FOSSA—Depression or groove in a bone.

GALACTOSEMIA—Genetic defect of metabolism involving the inability to convert galactose to glucose. A recessive trait, its signs may include mental retardation, cataracts, and digestive disorders.

GALAGO—A Lorisiforme, widely distributed in forest areas of Africa. Very small (three inches long), the galago is capable of astounding feats of bipedal leaping on the ground and in the trees.

GAMETES—Sex cells; male sperm or female ova.

GASTROENTERITIS—Inflammation of intestines or stomach.

GAZELLES—African ungulates.

GELADA—Species of cercopith monkey related to the baboon. Geladas are found in desert rocky environments. Their social system is based on a male plus female consorts, a "harem," as contrasted with the troop structure of the better-known savannah-living baboons.

GENOTYPE—Genetic constitution of an individual.

GENUS (plural "genera")—Taxon that includes a number of related species.

GIARDIASIS—Infection of intestinal tract by protozoan organisms. Symptoms may include nausea, loss of weigh, and diarrhea.

GIBBON—Small ape native to southeast Asia.

GLABROUS—Smooth, hairless.

GRACILE—Fine, linear, or small-scale.

GRANIVOROUS—Grain or seed-eating.

HALLUX—The big toe.

HAPALEMUR—Lemur of the East African rain forest.

HEMIACODON—Fossil of the Eocene prosimian family of the Omomyidae.

HEMOLYTIC ANEMIA—Deficiency of hemoglobin in the blood due to the breakdown of red corpuscles.

HETEROSIS—Pertaining to the greater fertility and vigor, for certain traits, of heterozygotes compared to homozygotes.

HEURISTIC—Useful for investigation or research.

HOMINIDAE—Subfamily that includes all hominoids not assignable to the Pongidae. It includes all forms of *Homo,* both living and extinct, the australopithecines, and *Ramapithecus.*

HOMINIZATION—Evolutionary process whereby populations of terrestrial apes have been transformed over geological periods of time into human populations.

HOMO SAPIENS—Species to which all modern human populations, as well as Upper Pleistocene hominid populations, are assigned.

HOMOLOGY—Structural similarity in organisms, due to common evolutionary descent (human hand and bat "wing").

HOMOZYGOUS—Possessing two identical alleles of a gene on a pair of chromosomes.

HYALINIZATION—Formation of horny tissue.

HYBRIDIZE (-ATION)—Genetically, the formation of a new population through the merging of two or more parent populations. Hybridization is also loosely used as a synonym for *gene flow* between adjacent populations.

HYLOBATIDAE—The smaller ape subfamily that includes the gibbon and the siamang.

HYPERTELORISM—Deformity characterized by abnormal width of the nose bridge.

HYPERTONICITY—Abnormal tension of tissue tone.

HYPOGONADISM—Underdevelopment of the gonads or reproductive organs.

HYPOPLASIA—Arrested development; failure to attain full size or growth.

HYPOTONIC—Showing less than normal tension of tissue tone.

HYPOTRICHOSIS CORPUS—Genetically caused absence, partial or complete, of body hair.

HYPOXIA—Oxygen deprivation.

IMMUNOCHEMICAL—Referring to the physiology of immunity and the study of chemical incompatibility.

INDRI—Modern arboreal prosimian, closely related to the lemur, found only on the island of Madagascar (Mali).

INFRAHOMINID—Animal forms "below" man in the evolutionary scale, especially monkeys and apes.

ISCHIAL CALLOSITIES—Naked, thickened skin areas on the buttocks, found in many anthropoid species, used in sleeping on tree branches in a sitting-up position.

ISCHIUM—The lower margin of the pelvic girdle.

ISOMALTASE—Enzyme involved in the digestive breakdown of maltose to a monosaccharide, glucose.

JEJUNAL—Pertaining to the jejunum, a portion of the small intestine.

KARYOTYPE—The characteristic number, shape, and arrangement of chromosomes of an individual, or of a species.

KERATINIZATION—Formation of horny tissue from the outside corneal layer of the skin.

KINESICS—The study of the biology of movement.

KLINEFELTER'S SYNDROME—A disorder characterized by underdevelopment of male reproductive organs and mental retardation, resulting from the abnormal karyotype XXY, instead of the normal male XY.

KWASHIORKOR—A disease of protein malnutrition in children and infants. Most commonly found among African children raised on cereal or manioc with lack of protein. Its symptoms include severe bloat of body tissues, "pot belly," reddening of hair, cracking of skin, diarrhea, and mental apathy.

LABILE—Changeable, flexible, unstable.

LACTASE—An enzyme capable of breaking down milk sugar. Some individuals lack the genetic ability either to produce any lactase, or in other cases, sufficient lactase.

LACTOSE—A disaccharide (sugar) present in milk.

LAGOTHRIX—Commonly known as the woolly monkey. Widely distributed throughout the Amazon basin rain forest, this ceboid has a thick prehensile tail.

LANGURS *(PRESBYTIS)*—Small Old World monkeys, some species of which are highly terrestrial, particularly those populations that "infest" villages and temple complexes in India.

LEMURS—Prosimian members of the order of primates.

LORIS—Very small prosimian, found in tropical rain forests of India and Ceylon; related to lemurs.

LUMEN—Tube, specifically the intestine.

LUTEINIZING—Process of chemical change in uterine lining during ovulation.

MACACCA—Most widely distributed monkey genus, macaques are found from North Africa all the way to the Philippine Islands and beyond. Macaques are equally at home in the trees and on the ground.

MACROEVOLUTION—Large-scale, long-term evolution resulting in the differentiation of large taxonomic categories such as genera, families, and orders.

MANDIBLE—Lower jaw.

MANDIBULAR SYMPHYSIS—Thickened area on the outer frontal surface of the mandible, or lower jaw.

MANDRILLUS—Monkey genus including the drills and the mandrills, found in Central Africa. Similar in many ways to the baboons, these animals are primarily terrestrial. They show spectacular coloration of face and buttocks.

MELANIN (-IZATION)—The brown or black granules of pigment in the malphigian layer of the epidermis, which impart skin color. Melanin is also found in other parts of the body. The ability to produce melanin is under genetic control. Differences in human skin color are due to differences in the density with which melanin granules are distributed within the skin cells, and between them.

MESOPITHECUS—Miocene fossil monkey.

MICROCEPHALY VERA—Abnormally small brain case, associated with severe mental defect.

MICROGNATHIA—Abnormal underdevelopment of the jaw.

MICROSYOPIDAE—Prosimian family from the Eocene.

MICROVILLI—Minute structures on the mucous membranes of the small intestine that absorb nutrients from food materials.

MIOCENE—Epoch of the Cenozoic Era, from 25 to 12 million years ago, which saw the appearance of such ancestors to modern pongids as the dryopithecines, as well as the earliest protohominid, *Ramapithecus*.

MONOSOMY—Karyotype anomaly in which a chromosome occurs singly, instead of in the normal pair.

MONTANE—Mountain slope region.

MORPHOGENESIS—Structural growth and development.

MORPHOLOGICAL—Pertaining to animal structure and form. Morphological study of a fossil includes the analysis of bones, muscles, and other anatomical features.

MOSAIC—Phenotypic result of nondisjunction in cell division, wherein some cells have a normal chromosome pair count, some cells lack a chromosome of a pair, and other cells have an extra chromosome of a pair.

MOSAIC EVOLUTION—Evolution of different parts of a system at different rates.

MUTATION—Term that covers the processes resulting in the appearance of new hereditary characteristics.

NECROLEMUR—Fossil of prosimian family of Tarsidae.

NONDISJUNCTION—Failure of chromosome pairs to separate normally into daughter cells in mitosis.

NOTHARCTUS—Extinct Eocene prosimian fossil, recovered from Wyoming.

OCCIPUT—Rear base bone of the skull.

OCCULUSAL—Biting or grinding surfaces of the teeth.

OLIGOCENE—Epoch of the Cenozoic Era, from about 35 million years ago to about 25 million years ago, characterized by the emergence of monkeys and apes, and decline of the prosimians.

OMNIVORES—Animals whose diet may consist of both plant and animal materials.

OMOMYDAE—Extinct Eocene prosimian family.

ONTOGENY—Development of the individual organism from conception to maturity.

OREOPITHECUS—Extinct Miocene genus of brachiating anthropoids, sometimes thought to be early hominids.

ORTHOGENESIS—Apparent evolution toward some climax form.

ORTHOGNATHOUS—Straight-faced, as modern man is, in contrast to the protruding angular faces of apes.

OSTEOLOGICAL—Referring to bones or skeletons.

PALEOANTHROPOLOGY—Specialty within anthropology that investigates human evolution through the study of the fossil record.

PAPIO—Genus to which baboons, mandrills, and drills belong.

PARALLELISM—Separate evolutionary development of similar structures in populations or taxa with common ancestry.

PARAPITHECUS—Oligocene fossil from the Fayum, in Egypt, considered ancestral to modern pongids.

PARAPHILIA—Any genetic behavioral abnormality that interferes with reproduction in its carrier.

PARASITOSIS—Parasitic disease.

PATAS—African monkey.

PES PLANUS—Flatfootedness.

PHENOTYPE—The structural or behavioral expression of the genotype.

PHEREMONE—Chemical odor-bearing substance secreted by an animal that can influence behavior of other animals of the same species.

PHONOLOGY—System of sounds and sound combinations that characterize a language.

PHYLETIC—Evolutionary transformation of an array of interbreeding populations to a new grade or stage.

PHYLOGENESIS—Origin, descent, and evolutionary succession.

PHYLOGENY—Evolutionary history of populations.

PHYTOAGGLUTININ—Chemical substance in plants, such as beans, used to stimulate cell division in chromosome study.

PHYTOGEOGRAPHY—Study of the geographical distribution and relationships of plants.

PLASMODIUM ORGANISM OF FALCIPARUM MALARIA—Microorganism that causes malaria.

PLEIOTROPIC—Pertaining to the multiple phenotypic effects of a single gene.

PLIOCENE—The geologic period immediately preceding the Pleistocene; approximately 2 to 12 million years ago.

PLIOPITHECUS—Miocene fossil, frequently claimed as ancestral to modern hylobatids.

POLLEX—Thumb.

POLYDIPSIA—Abnormal intake of water.

POLYGENES—Different multiple genes; genes at different loci that act in concert to determine a trait, any one of which may be shown to make only a small contribution to the phenotype. Human skin pigmentation is an example of a trait determined by polygenes.

POLYMORPHIC—Maintenance of genetic differences, with regard to a given trait, within a population, by natural selection. Almost all human populations are polymorphic at the ABO blood group locus.

POLYPLOIDY—Characteristic of cells in which chromosomes occur in fives, rather than pairs (diploidy) or singly (haploidy).

POLYTENE—Giant chromosomes in the somatic cells.

POLYTYPIC (SPECIES)—Species that is widely distributed in space and is characterized by considerable genetic diversity among the populations that compose it.

PONGIDAE—Subfamily that includes the chimpanzees, gorillas, and orang-utans.

PONGIDS—Apes, including the chimpanzees, gorillas, and orang-utans as well as gibbons.

PORPHYRIA—Inherited disease characterized by abnormal excretion of certain blood pigment products, with such signs as cramps, malfunction of the digestive tract, and motor disturbances. It is inherited as a dominant trait.

POTTO—Largest of the Lorisiformes found in the West African forests. The hands are prehensile, and the potto is sometimes seen on the ground.

PRECIPITIN—Antibody produced in the blood serum by an antigen.

PREDACEOUS—Living by preying upon other animals.

PRESBYTIS ENTELLUS—Langur monkey of India.

PRIMATOLOGIST—Specialist in the study of primates.

PROCLIVITIES—Tendencies, natural inclinations.

PROCONSUL AFRICANUS—Fossil ape from East Africa, Miocene in date, widely believed to be ancestral to modern pongids, and close to the divergence of the earliest hominids.

PROPITHECUS—Modern arboreal prosimian, related to the lemur, found in Madagascar.

PROSTHION—Anthropometric landmark on the facial aspect of the skull; that point of the center of the upper jaw at the tooth margin.

PROTOHOMINID—Pertaining to the earliest hominids.

PROXEMICS—An anthropological subdiscipline, pioneered by Edward T. Hall, that deals with culture patterns and cultural differences as expressed by the ways in which people maintain, reduce, or increase the physical distance between them in social interaction.

PTILOCERCUS LOWII—Small tree shrew found in Malaya and Sumatra. The tree shrews are considered by any primatologists not to be primates at all.

PULMONARY EDEMA—Fluid in the lungs.

PUTATIVE—Usually regarded as; supposed.

QUADRUPEDAL—Four-footed locomotion (in contrast to bipedal or two-footed hominid locomotion).

RAMUS—The ascending side-portion of the lower jaw (plural "rami").

RETROGNATHISM—Developmental jaw defect which results in an under-slung lower jaw.

RHIZOMES—Underground rootlike stems.

SACRUM—Wedge-shaped bone, component of the fused lower vertebrae making up the rear wall of the pelvis.

SAGITTAL CREST—Thickened ridge of bone running sagittally, or in the middle from front to rear on the skull, as in the gorilla.

SAGUINUS—South American monkey without prehensile tail, found in forest areas, both in low altitude and mountain regions.

SAHAEL or SAHEL—Dry woodland area with thorny, tussock-shaped grasses, transitional from savannah to desert; borderland.

SALTATION—Rapid, large-scale evolutionary change that may result in the appearance of new species.

SEGREGANTS—Results of the sorting of chromosome material which occurs during cell division.

SEROLOGY—Study of chemical properties of the blood.

SICKLER—Individual whose blood demonstrates the presence of sickle-shaped red cells.

SIFAKA—Lemuriforme prosimian of medium size, found on the island of Madagascar.

SIMIAN—Monkey- or apelike.

SMILODECTES—Extinct Eocene prosimian, known from fossils recovered from Wyoming.

SOMATOTYPES—Categories of morphological classification devised by Sheldon, for identifying and describing human body shapes.

STRATUM CORNEUM—Outermost, or horny, layer of the epidermis. It is transparent and varies in thickness on different skin areas.

STRATUM GRANULOSUM—Second layer of the epidermis.

STOCHASTIC—Conjectual, or, in statistics, based on a single item in a probability distribution.

SUBLETHAL GENE—A gene that may lead to death at some stage in the individual's life, perhaps long after conception.

SUCRASE—Enzyme that breaks down cane sugar to invert sugar.

SUPRAORBITAL TORUS—Thick bar of bone projecting visor-like over the eye orbits. Found in hominid fossils (compare Broken Hill, Peking) as well as in pongids, such as the gorilla.

SYMPATRIC—Populations inhabiting different ecological niches in the same zone or territory.

SYNCHRONOUS—Occurring at the same time.

TARSIOIDEA—Family of arboreal prosimians represented by a single genus, found in the Far East. Tarsiers most closely resemble fossil prosimian forms from which later monkeys and apes evolved.

TARSIER, TARSIUS—Prosimian member of the order of primates, highly arboreal, notable for its leaping gait in locomotion.

TAUTOLOGY—Unnecessary repetition of an idea.

TAXON (plural "taxa")—Zoological or botanical category, such as genus, family, and order.

TAXONOMIC—Pertaining to a system of classification of plants and animal forms that demonstrates the evolutionary relationships among them. The taxon in which modern humans are grouped is the species *Homo sapiens*; the taxon genus *Homo* includes *H. sapiens* and an extinct taxon, *H. erectus*; thus it includes two *taxa*.

TAY-SACHS DISEASE (INFANTILE AMAUROTIC IDIOCY)—Degenerative disorder of the nervous system inherited as a recessive trait. The disease shows as a progressive diminution of vision, severe dementia, convulsions, paralysis, and death usually by the age of three years. Found in high frequency among East European Jews. The Tay-Sachs gene apparently confers resistance to tuberculosis in the heterozygote.

TERRESTRIAL—Ground-living, as contrasted with "arboreal" (tree-living).

TETONIUS—Fossil of extinct prosimisan family (Tarsioidea) dating from the Eocene.

TETRAPLOIDY—Characteristic of cells wherein the chromosomes occur in groups of four, rather than in pairs (diploidy) or singly (haploidy).

THALASSEMIA—Kind of anemia found primarily among populations of the Mediterranean region.

THEROPITHECUS—Known as the gelada, this monkey is found in Ethiopia and is probably the most terrestrial of African cercopithicines. The gelada is closely related to the baboons.

TORUS—Bony bar or shelf, as over the eye.

TRANSFERRINS—Blood proteins that may be significant in maintaining body resistance against infection.

TREHALASE—Digestive enzyme that breaks down one of the disaccharides, trehalose.

TRENCHANT—Incisive; sharply expressed or conceived.

TRIPLOIDY—Characteristic of cells in which chromosomes occur in triplets, rather than in pairs (diploidy) or singly (haploidy).

TRISOMY—The cell condition wherein chromosome replication results in three chromosomes rather than the normal two.

TROPICAL SPRUE—A chronic disease characterized by diarrhea, anemia, mouth sores, and indigestion.

TYROSINEMIA—Inherited defect in the metabolism of tyrosine.

TRYPANOSOMIASIS—Infection of the nervous system commonly known as sleeping sickness.

TRYPTOPHANE METABOLITE—Amino acid released from protein during digestion.

TYPOLOGY—Concept of a set of ideal features which characterize a kind of organism. The typological approach to the study of evolution has given way to the population approach.

UNGULATES—Herbivorous mammals, such as deer, sheep, and goats.

VELD—Open plain or savannah.

VON WILLEBRAND'S DISEASE (VASCULAR HEMOPHILIA)—Hereditary disorder of the blood and circulatory system. Inherited as a dominant, the disorder includes the lack of an antibleeding factor in the blood chemistry, and a vascular (blood vessel) defect. The tendency to bleeding may decrease as the victim grows older.

ZYGOTE—An ovum that has been fertilized by a sperm.

INDEX